ROUTLEDGE LIBRARY EDITIONS:
20TH CENTURY SCIENCE

Volume 3

MY LIFE

MY LIFE
Recollections of a Nobel Laureate

MAX BORN

Routledge
Taylor & Francis Group

LONDON AND NEW YORK

First published in 1975 as *Mein Leben: Die Erinnerungen des Nobelpreisträgers* by Nymphenburger Verlagshandlung, Munich.

First published in English in 1978.

This edition published in 2014
by Routledge
2 Park Square, Milton Park, Abingdon, Oxfordshire OX14 4RN

and by Routledge
711 Third Avenue, New York, NY 10017

Routledge is an imprint of the Taylor and Francis Group, an informa business

First issued in paperback 2015

British Library Cataloguing in Publication Data
A catalogue record for this book is available from the British Library

ISBN 978-0-415-73519-3 (Set)
eISBN 978-1-315-77941-6 (Set)
ISBN 978-1-138-01349-0 (hbk) (Volume 3)
ISBN 978-1-138-97666-5 (pbk) (Volume 3)
ISBN 978-1-315-77937-9 (ebk) (Volume 3)

Publisher's Note
The publisher has gone to great lengths to ensure the quality of this book but points out that some imperfections from the original may be apparent.

Disclaimer
The publisher has made every effort to trace copyright holders and would welcome correspondence from those they have been unable to trace.

MY LIFE

Recollections of a Nobel Laureate

Max Born

Taylor & Francis Ltd
London
1978

First published in 1975 by Nymphenburger Verlagshandlung,
Munich, under the title
Mein Leben: Die Erinnerungen des Nobelpreisträgers.

This original English version first published in 1978 by Taylor & Francis Ltd,
10–14 Macklin Street, London WC2B 5NF,
and Charles Scribner's Sons, New York.

Text set in 10/12 pt VIP Bembo, printed by photolithography, and
bound in Great Britain at The Pitman Press, Bath

ISBN 0 85066 174 9

Distributed in the United States of America and its territories by
Charles Scribner's Sons, 597 Fifth Avenue,
New York, N.Y. 10017, U.S.A.

Contents

Preface

Gustav Born

These 'recollections' by my father were written for his own family, starting in about 1940, and continuing, in his spare time, until about 1946. This took his account up to the discovery of quantum mechanics (1925), and the work of writing was then interrupted by other activities until about 1961, after he had retired and moved back to Germany.

As the manuscript was written for us, his children, and for his grandchildren, the style may be found rather different from autobiographies intended for publication. It has in fact been edited to make it more suitable for English-speaking readers; although my father's English was good, minor lapses of idiom have been anglicized.

The story continues up to about the beginning of the second world war, with occasional mentions of later periods. I have added a Postscript to provide a general account of my father's later years, which I hope will put into perspective the references to this period and round off a story which will be of interest to scientists, social historians and general readers alike.

The illustrations are mainly from family collections, but we are grateful to Professor N. Thompson of the University of Bristol for the photograph on page 245, and to the Institute of Physics for the photograph of the Max Born medal which is reproduced on the dustjacket and which was made by my artist sister Gritli (Mrs. Margaret Farley). The map of Breslau is taken from the 1910 edition of Baedeker's Guidebook to Northern Germany.

Cambridge, July 1978

Max Born in the 1920s

The Scientific Work of Max Born

Nevill Mott

Max Born was a scientist of the greatest distinction who presided over the Göttingen school of theoretical physics in its greatest days, from 1925 when a brilliant group there suddenly saw its way to formulate a set of laws for the behaviour of inanimate matter which has served us well up to the present time. This set of laws, which from the very beginning was called quantum mechanics and for part of which Born was awarded the Nobel prize much later, in 1954, is what he will be remembered for and what he would probably have wanted to be remembered for. But it was by no means the only part of his achievement. Perhaps equally important in its effects on chemistry as well as on physics was his earlier work on lattice dynamics, that is the theory of how atoms in solids stick together and vibrate. This achievement was substantially his own; on quantum mechanics he was the centre of a very active team.

Born was born in 1882 and so was already forty-six when quantum mechanics was invented. The account which he gives of his education shows the very thorough grounding in mathematics that he received, and this was certainly one of his strengths. Some of his early work was on Einstein's relativity theory. But perhaps the first paper which has made a lasting mark was—in collaboration with von Kármán—published in 1912 on lattice vibrations. This also was stimulated by Einstein's theory of specific heats (1907), in which he 'quantized' the vibrations of an atom in a solid. At about the same time Debye's work was published on the same subject, perhaps better known to students than Born's today because it is so much simpler. To quote from the *Biographical Memoir* published by the Royal Society, Born's approach here, as in most of his other work, was to face the problem in all its complexity, to devise a mathematical formulation of appropriate generality and then to descend to the simpler, more tractable and often physically more interesting cases. It was in fact many years before the need for his more general approach was appreciated.

Born continued to make contributions to crystal physics throughout his career. Very well known is his work on cohesion in ionic crystals, with the bridge it introduced between the physicist's and chemist's ways of studying these materials. Physics was beginning to produce quantitative values for the lattice energies of crystals, while for the chemists heats of reaction were of central interest. Born showed that the ionization potentials of the atoms concerned could be used to compare the chemical and physical concepts. This was a landmark, and led to the so-called Born–Haber cycle, in which the distinguished chemist Haber presented Born's ideas in a way which was widely recognized in chemistry.

Born returned to crystal theory in his Edinburgh days before and after the second world war, contributing among other things to the theory of liquids. Certainly as the founder and undisputed master of lattice dynamics he would rank as one of the great physicists of the century. In addition to this, however, there are his contributions to quantum theory. A paper with Pauli dated 1922 deals with the quantization of the mechanical vibrations with which his earlier work was concerned. At this time it was understood, since Planck's and Einstein's contributions fifteen years earlier, that the energy of vibrations was quantized, and since the brilliant work of Bohr in 1913, the same was true of angular momentum. But soon after physicists got to work after the war of 1914–18, it became clear that the extension of Bohr's theory to anything more complicated than the hydrogen atom simply would not work. Something quite new was needed. This was to be 'quantum mechanics'. Born's first paper, using this term, was *Über Quanten Mechanik*, published in 1924. He was now clearly near to discerning what the shape of the future theory had to be, but it is equally evident that the decisive breakthrough eluded him. It is on record that the ideas that made Born's dream of a consistent quantum mechanics a reality are due to Heisenberg, and he discussed them with Born very early on. The very important paper by Born, Heisenberg and Jordan was the result. When Heisenberg was shortly afterwards awarded the Nobel prize, he expressed to Born his regret that he was not included, as this book makes clear.

Meanwhile Schrödinger produced his wave equation, and this was shown to be formally equivalent to the theory of Born and his colleagues. De Broglie had earlier speculated that electronic waves might exist, and the experimental work of Davisson and Germer and G. P. Thomson showed that this was so. The question was, what was the interpretation to be of Schrödinger's 'wave function'? Born, in a classic paper on collision problems published in 1926, first said—in disagreement with Schrödinger himself—that the value of a wave function at a point gave, when squared, the probability that an electron was to be found at that point in space. This immensely important hypothesis, accepted by nearly everyone (with the very important exception of Einstein), gained Born his Nobel prize.

In any consideration of Born's work, his massive contributions to crystal physics, in which he stood alone, are in my view themselves of Nobel prize calibre and represent work which—but for Born—might have waited a decade or more. As regards quantum mechanics, the importance of this was greater still. Born was there with his brilliant young men, particularly Heisenberg, and what he contributed is well told in this book. Perhaps the probability interpretation of the wave function was the most important of all, but given Schrödinger, de Broglie and the experimental results, this must have been very quickly apparent to everyone, and in fact when I worked in Copenhagen in 1928 it was already called 'the Copenhagen interpretation'—I do not think I ever realized that Born was the first to put it forward. This is the reason for his rather sad remark that the book I wrote with Massey in 1933 on atomic collisions did not give him credit for this, and that this did him harm. I wish he had told me, so that we could have set it right in the second edition (1949). But the Stockholm committee at last recognized it, and awarded him the Nobel prize at the age of seventy-two.

After Born retired from the chair at Edinburgh, he and Mrs Born went to live at Bad Pyrmont in Germany. Here he became increasingly active in seeking to influence German public opinion about the dangers of nuclear weapons. In 1955 a group of eighteen Nobel prize-winners issued a statement, drafted by Born, Hahn and Heisenberg, which after the Russell–Einstein manifesto marked the beginnings of the Pugwash movement. Two years later, when the nuclear policy of the Federal Republic of Germany was under debate, Born was one of the leaders of the 'Göttingen 18' who declared that they would in no circumstances collaborate with government in any work associated with the development by Germany of these weapons.

He died in 1970 at the age of 87.

Part 1

The Good Old Days

Above:
My birthplace: the house on the right, in front of the church. To its left is the King's Palace

Below:
Myself as a baby

I. Childhood

The house in which I was born, Wallstrasse 8 in Breslau,* was in a rather special situation. On the left, if you looked out of the front windows, was the big women's prison, a bleak, ugly building; later it was pulled down and replaced by a modern block of shops. On the right, adjoining our small garden, which contained hardly any lawn but some tall chestnut trees, was a 'café-restaurant', with a similar but larger garden, filled with tables and chairs, which in the afternoon were occupied by elderly ladies enjoying a chat and a cup of coffee with plenty of cake. Then came the royal palace, a big and stately, but extremely plain building. The opposite side of the street was not built up; it was taken up by the *Exerzierplatz*, the parade-ground of the garrison, mainly of the 11th Infantry Regiment, which enjoyed a special fame for heroism from the war of 1870/71. The view from our windows was onto the vast extension of this military ground, bordered on the opposite side by the 'Promenade', with a double row of magnificent chestnut and lime trees. Above these you could see the roofs of the red-brick law courts, the synagogue, and in the distance the cupola of the Arts Museum. In this way my first impressions of the external world were saturated by things typically Prussian; the royal palace, the prison, the courts, soldiers practising the goose-step from morning to night, and in the distance the temples of the Jews and of the arts.

Breslau was still a fortress during the Napoleonic wars; but after Waterloo the walls were pulled down and transformed into the 'Promenade' already mentioned, a narrow strip of lawn and walks under beautiful trees. It widened to little parks, on small hillocks, wherever one of the old bastions had been. One of these, the Liebichshöhe, was crowned by another café-restaurant and a tower, commanding a lovely view over the town. The most beautiful spot was the Holtei-Höhe (named after the local poet Holtei, who wrote mainly in the Silesian dialect), just in the eastern corner where the old fortifications surrounding the city in a semicircle met the Oder river; from the top of this little hill you had a marvellous view, across the wide river, of the oldest parts

---

* A plan of Breslau in 1910 appears on page 5.

3

of the city, built on two islands, the Kreuzinsel and the Dominsel, a maze of old buildings, roofs and towers, partly in the typical Silesian red-brick Gothic, including the Kreuzkirche and the cathedral, partly in the noble Silesian baroque, such as the archbishop's palace. We children loved this view. It is impressed in my soul deeper than anything else I have seen since. But when I returned to Breslau in 1937, by then under the Nazi régime, for one day in order to visit my old aunt, Gertrud Schäfer, I was disappointed in some way: the grand view was there still, lovely and romantic as ever, but it had shrunk in size. The Holtei-Höhe was hardly more than a molehill, the river rather narrow and the towers small, compared with the image which my mind had preserved from the days of my childhood.

The Promenade, over its whole extent around the inner city, ran alongside the *Stadtgraben*, the moat of the old fortress. Parts of it formed little lakes where you could hire boats in summer; and in the winter it was the most splendid skating-rink. For two or even three months the ice was thick enough to skate on, and the whole population would enjoy the sport, day and night; yes, even at night, when the ice was illuminated by colourful Chinese lanterns and a band played on a wooden platform, on which a red-hot iron stove was placed to protect the musicians from freezing to death. It was on the ice that my father and mother used to meet when they were engaged. And my own first love-affair is also inseparable from the ice, which offered the most natural meeting-place for young people in a period when there were still many 'Victorian' or 'Wilhelminian' prejudices restricting the movements of well-bred youth.

In the nineteenth century the city had grown far beyond the ring of the old fortress walls. Modern suburbs stretched in all directions. But the inner city preserved much of its ancient beauty, as in many other German towns. I never quite understand why the British have sacrificed so much of their architectural heritage, which they once certainly possessed, to ugly utilitarianism.

Our house belonged to my grandfather Kauffmann. He was a great textile industrialist and owned several houses, mostly apartment houses in not very attractive districts. The Wallstrasse house was of a better type and well situated. It had only two flats of which we occupied the lower one. My father was not a rich man, a young lecturer at the medical faculty of the University, with a small salary. The Kauffmann family did not accept him without some opposition, but once my mother had got her way, it was considered necessary to fit the young couple up in a way worthy of the wealth and importance of great merchants. This was accomplished with the help of our nice flat, filled with rather pompous and heavy furniture of the 'Victorian' type. The house of my grandparents, in the Tauentzienplatz, was only five minutes' walk away.

Oddly enough, I remember the names of the people who occupied the flat above us. The first tenant was a titled old lady, a Countess von Rothkirch, with her elderly daughter. At that time one of the most popular novelists was Gustav Freytag; my grandmother Born told me many a time that she knew him well as a child when they lived in the same little town (Kreuzburg). Freytag's most popular novel, *Soll und Haben*, is set in Breslau and the

Plan of Breslau in 1910

5

surrounding country, and the principal figures were based on real people: the princely merchant Schröter of the novel on the former head of the well-known Silesian firm of Molinari (an Italian word meaning miller = *Müller* or *Schröter*), and the von Rothsattel family on an actual family called von Rothkirch. The old lady living in our house belonged to this clan, so my father assured me. When I re-read *Soll und Haben* I could not but marvel at the boundless admiration which was given to it in the circle of our Jewish families, for it is violently anti-semitic and emerges today as one of the roots of the Nazi catastrophe. I shall be returning to this point later on.

When the old Countess died, a Jewish family called Kolker occupied the flat; they had some boys about our age, but we did not take to them, and our only relations consisted of more often than not rough encounters in the garden.

My mother died when I was only four years old. I do not think that I have any direct recollection of her; the image of her which I carry in my soul is very likely the product of what I have learned about her from my father and other relatives, together with pictures and photos, and last but not least by her living image, my sister. I am sure that the lovely features and the character of mother and daughter were similar to an astonishing degree. But my mother was, according to all accounts, more brilliant. She was very musical, an excellent pianist. Young ladies of that period used to have 'Poesie-Albums', little books in which they asked their friends to write a few lines, a poem or sentimental quotation. I possess my mother's album, which contains few poems, but very many musical entries, and not only from 'ordinary' friends but from a great many famous musicians, among them Sarasate, Joachim, Brahms, Clara Schumann, Max Bruch (who was at one time conductor of the Breslau Symphony Orchestra), and several of the celebrated Wagner singers from Bayreuth. Brahms, for instance, wrote a few lines of his Academic Festival Overture, which was first performed in Breslau when he received an honorary doctor's degree from the University. Many artists who gave concerts in our town stayed at my grandfather's house and my mother, who alone of the three daughters was musical, did not waste this opportunity. There is very little else I know about her. She must have been exceedingly charming and graceful. But shortly after her marriage she began to suffer from gall-stones and she died when she was pregnant with her third child. At that time the operation which saved me a few years ago from a similar fate had not been invented. But as I have had plenty of gall colic, I know how she must have suffered.

My sister and I were then left to the care of servants. The first one I remember is my little sister's wet-nurse, a Polish woman named Valeska. There is still in my mind a faint image of this sweet and kind creature, in the colourful costume of Polish peasant women. She was a faithful Catholic and taught us the word 'God' and the Lord's prayer, and whatever her simple mind knew. Most children's nurses and maids in Breslau's well-to-do families were Polish at that time, and this fact was so deeply embedded in my mind that the word 'Pole' was almost synonymous with servant girl; and when later I first met an educated Pole, proud of his nationality, it gave me a kind of shock.

Then came Fräulein Weissenborn, an elderly spinster who took care of the household and our education until my father's second marriage. I still remember that at her first appearance my sister and I thought it extremely funny that the second half of her name was identical with our own. She was kind enough, but rather pedantic and dull, and a substitute for a mother only with respect to the material side of life. There was never anybody with whom we children could take refuge with our little sorrows and joys, as other children did with their mothers. This made us all the closer. We shared our games and play, and we developed a kind of private language, distorted expressions for everything, with which we could converse in the presence of grown-ups without being understood.

I suppose my father suffered not a little from Fräulein Weissenborn's loquacity, which she gave full rein to at dinner and supper. There is a German proverb, '*Nichts ist schwerer zu ertragen, als eine Reihe von guten Tagen*' (there is nothing harder to bear than a series of good days on end). Now Fräulein Weissenborn had been lady companion to an old Frau Guttentag, who was pushed about in a bath-chair; and whenever we met her on our walks with Fräulein Weissenborn on the Promenade, she entertained my father at supper with endless accounts of the old lady and all her relatives, until the sufferer once deeply offended her by interrupting her with the words: '*Nichts ist schwerer zu ertragen, als eine Reihe von Guttentagen.*' I remember very little of these early years. We rather kept to ourselves, and our only society was our cousins; and we were never out of range of one of the Fräuleins, either our own or that of our cousins. Behind our house was a rather picturesque courtyard, surrounded by old buildings in traditional timber framework, one of which was an elementary school. From our kitchen window my sister Käthe and I used to watch the hordes of children playing in the courtyard during break and after school, and to envy their freedom of movement. We spent long hours watching the drilling and parading soldiers from the front windows, and we invented nicknames for the officers and N.C.O.s, who behaved in a particularly grotesque way. But there were also great events on the parade-ground: every year the military celebration of the king-emperor's birthday, when a battery of the 6th Artillery Regiment fired a thundering salute just in front of our house; and occasionally even bigger parades, when the emperor visited the city and stayed for a few days in his palace, as our 'neighbour'. The biggest event of this kind was the Three Emperors' Conference: Wilhelm II of Germany, Franz Joseph I of Austria, and Nicholas II of Russia. We children did not know at that time the political significance of it—and I confess that I cannot remember it now; it was some last attempt to keep Russia from sliding slowly into the anti-German entente. We took it as a marvellous spectacle which we could comfortably watch from our windows, while other people had to pay considerable sums for such opportunities at hired windows or the wooden grandstands erected around the ground. There was, of course, always a crowd of uncles, aunts, cousins and friends in our rooms on such occasions, and there was a lot of fun.

We were certainly rather spoilt children. Grandmother Kauffmann was a small, but extremely brisk lady, who governed her husband partly by her energy and partly by her poor health. My father at least did not take her frequent complaints very seriously; this we found out in later years when we began to understand the rather delicate relation between the Kauffmann and Born houses. After our mother's death, grandmother Kauffmann took over supreme command, Fräulein Weissenborn acting as her lieutenant. There was hardly a day that her carriage did not stop at our door and she looked in, accompanied by Fräulein Wedekind, her lady companion, a charming, very musical and pretty woman, with an incredibly thin waist, always dressed in the most modern style, and full of life and fun. Grandmother exercised a strict control over all our doings and supervised our health with particular care, though I think with very little medical knowledge. There were frequent clashes with my father about such questions. Later he told us how furious he sometimes was when 'those womenfolk' countermanded his sound medical orders with silly popular cures. He did his best, but he was working hard in his Anatomical Institute, and came home at night too tired to argue with Fräulein, who was certainly more afraid of grandmother than of him. As he did not see us much, he was easily scared when he was told that we had a cold or some other little ailment, and in this way he did not prohibit 'those women' from treating us as milksops, shielding us from any 'draught' or strenuous exercise. But on the other hand grandmother was very loving and kind, and she always had some little present or surprise. Our birthdays were certainly glorious, with heaps of toys and sweets from the grandparents, uncles and aunts, and a children's party. Yet I am more than doubtful whether a child who is accustomed to have every wish fulfilled, that can be done with money, is happier than one of a poorer family. One little incident has stuck in my memory which was a consequence of this spoiling, and very properly led to a shameful disappointment. My mother had, apart from two sisters, one brother who had the same name as I—though I was not called after him. My father wanted to call me Marcus, after his father, a name which my mother did not like, perhaps because such classical names (Julius, Hector, Marcus, etc.) and Teutonic names as well (Siegfried, Siegmund, Gunther), were the fashion in Jewish families and became indeed specifically 'Jewish'. So Max was considered as a 'condensed' form of Marcus and chosen as a compromise. Well, there was this Uncle Max, a nice, sturdy little man with a beautiful wife about whom I shall have more to say later. One day the door of our nursery was ajar when grandmother was discussing things with Fräulein Weissenborn and Fräulein Wedekind in the drawing room. Suddenly I heard the latter say: 'I got the couple of love-birds yesterday for Max, I hope he will like them.' Now my birthday is shortly before Christmas, and I wanted passionately to have such a pair of little parrots; but my father, afraid of the noise and smell, objected to it. So grandmother had overcome his will—how marvellous! I looked forward to the morning of my birthday with greater expectation and excitement than ever before. But when I saw the table with the presents, wonderful toys and books, cakes and sweets, but no love-birds, I broke into quiet crying, and nobody could understand it. Then I waited for grandmother's appearance, hoping she would bring the birds; but again a disappointment and a new outbreak of tears instead of joy about the presents. At last Fräulein Wedekind, clever and kind, took me into a corner and succeeded in getting the words out of me: 'But you said the other day you had got a pair of love-birds for me,' which provoked an explosion of laughter from her and subsequently from the whole family.

When I learned that the birds were a Christmas present for Uncle Max, I was overwhelmed with shame: my eavesdropping and greed and discontent had been disclosed to the whole family, and I was the laughing-stock of all of them. The others forgot it quickly—I was not teased. But I did not forget, and it was the first experience which taught me to be reserved. Many others followed. There was no mother in whom I could confide, and who could restore my confidence. So it happened that I became a somewhat odd fellow. I did not know this myself, but after I was married and, for the first time, became intimate with another human being, I learned from my wife what was wrong with me. Well, that was twenty-five years later, and until my narrative reaches that period I shall have to refer to this point more than once. I was never deeply interested in characters and in the art of describing them, and I find it an awkward business, especially in respect of my own. But it has to be done. I shall however feel happier when I come to describe why I became interested in crystals and relativity and quantum theory.

But to return to our childhood. It might be thought that we were genuine big city children, living in apartment houses, accustomed to the traffic of busy streets and cut off from the natural life and beauty of the countryside. This is, however, not quite true. There was in the first place Kleinburg. My grandfather Kauffmann had bought a big house in a magnificent park-like garden, in the village of Kleinburg. At that time Kleinburg was still separated from the town by a strip of open country, but was already becoming transformed into a suburban agglomeration of villas, used as summer quarters by well-to-do merchants and civil servants. It was connected with the city by tram, pulled at that period by horses. During the summer months, we spent nearly every afternoon in the Kleinburg garden; sometimes the journey was made by tram, but very often in grandfather's carriage, which collected the different groups of grandchildren, with their nurses or Fräuleins, and took them back in the evening. The garden contained an enormous expanse of lawn, a double row of old chestnut trees, a large orchard with fruit trees and vegetable beds, two hard tennis courts, a little 'forest' of pine trees, and a playing-ground for children with swings, ladders and other gymnastic apparatus. There were always a considerable number of children, mostly relatives: the two daughters Alice and Claire of Uncle Max Kauffmann; Hans, Lene and Grete Schäfer, the children of my mother's sister Gertrud, and a lot belonging to second cousins, mostly the sons and daughters of grandfather's brothers. As the garden was well fenced in, we children were more or less left to ourselves, while our nurses sat chatting in the shade of a group of lovely trees. So we could run about and play in almost complete freedom. The only source of mischief was the fruit garden. There were stretches of it allowed to us children; but as we never could wait until the strawberries, peaches, pears, cherries and plums were really ripe, we ate plenty of them in a state when they were apt to produce indigestion. Later, when our own harvest was eaten up, there was the marvellous crop of grandfather's own beds and trees growing under our noses, and although we certainly got our proper share, we were exposed to a temptation which led to regrettable raids and sometimes to sad consequences.

This was our little paradise. But then there was the really great paradise, the summer holidays in Tannhausen. Here I must tell you a little more about the Kauffmann family, and about the textile firm which was the core of it. The founder of this firm was my grandfather's father, Meyer Kauffmann, a small Jewish trader in the little Silesian town of Schweidnitz. He bought textiles woven on hand-looms by poor people in the lovely valleys of the Silesian hills (Sudeten), sold them at the big fairs in Breslau, Frankfurt on the Oder and Leipzig, and supplied the home industry with raw material, wool and cotton. He was one of the first to have the idea of collecting a group of workmen in a house where the looms were driven by the power of a simple water-wheel. In this way the first textile factories came into existence. He had five sons: Salomon (my grandfather), Julius, Robert, Wilhelm and Adolf; the first four of them joined the firm and all became brilliant industrialists and merchants. Only Adolf went his own way; he became a physician but, as far as I know, never actually practised. He was deeply interested in music, and it was mainly due to his initiative, and his brother's money, that the Breslau Orchestral Association was founded. He seems to have been a great friend of my mother, who shared his musical enthusiasms. I cannot describe the growth of the firm Meyer Kauffmann. They bought or built a considerable number of factories; at the height of their prosperity they had a spinning-mill in Breslau and five or six weaving-mills in different valleys among the mountains. Today, less than half a century later, all this splendour has disappeared. There was only one generation of clever, humane, energetic and prosperous men. The next generation was still clever and humane, but not energetic or prosperous. The firm was already declining when I was a young man; it became a limited liability company, later a joint-stock company. The shares were not kept by the family, who slowly became impoverished, and the end came when the Nazis confiscated all Jewish property and drove the last directors of the family stock, my cousin Hans Schäfer and my second cousin Otto Kauffmann (grandson of Robert), out of the firm and indeed out of the country.

But at this point I have still to tell of those happy days when the firm and the family were flourishing.

The Weistritz is a tributary of the great river Oder; its source is on the ridge of the Sudeten mountains, not far from the German–Austrian (now Polish–Czech) frontier, and its valley winds through lovely hills, partly covered with dense forest. Here the Kauffmanns owned two big factories in the villages of Wüste-Giersdorf and Tannhausen. Both names are characteristic of the country; Giersdorf is quite a common name in these parts, but the epithet 'Wüste' (wilderness) indicates that it was, in mediaeval times, a particularly wild and lonely stretch of the valley. And Tannhausen means a place amidst spruce or pine woods (*Tannen-Wälder*). The valley is so narrow that there is seldom more than one row of houses on either side of the road, which follows the stream; but the length of all these hill villages is tremendous, and very often one village begins where the last one leaves off, so that there are miles and miles of picturesque cottages, smithies, little shops, each set in a tiny garden full of colourful flowers. On the slopes of the hills are fields and meadows, and above them the dark, mysterious spruce forests. Where the valley widens, you would often find a factory, mostly for textiles, with the owner's house and garden.

The Kauffmann factory in Wüste-Giersdorf was on one such wide section of the valley, a little above the valley floor; a vast expense of buildings and sheds from which the clattering of the looms was audible, like a permanent distant thunder. Between the works and the river was a stretch of park-like garden in which two of the brothers had their stately homes. That of Julius was often referred to as the 'Schloss'; it was indeed more than an ordinary villa—a rather monstrous brick building with quite a considerable tower. Julius's taste was altogether in the direction of the magnificent. He was the only one of the brothers who accepted the Prussian title Kommerzienrat (commercial counsellor), while my grandfather Salomon, the head of the firm, refused it, as a convinced liberal and democrat. The other house, also big but within the dimensions of an ordinary country house, was Wilhelm's, who was the technical expert among the brothers on the wool products manufactured at Giersdorf.

Half a mile downstream the houses along the road and river changed their name into Tannhausen; here was the firm's main cotton-weaving mill, housed in a long old building of generous proportions, situated on high ground above the river. And here were two more abodes of the Kauffmanns', known as 'the big house' and 'the little house'. But it was not in the big house that the head of the firm lived; his permanent residence was in Breslau, where the central offices were, and he used his house in Tannhausen only as summer quarters for his family. The big house was inhabited by Robert, the cotton expert, and after his early death by his son Franz, who married an English girl, Violet Jones.

The little house was the centre of our summer paradise. When you passed the bridge which led from the road over the Weistritz to the factory you had on your left the stables and agricultural buildings; and passing these you entered the lovely park which covered a large area on the hillside and merged in its upper part with the endless woods. It was an old-fashioned house; a basement containing an enormous kitchen and servants' quarters, a main storey with a central dining room, a drawing room and several bedrooms, and on top an attic with several strange rooms, which revealed the primitive wood construction of the roof. The main entrance was from stairs leading to the veranda in front of the dining room, and there you had a grand view, over the flower garden with a fountain and a row of big trees along the river on a gently sloping plain bordered by a range of beautiful mountains whose silhouette I shall never forget, no less than their names: Hornschloss, Langenberg and Schindelberg.

A hundred yards farther into the park was the big house, corresponding to its name in size, as well as in splendour, built in a not too tasteless nineteenth-century Gothic.

11

Beyond this building were large lawns and a pond and shrubberies, little woods of beautiful old beeches, oaks and pines, and of course, a big vegetable garden and orchard. Higher up the slope, where the park ended and the forest began, was the forester's cottage, called '*das Witwer-Häusel*'; not that the man was a widower—he had a wife and a numerous family; Witwer was his name. That cottage was a great attraction for us children. There were not only many dogs and cats, goats and sheep, beehives, and birds in large cages, but very often wild animals: a mother roe with her fawn, a young fox or a hedgehog. My cousin Hans Schäfer and I had a little cart, and Witwer taught us how to tame a goat and harness it to the cart. But I do not think this mode of transport was very successful, and after having been knocked down by the goat and suffered other mishaps, we gave it up and preferred to use the cart for tearing down the slopes at record speeds.

How could I describe the happy life in those summer days? My memory is filled with pictures, noises, smells which are all indispensable to give you an idea of what Tannhausen was for us. It began in the morning with a delicious breakfast, the main attractions of which were a kind of white crescent-shaped rolls, called '*Hörnchen*', which were better than anything of the sort I had later in life. Then we all moved through the park up to the woods and settled either on the Sylphenplatz, a circular arrangement of wooden benches in a lovely valley surrounded by beech and pine woods, next to a little pond formed by a clear stream tumbling over rocks in miniature cataracts; or higher up on the Haunstein, a steep rock emerging out of the forest, from which you had a broad view over the Weistritz valley and the hills. Later, when the wave of nationalism was spreading over the country, the old Haunstein was rechristened 'Kaiser Wilhelm Fels' and was fitted with a memorial tablet on some patriotic celebration, and a flagpole. But we did not use this new name. We were either Romans and Teutons, or trappers and Red Indians, roaming through the forest and fighting with swords made of hazel twigs and using pine cones as missiles. The girls meanwhile played more peacefully with their dolls. Sometimes they joined us in our games—for instance, if we sailed our flotilla of little ships on the pond and stream.

The afternoons were given over to bigger excursions, often beginning with a carriage-drive up one of the branch valleys, the Reimsbach Tal or the Donnerauer Tal, to a village inn, where coffee and cake were consumed before the walk in the hills and woods began. My Aunt Gertrud Schäfer, 'Tante Trude', was always the guide. She knew the topography of this country by heart and constantly led us on new and surprising paths. She also taught us the lovely old German folk-songs, and so we walked up the narrow winding tracks through meadows and woods, singing '*Wer hat Dich, Du schöner Wald, aufgebaut so hoch da droben*', or when the evening came, '*Der Mond ist aufgegangen, die goldnen Sternlein prangen am Himmel hell und klar*', and many others.

There were special joys for us boys when we drove back and the horses were sufficiently tired to be trusted to our hands. We were, of course, quite at home in the stables and great friends with the horses, and with the coachmen as well. Our seat was always on the box beside the driver, and we learned to handle the reins, although the grown-up passengers behind us did not like this particularly. There are very few incidents of these walks which I remember; one was an encounter with an adder which I killed with a stick, gaining a reputation for heroism among our gang. But there was another incident concerning a species belonging to the lower animals, a salamander, and its influence on my mind was much deeper than, and the reverse of, that heroic snake-killing. My father did not like to spend all his holidays with the family; he used to go for two or three weeks to the Tyrol or Switzerland with his brother Hermann, or with one of his intimate friends, the anatomist Professor Grützner, of Tübingen, or the physician Dr Steiner, of Cologne. They did not climb peaks but walked over the high passes, and their pleasure in the views, the clean air and the physical exercise was doubtless much enhanced by the exchange of ideas, on science, medicine, philosophy and politics, in that intense and passionate way of the old German liberals. My father always returned in high spirits, and when he appeared in Tannhausen life became intensified and more interesting altogether, in particular for us boys. For he was a real naturalist, an intimate friend of all living creatures; he knew the flora and fauna of his home province in every detail, and whenever he found a plant unknown to him, he took a little book out of his pocket with which he could determine the species. He made us keep aquariums and terrariums and taught us to feed the animals properly and to observe their behaviour. I must confess that I was never very good at this kind of occupation. I could not easily remember Latin names of species, orders and genera, and my interest was rather early directed more to the phenomena of inanimate matters, to crystals and colours and motions, than to biological things. Yet I loved to listen to father's fascinating tales about the marvels of life, and to watch tiny creatures in a drop of dirty water out of a pond which he showed us in his microscope.

Once, on his return from the Tyrol, he brought back a little alpine salamander which he had carried all the way with much trouble, and gave it to me with the warning that the animal needed special care and must have its food and fresh water every night. I was proud of having such a rare specimen, and I supplied it with flies and fresh water for a week or two. But one night there was a children's party or some other distraction, and I forgot the water. Next morning the little creature was dead—whether it was really caused by the lack of fresh water or simply by accident is beyond my knowledge. My father did not blame me, but said a few words about life and death the full meaning of which I could hardly understand, but which impressed me deeply. From that moment I suddenly knew what it meant to make another creature suffer and die.

I think this little experience taught me much more than religious lessons about Christ's or the martyrs' sufferings. In any case, it was inextinguishably fixed in my memory, together with several other examples of my father's method of teaching practical ethics. When I grew older, I understood him better, and I am certain that his deepest conviction was that the most fundamental rules of ethics can just as well, or even better, be derived from a

13

study of nature, including man, than from religious tradition. For he believed that these rules, during the progress of evolution which culminates in man's civilization, become intrinsic parts of our soul, and that education has not so much to teach them in one or the other traditional form, but to draw them out of the recesses of the childish mind into the light of consciousness. Concerning the question of suffering and death, he never told me anything about their mysterious meaning, but about their obvious place in the natural order; he did not say 'never kill'—he killed many animals in his biological research, and he despised the anti-vivisectionists as idiots—but nevertheless he imbued in me a deep respect for the right of every living creature to enjoy its span in the light of the sun. From him I learned to hate killing for pleasure, hunting and fishing as a sport; but he was not a vegetarian, nor made me one, because he knew man to be a carnivorous animal and likely to remain one in these northern latitudes.

You may get the impression that these grandparents on my mother's side were all that mattered to me. But that would be quite wrong. My father's ancestry was humbler in regard to all things of property and consequence in the world, but I believe that the essence of my nature is due to the genius of the Born clan.

My grandmother, Fanny Born, was a small and fragile lady in a stiff old-fashioned dress with a black bonnet. She was very quiet and rather stern, but never brisk and haughty like my other grandmother. She seemed to live in the past, in the recollection of her father and her husband. When my wife and I were a newly married couple she once visited my stepmother in Breslau, who showed her some large bundles of old letters which she intended to destroy. But Hedi began to read them and was so fascinated that she copied a great number of letters, beginning with some from my great-grandfather, and combined them with some poems by my father and a short biographical sketch of him, written by his sister Selma, into a volume which was the most precious birthday gift I ever received. There it was recounted that grandfather Marcus Born was a physician who practised in Kempen, a little town in the Polish–Prussian province of Posen (Poznan), when he met Fanny, the daughter of the rich Silesian industrialist Ebstein. My sister and I knew her only in her sad and lonely widowhood, and she seemed to us very old indeed, venerable and dignified. My grandparents lived a few years in Kempen, but later moved to Görlitz, a bigger town in the Lauzitz, a part of Lower Silesia near the Saxon border. Here Marcus Born became a well-known physician, practising not only in the town but also at the big estates of the aristocracy, who took very much to the Jewish doctor. There were very few Jews at this time in Görlitz, and practically no anti-semitism. My grandparents moved in the best society, and Marcus was appointed a district medical officer, the second Jewish doctor to be entrusted with this office by the Prussian government. My father was the eldest and the most brilliant of his children.

Marcus Born died rather young. At that time my father had ended his medical studies and accepted a position as assistant at the Anatomical Institute of Breslau University. He persuaded his mother to move to Breslau too, but she could not settle down there and returned to Görlitz. There however the

My father, Gustav Born

same happened again, she was estranged and unhappy. So she came back to Breslau, where my father had meanwhile been appointed principal dissector at the Anatomical Institute and lecturer at the University. He had found a pleasant home for her, a flat in Palmstrasse, one of the then new streets just outside the old fortifications. There she lived until a short time before her death, with her youngest daughter Henriette, known as Jettchen. I presume that her second son, Hermann, also stayed with her when he studied medicine at Breslau University. The elder daughter Selma had married a young physician and friend of my father, Dr Joseph Jacobi. I remember my grandmother's home quite well, for we had to visit her regularly at least once a week, and we loved to do so. Her rooms were full of lovely furniture, in the style of the 1840s called 'Biedermeier', of a similar type and just as noble as those in the little house in Tannhausen; and in the cupboards was a strange collection of china ware, old books and many other things which seemed to us 'modern children' out of another world.

There we regularly met our Jacobi cousins: Reinhard, who was about my age; Fritz, a year younger, and the little Walter. Behind the house was a long and narrow garden with many high beech and chestnut trees and rather poor lawns and flower beds, where we could play and roam about. On rainy days Aunt Jettchen was very good at entertaining us with table games. And there was of course always a splendid meal which you would call a tea, or high tea, except that the beverage was very milky coffee. Grandmother Born was always quietly sitting in a corner of her sofa, or in her easy chair at the window, and said very little.

There was hardly any communication between the two sets of grandparents. Grandmother Born, and still more her two daughters, resented the attitude of the Kauffmann family right from the beginning: their resistance to my mother's wish to marry a young and obscure doctor, the reception my parents found at their first visit after the official engagement, when grandmother Kauffmann appeared only for a few minutes in the drawing room and left the callers to her daughters and Fräulein Wedekind, and other happenings which appeared to them as the expression of conceit based on wealth. And in grandmother Born's mind must have been still the recollection of her own parental home which was also that of well-to-do industrialists.

In this way our childhood moved in two distinct spheres, that rich, dignified and proud community of the Kauffmann family and the simple, but no less proud Borns. An essential part of the latter were the Jacobis. My Uncle Joseph Jacobi, like my father, was a doctor, and a great friend. They were men of a very similar type, old German liberals and democrats—I say German, although they were Jewish for in those days the race theory had not been invented; the emancipation of the Jews dated far enough back in the past to allow a cultivated man to consider himself a German independent of his Jewish origin, which was no greater difference from the norm than that of being a Roman Catholic in a Protestant region. Often they were great German patriots who had fought and suffered for freedom and constitutional rights in the revolution of 1848. A brother of my grandfather, for instance, named Stephan Born, was involved in the revolt in Baden and escaped to Switzerland, where he later became a professor of literature at the University of Basle and editor of the most important Swiss newspaper, the *Basler Nachrichten*. I have a most interesting letter written by him to his brother Marcus, my grandfather, in which he expresses his hatred of Prussian politics, and of Bismarck's methods in particular, in the most violent terms, and outlines in prophetic words the future of Germany—alas, quite wrongly. For he predicted the collapse of Prussian military and reactionary politics in a war with Napoleon III, and he hoped after that for a splendid development of German democracy. How sad he must have been, disappointed by the developments which he later witnessed: the defeat of Napoleon and the triumph of Bismarck. And if he had been looking down from Elysium on subsequent events, what would he have said to the Germany of Wilhelm II and Adolf Hitler!

To return to my Uncle Joseph, he was not a liberal of this pugnacious type, but made his peace with those in power, as did my father and most of his generation. Uncle Joseph took part in the wars of 1866 against Austria, and of 1870/71 against France, as an Army surgeon and came back with the Iron Cross and the rank of 'Ober-Stabsarzt' (corresponding to a Colonel in the R.A.M.C.). He settled in Breslau as a physician and soon had an enormous practice, to which we belonged as a matter of course. And he was indeed an excellent doctor; his quiet, friendly and positive behaviour inspired infinite confidence. He became royal district medical officer, like grandfather Born, lecturer in hygiene at the University, and was honoured by the title *Geheimrat* (privy councillor).

My Aunt Selma was his second wife. There were two children of the first marriage, a son Siegmund and a daughter Helene. This Siegmund was a devil of a fellow, clever but lazy, difficult at school and even more so later in life. I do not know all the different professions he tried; but I remember that he changed his name to Hans, as the Teutonic name Siegmund at that time (i.e. twenty years after the period which I am describing) had 'degenerated' into a 'Jewish' one. He looked absolutely 'Aryan', as one would say today, and became an officer of the army reserve and, as such, was killed in the first world war. His sister Helene was also a most temperamental young lady, but settled down after her marriage to a doctor, Arthur Brieger; some of their children are now in America, some in England.

My Aunt Selma herself had the three sons whom I have mentioned already, Reinhard, Fritz and Walter. The elder two sons were great pals of my sister and myself. They lived rather far from us, in a suburb on the right bank of the Oder, a rather poor and slummy district where Uncle Joseph had a big practice among the poor, so we could not see each other every day; but we made the journey by tram often enough. However, the Jacobi boys hardly ever came to Kleinburg. My aunt taught them to dislike the Kauffmann atmosphere, and this made them shy and awkward when they were persuaded to visit the Kleinburg garden. They had a garden of their own, namely that of a big hospital for poor elderly people (*Siechenhaus*), where their father was head physician. There we were their guests very often, and we liked it, as no 'Fräuleins' were present and our freedom was even greater than in Kleinburg. My cousin Reinhard studied law and became in the end a judge at the Prussian High Court in Berlin. When Hitler came into power he was dismissed and died a short time later. I always suspected that he committed suicide, and that has recently been confirmed.

His brother Fritz became a dentist and lived in a small town with his (Aryan) wife and two children. He was very badly treated by the Nazis, and in spite of his being a veteran of the war of 1914–18 he was in a concentration camp for a considerable time. Whether he is still alive or has shared the dreadful fate of the German Jews who perished in Poland, I do not know. His son Hans escaped in time and became a press photographer in London.

The youngest brother Walter was extremely gifted, especially musically. He was a brilliant pianist even as a small boy, and his early compositions were considered by professionals as highly promising. But he was already strange when he was only about twelve or thirteen years old, and this became more pronounced in the following years; he became insane and died quite young in a mental asylum. My Uncle Joseph had died before this catastrophe. Aunt Selma was still living in Berlin when war broke out; she must then have been well over eighty. I learned from my sister in America that she seems to have disappeared, together with many other old relatives still living in Berlin. It is more than likely that they all perished with the rest of the German Jews in the death-chambers of the concentration camps in Poland.

II. Schooling. My Father's Second Marriage

As I have already mentioned, I was a rather delicate child, suffering very much from bad colds and coughs. This was not improved by the methods of grandmother Kauffmann who was afraid of fresh air, which she called 'draughts', and disliked the ordinary wild games of children because they made them get hot and perspire—apparently a very dangerous condition. After I had suffered from a bad attack of bronchitis—it must have been about a year after my mother's death—my father decided to take us to the seaside, in order to harden our constitution and to give us good exercise beyond the clutches of his mother-in-law. The place chosen was the Isle of Norderney, in the North Sea, where we were to spend four weeks under Fräulein Weissenborn's care. If I remember rightly, my father made the long journey with us and stayed a few days until we had settled down to a life according to his prescription, with plenty of fresh air and bathing.

But fate was against him. Hardly had he returned to Breslau when he got a telegram saying I had become very ill with strange symptoms, and the poor man had to hurry back again by the night train. He found me indeed in a bad state; for I had a sharp attack of bronchitis with asthma, so severe that the local doctor advised him to take me away at once. So I was brought back to the mainland, where the asthmatic trouble abated.

We returned to Breslau, a beaten army, and grandmother triumphed. How my father must have suffered under her 'Didn't I tell you? To expose this fragile child to the gales of the North Sea!'

And I suffered too, not only because I was now under still stricter control, forbidden to do many things which little boys enjoy, but because this asthma has accompanied me constantly all through my life. Heavy attacks have been rare, but they are terrible experiences. Even worse than the actual attacks was the permanent menace of their occurrence and the very frequent attacks of heavy breathing which, while not an illness, were just bad enough to spoil many a pleasure. On the other hand, this indisposition was not altogether a curse. It taught me endurance in spite of suffering, and the understanding of suffering in others. I think I can now, approaching the end of my life, say that at no time have I yielded to this scourge and have never behaved like an invalid.

The first serious consequence of my bronchial weakness was the decision that I should not go to school like my cousins Hans and Reinhard (who were a little my senior) but should have private lessons at home. A schoolmaster of an elementary school, Herr Böhr, was engaged for this purpose, and I still remember his face, surrounded by a blond Teutonic beard, his friendly but penetrating look and his full resounding voice. I also remember that at the beginning I was extremely shy and refused to answer, or did so with such a low voice that nobody could understand me. It is fashionable today to call this a complex, produced by the sheltered and isolated life which we children led. So the master's first task was to make me speak, and Herr Böhr succeeded. I had to shout my answers, and after a while I enjoyed it. I must also have enjoyed the lessons and my own progress, for never again in my life have I felt such an awe and veneration for a teacher, not even for the great Hilbert in Göttingen, twenty years later. One day we gave a party for the children of Herr Böhr, and we looked forward to it; but it was a disappointment, for we expected the offspring of such an extraordinary man to be themselves extraordinary creatures. Yet they turned out to be fat, pink boys without any particular merits, who were rather shy to begin with and developed later into ruffians.

After a year of these private lessons, I was considered fit enough to be sent to school. There were several old and famous high schools in Breslau, with resounding names, like Magdaleneum, Elisabethanum, etc., as they had sprung from monastery schools attached to the churches of St Magdalen, St Elisabeth, and so on. These had preparatory classes attached to them. But (in consequence of their origin) they were all at that time in their old buildings, near the centre of the city, with dark and badly ventilated rooms. They were also reported to have preserved to some degree mediaeval methods of education and punishment, combined with a system of bullying among the boys. Whether this was true or not, all 'good' families preferred to send their sons not to the prep-schools of these high schools, but to private schools of which there were two or three renowned ones. All my cousins went to *Wankel's Knabenschule*, situated on the 'Ring', the central square of the city, so I was sent there too. Each Silesian town has such a 'Ring', which does not look like a square because the centre of the enormous quadrangle is covered by a block of buildings, which consist of the guildhall and other official buildings, a chapel or church and some stately merchants' houses; the rest of the square surrounds this central block like a ring, though not a circular one. The Wankel school was just opposite the old *Rathaus*, and for two years my eyes, wandering to the window during a dull lesson, fell on this beautiful Gothic structure; again an image which is indelibly impressed on my mind. I can still see in my imagination this lovely picture on a sharp, cold, eastern winter day, above the snow-covered pavement the black walls, with the irregularly placed windows of noble proportions, the high roof of old coloured tiles, the graceful tower silhouetted against the blue sky.

The school itself was on the first floor of a bleak old building surrounding a narrow courtyard, partly covered by glass and called '*die Passage*' (arcades).

19

There we had to spend our break. Yet there were some attractive features; for among the shops of the 'Passage' was the biggest toy shop of the town, owned by Gerson Fränkel, the windows of which displayed the most splendid things a boy's heart could desire, tin soldiers in the uniforms of all periods of history, boxes of bricks, toy railways, etc. The classrooms were of the same low standard of lighting and hygiene as the courtyard. Yet the classes were small and the teaching not bad. I suffered from the fact that I had missed the first year, called Nona (the ninth form), during which the little boys had learned the elements of community life. I never succeeded in becoming quite one of them, though half a dozen of us remained together for the whole twelve-year period of our school life. The barrier was increased by my own shyness and sensitiveness, for which there was no outlet through the loving interest of a mother.

One little incident has stayed in my memory. There was no Jewish religious teaching at the little school. My father decided that I should take part in the evangelical (Lutheran) instruction. He had no particular preference for the Jewish religion, but none for the Christian either. He remained a Jew mainly from the standpoint of strict liberalism, which considered religious questions as purely private; what was good enough for his ancestors he thought good enough for himself. His remarks on religion were often critical, yet never hostile, and he was quite aware that the whole of European civilization was saturated with ideas, pictures and anecdotes taken from the Scriptures. Therefore he wished that we should be acquainted with the biblical stories, and he did not mind in which particular form they were presented. So I had to attend the class of an evangelical parson. As I was a newcomer, he tried to find out what I knew about religious things and asked me whether I could say the Lord's prayer. This I did fluently and not without pride. So I was very astonished when he corrected my wording several times. I insisted that my version was right. Then he said: 'Yes, my boy, that is the Roman Catholic version, but here you are in a Protestant community.' In fact, I had learned the prayer from my sister's Polish Catholic wet-nurse, Valeska, who was in charge of our nursery after my mother's death. The boys found the teacher's remark very funny and began to tease me. The Jewish intruder they might have accepted; but this Roman Catholic business was too much for them. The effect on me was an increase in loneliness and isolation, and the beginning of a deep resentment against religion. As religious questions had never played a great role at home, I had no instinctive veneration to overcome. It must have appeared strange to my childish mind that some tiny modifications in the wording should be considered as important distinctions between two such groups of people as the Protestants and Roman Catholics. In this way I early became an outsider in a matter which for many people is of fundamental importance. However, I was aware of this only at school. Religions, churches, sects, were hardly touched on in our house, or in the wider family.

In spite of this incident, I settled down in my form as an inconspicuous and almost normal member. Whether I enjoyed this life and the teaching I cannot remember. There are a few incidents which stick in my memory, though I

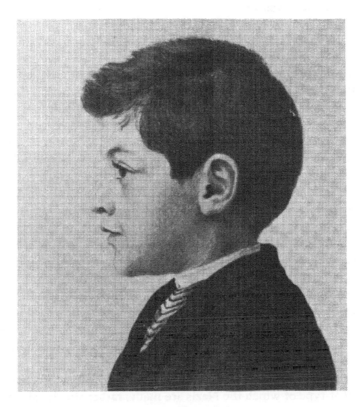

Myself as a boy

rather think they were put there later by the tales of grown-ups than by my own direct recollection. This is one: I had to stay at home for a few days because of one of my frequent attacks of asthmatic bronchitis. When I was well enough to go to school again, there was a period of very severe frost and snow. So my father took a cab for both of us and dropped me at my school; he gave me a coin and told me to take another cab for my return home. But it appeared that our prep-school had that very day been challenged by the prep-classes of the Magdaleneum, which was situated in the neighbourhood, to a snowball fight. This battle went on during the breaks and after school. I took part in spite of my state of convalescence and fought with devotion and courage until I looked more like a snowman than a living boy. When it was over—I have forgotten whether we were victorious or not—I remembered the mark piece in my pocket, went to a cab-stand in front of the guildhall and gave the driver my address. When I arrived home, very late and anxiously awaited, I quite innocently and proudly reported our marvellous snow battle and was astonished when my father called me an idiot and other names, and had me put to bed at once. He was always mild and kind, and he used thrashings very sparingly. We were all the more impressed when he became really angry. Therefore I remember this case, although I daresay his anger was not so much directed against me than himself for not having instructed me properly. As far as I remember, the adventure had no bad consequences on my health.

After two years of prep-school the question arose of which of the numerous high schools I should be transferred to. As I have said already, those in the city with monastic names were considered by the progressive part of society to be old-fashioned and not up to the standard of modern buildings and hygiene. But there were a few 'satisfactory' schools, one of them in the fashionable West of the town, the Königliches König Wilhelms-Gymnasium. This was chosen. It was housed in a modern building with ample space and light and was considered up-to-date in every respect. Not only I, but quite a group of boys in my form were transferred to the Sexta of the K.W.G. This school had its own prep-classes (Nona, Octava, Septima); so we entered a well established community of about forty boys, and for a time we Wankelians were all outsiders and therefore compelled to keep together.

I have forgotten the names of our little group, apart from those of two boys who were my classmates throughout the whole time I was at school. Their fathers were colleagues of my father, professors at Breslau University; a theologian and a pathological anatomist, hence a scientific associate of my father. These two boys were inseparable pals, always alternating in being the *primus* and *secundus* of the form. For some time I was somewhat loosely allied to them; but it was more in common defence against the rest of the form than in sympathy. They were an odd pair indeed. K. was thin, quiet, sedate, taciturn, matter of fact, and would have been quite a decent fellow if he had not been attached in an exclusive way to the other one. This boy, P., was from the beginning that type of which the Nazis are made: rather sturdy, almost fat, yet energetic, loquacious and a perfect brute—at least, when he was certain not to be seen by one of the masters, to whom he was most deferential. My father wished to establish some friendship or at least social intercourse between me and the sons of his colleagues. So they were invited to a children's party, and in turn I was invited to the P.s'; they lived in a big villa in a spacious garden at the extreme other end of the town, near the Scheitnig Park. There we played trappers and Red Indians in a most realistic fashion: the prisoners at the stake were really tortured under the expert direction of Hans P. I escaped this fate but watched with indignation how a boy was tortured until he quietly cried, and was threatened with more of the medicine if he told the grown-ups. He was saved in the end by one of P.'s sisters; about twelve years later I met her again in Göttingen; she was married to a lecturer in anatomy, Stolper, who died shortly afterwards, and I remember quite well her saying to me at once: 'Well, Dr Born, I know you dislike my brother, but let us be friendly', and then we recalled that adventure and other similar affairs and laughed about it. Yet at the time it happened I did not think it funny. Maybe I was a bit of a weakling and not accustomed to the ordinary brutality of boys, but my cousin Reinhard Jacobi, who was also at that children's party, shared my impression and distaste, although he was a year older and a tough boy.

Similar things happened at school, though the sharp discipline prohibited extensive bullying. Our form contained a considerable number of Jewish boys, perhaps one third, and there were already the first signs of anti-semitism. In defence of the staff I must say that there was hardly any racial or religious prejudice on the side of the masters. But P. and a few others were more or less openly anti-semites and enjoyed a little Jew-baiting whenever it could be done unnoticed by the liberal-minded masters. In the upper forms P. developed to an extreme degree the kind of nationalism which was later to infect the whole country. In the geography lessons he liked to draw new frontiers of a future Greater Germany exactly in the Nazi style. He studied law and became a Prussian civil servant; and I was not surprised to learn from the newspapers that he became a leading member of the Prussian administration set up in the former Polish province of Posen (Poznan) in order to germanize it. When I had to leave Germany in 1933 and took refuge with my family in Val Gardena, South Tyrol, I met Frau Stolper again, who was as nice and human as before; she told me that my 'intimate enemy', her brother, had risen to a high rank in the Nazi hierarchy.

In this way I had a foretaste during my early school-days of what was to come in Germany.

Amongst the reasons favouring the choice of that particular school was our removal from our old house on Wallstrasse to a flat situated only five minutes' walk from the K.W.G. This removal was itself the consequence of my father's second marriage, about which I now have to speak in so far as it concerned my own life and development.

My mother's death had been a terrible blow to my father. In some of his letters written at the time she was already suffering from gall-stone attacks and was sent to Karlsbad in Bohemia for a cure, his deep anxiety for her health reverberates, and his deep love. At her death he was not broken, but sad and depressed. I have a vivid recollection of the impression on my sister Käthe and me one Christmas Eve when our childish joy was choked by seeing him sitting in tears. However, he was a young and handsome man, and after a few years he was susceptible to the attraction of women again. Later, a big box full of old letters came in my possession, and a short time before the Nazi catastrophe I decided to destroy them. But when I took the bundles of letters out of the box I began to read, and I continued reading the whole night. It was exciting, exhilarating and depressing. Many things which I did not understand about the relation of my father to other people, about his professional successes and failures, about his ideas, hopes and dreams, became clear. There were also love-letters (apparently returned to him after the end of the affair) and a very frank correspondence with intimate friends about a second marriage. When I had read them all, I burnt them.

In this way I learned that there was a passionate affair with a cousin of my mother, who seems to have played with him in an exasperating way. I found a bitter and ironical poem by means of which he apparently tried to free himself of this fascination. Then he fell deeply in love with Hermine Spiess, who at that time was considered the most brilliant singer in Germany. They met in grandfather Kauffmann's house, where all musicians used to stay when giving recitals at the Breslauer Orchester-Verein. But although she seems to have been much attracted by the young scientist, there was no question of her giving up her musical career and settling down as a professor's wife. So it came to nothing.

This series of disappointments hit him all the more as there were many strong reasons for entering a second marriage. I have tried to picture the kind of nursing and education to which we children were subject from the hands of Fräulein Weissenborn, under the powerful direction of grandmother Kauffmann. Both were certainly animated by the best intentions, yet hardly by the necessary knowledge and insight, measured by my father's standards. Grandmother kept to her old methods which she had successfully applied in bringing up her own four children. Yet my father had other ideas and clashes were frequent. I presume that the main differences of opinion, apart from health and hygiene, were concerned with the value attributed to money, prosperity and social consequence. He did not want to have the attitude and outlook of a wealthy merchant family implanted in his children. For he himself suffered from this attitude. His parents-in-law expected him to live on a social level which was far above his own income, and they compelled him to accept financial assistance to keep in step with them. As long as my mother lived he seems to have considered this state of affairs as natural, and a due consequence of marrying an heiress and a girl of delicate health who needed a fair amount of luxury. But he suffered from this dependence; whenever a promotion in his profession was in sight or had vanished out of sight, his hopes and his disappointments were not only the expression of his scientific ambition and of his wish for a wider field for his scientific activity, but just as much of his longing for financial independence and for escape from the perpetual humiliations inflicted by the Kauffmanns. If this was already the case during my mother's lifetime, how much stronger it must have grown after her death, when her mitigating influence had gone.

This desperate feeling of financial dependence must have become an obsession during the following years. Several times he had a hope of escape by being called to a professorship of another university. But he was always disappointed; often by a small margin. If I remember rightly, there were in particular two cases, where he competed with intimate friends of his: for Tübingen with Grützner, and for Bern with Strasser. In both cases the other man was chosen, and obviously because he was not a Jew. I remember a letter from Grützner himself in which he informed my father not only about his appointment, but about the preceding discussions and intrigues, sparing no words in condemning the anti-semitic motive which led to the final decision. But could Grützner, who was not in a better position than Born, be expected to refuse because his friend was unjustly treated? My father's disappointment about Bern was still greater; for he would have considered a call to Switzerland as a double liberation, not only from the family bonds, but also from Prussian servitude. He used frequently to tell me later: 'When I was young and was asked which profession I would choose, I used to say: I do not know what I shall become, but two things I shall avoid with all my strength: to be a teacher or a Prussian civil servant. And now I am both, a Prussian university teacher.'

How different his and his childrens' lives would have been if we could have become Swiss citizens!

I learned from my stepmother herself how it came to pass that she married him. Her family had close friends in Breslau, or to be more exact, near Breslau, for Herr Schottländer, though a Jew, was a big landowner and lived on his estate some miles from the city. My father was also friendly with them, and as they watched his struggle for independence, it occurred to them that a marriage between Gustav Born and Bertha Lipstein would solve the problem, for the Lipsteins were very well-to-do. So the Schottländers managed to start a correspondence between these two—how this was possible is beyond my understanding. One has to consider that at my grandfather's time marriages in Jewish families were still regularly arranged by the parents, with the assistance of common friends, or even of matrimonial agents (called, in Yiddish, 'Schadche'). Now my stepmother wrote very nice letters, and my father was so attracted by these that he decided to meet her. The result was an engagement, and soon afterwards a wedding.

Kindness and honesty were the main features of my stepmother. She had a simple nature, growing up to a very definite set of rules of behaviour and keeping strictly to them. She had not much charm, either in external appearance or in her manners. When she was introduced to the Kauffmann clan, she did not have a very friendly reception. How could Gustav, the lover of the fair and charming Margaret Kauffmann, marry this unattractive, shy and reserved spinster! That was certainly the reaction. But the Borns and Jacobis received her with great warmth, and many of my father's friends too. I think that if she did not give him a second great happiness, she gave him much nevertheless: a comfortable home, meticulous care for his health and well-being, and for that of his children as well, and financial independence, which allowed him to spend his time on research work of his own choosing.

Here I must add a few words about the Lipsteins. My stepmother's father had been a merchant in Russia, dealing I think in timber, and had become rich. Now, in the Russia of the Tsars there was violent anti-semitism. Jews were restricted to definite regions and not permitted to travel freely, except when they were wealthy 'merchants of the first guild'. Occasionally there were also pogroms with pillage and bloodshed. Old Lipstein had risen to be a merchant of the first guild, yet he detested the Tsarist administration and longed for higher culture, not to be found amongst the Muscovites. Therefore he sent his daughters to boarding-schools in Germany or Switzerland (my stepmother was at Lausanne) and his sons to German universities. Finally he decided to emigrate and settled in Königsberg, in East Prussia, but died soon afterwards. His wife, Ida, who became our grandmother Lipstein, lived a few years in Königsberg, but removed later to Berlin, where two of her daughters were married. The eldest, Paula, was, like her mother, a handsome woman, but her husband, Julius Hirschfeld, a merchant trading mainly with Russia, was one of the ugliest men I have ever seen, almost like a gorilla. The second daughter, Helene, was no beauty herself, but married a distinguished, almost aristocratic-looking man, Ismar Mühsam, who together with some equally good-looking brothers, owned a big business.

Uncle Ismar was highly cultured; his lovely house in Victoria Strasse, near the Zoo, contained a collection of good modern pictures. He was an accomplished piano-player and had regular musical evenings where excellent chamber music was performed, by high-ranking artists, such as the brothers Klingler and Alfred Wittenberg. The players were of course well paid, but obviously enjoyed these evenings not only because of the charming enthusiasm of the host, but also because of the excellent dinners offered by Aunt Helene. In later years I was a frequent guest on these occasions and acquired in this way a fair knowledge of classical chamber music. When Ismar died, his wife gave me his violin, viola and cello and all his volumes of trios, quartets and quintets, which I still possess.

Rosa, my stepmother's youngest sister, was married to a merchant, Siegfried Jaffe, in Leipzig. She was rather pretty and sweet, but delicate in health, and she died very young.

There were also six brothers. Only one of them, Alfred Lipstein, was of a type which attracted me. He studied medicine and was for a long time Paul Ehrlich's assistant in Frankfurt. When I later studied in Heidelberg we often met and became good friends, and this friendship lasted. He did not succeed in leaving Germany and I am afraid that he and his charming wife perished under the Nazi terror, just as did his brothers and sisters.

All the Lipsteins received us children with the greatest kindness and hospitality. They spoiled us in every possible way by giving us expensive presents and tickets for the circus or theatre. In later years when we came to Berlin there was always a spare room for us, either in the old lady's home in Alsenstrasse, or in the Mühsam's house in Victoria Strasse.

Yet there came a time when I resented the obligation which I owed to the new relatives. It was not only the feeling of a burden produced by the considerable number of calls I had to make whenever I came to the capital, and of the restriction of my freedom imposed by the lavish invitations. There was more than that. The Born and Kauffmann families were Jews emancipated some generations earlier and assimilated to their gentile surroundings. They believed they differed from their neighbours only in respect of religion, for which they had hardly any use, and that they were Germans by culture and custom. I was reminded of this attitude recently when a volume of Gustav Freytag's novel *Soll und Haben* fell into my hands, which, as I mentioned before, was then, in the years before 1900, considered a classic. Reading it again, I must admit that it is well written and has a fascinating plot. But it is extremely anti-semitic. The heroes are 'Aryan' noblemen and merchants, the villains Jewish financiers and traders. There is one exception: the son of one of these evil money-makers is a quiet and noble scholar. The Jews as pictured in this book are essentially the same as those portrayed in *Stürmer*, the infamous newspaper of the Nazi chief Streicher. Yet this book was enthusiastically read and praised in our families. They obviously considered themselves as belonging to the class of noble German merchants, and having nothing to do with the 'Galician' small traders with their 'Yiddish' dialect and low standard of education. History has proved that this view was not shared by the majority of the Aryan Germans.

Now the Lipsteins had escaped only recently from the Polish ghetto. Old Mrs Lipstein never learned to speak the German language properly. Some of the sons retained a trace of Jewish accent—or should I say, cultivated it purposely, as it suited their mental attitude, which was the well-known mixture of irony, frivolity and melancholy of the educated Polish Jews, expressed in so many brilliant, untranslatable jokes and puns. The daughters, and in particular my stepmother, had much less of that specific taint, but just enough to create a gulf between them and the germanized Kauffmann family. I cannot deny that I was affected by this state of affairs.

At first, the new regime under my stepmother was certainly beneficial for sister Käthe and myself. I do not think we regretted the departure of Fräulein Weissenborn, and still less the end of grandmother Kauffmann's persistent interference which, with his wife's support, my father put an end to. The only concession he made was the preservation of the custom that Käthe and I had our dinner (or lunch—it was the chief meal served between 1 and 2 p.m.) with the grandparents every Saturday. We did not like it very much, although we had all our favourite dishes and little presents in addition. For we were subject to sharp criticism not only for our own behaviour, but for the educational methods of our parents, and we could not refrain from reporting these things at home. That led to new conflict. A sharp letter went from my father to the grandparents in which he described this impossible situation and rejected any interference in his way of bringing us up. This letter shows what the main point of disagreement was. The grandparents wanted our participation in all kinds of social functions and amusements: parties, theatre performances, concerts; the latter in particular were considered essential for 'cultivating the mind'. My father, however, knowing how occupied our time was with school, homework, music lessons and sport, was eager to keep us quiet and to protect us from nervous strain. He already knew of his youngest sister's mental disease, and being a scientist influenced by Darwin he feared the hereditary characteristics of such a curse. He did not give in, and he insisted that our amusement was strictly rationed according to his prescription. We did not object to this, as we loved him and had absolute trust in him; and when he explained his reasons to us we understood. Even when I was almost grown up, and in the top form of my school (Prima), punctually at nine o'clock, and without the least embarrassment, I left my grandparents' box where I attended the regular orchestral concerts with them, although some of my younger cousins stayed to the end—even with some amusement—just to show that I was on my father's side. My sister, it is true, was not always so obedient.

Our new flat, Zimmerstrasse 5, was in the western suburb just beyond the Stadtgraben and Promenade. The rooms were big and well proportioned, but the street and the whole district were ugly and depressing, by modern standards: rows of shapeless grey apartment houses, becoming darker day by day through the smoke from the numerous workshops and factories; hardly a tree or lawn; a great deal of noise from traffic, tramcars, cabs, landaus, later

27

intermingled with automobiles. A few steps round the corner was the Sonnenplatz, not a real square, but just the crossing of half a dozen streets and tramway lines; not at all a 'sunny place', but extremely drab and suburban. I had to cross this square on my way to school which was situated in a side street of particularly depressing character; but the building itself was modern (at that time) and bright, and it had a large yard, which was even adorned with a number of trees, for recreation in the intervals.

My father had a longer journey to his Anatomical Institute, which was on the periphery of the inner town, where the Stadtgraben approached the Oder river. A few years later it was transferred to a new and modern building still further away, on the opposite side of the town on the outskirts (Scheitnig). Then he had a long and fatiguing journey two or even four times a day. He made several attempts to get a house or flat near the new institute, but as that suburb was not yet developed no suitable place was found.

The old 'Anatomy' is still vividly in my mind. It was housed in a secularized monastery in Katharinenstrasse, a narrow lane of mediaeval appearance. Enclosed by high brick walls were lawns with tremendous old trees, and old monastic buildings with narrow but high windows and gigantic steep roofs of red tiles. It was just the place to impress a child's fantasy with pictures of mystery and horror. For somewhere in the cellars were stored the corpses—which we never saw. And higher up, in the Gothic refectory, was an enormous collection of skeletons which we were allowed to inspect and admire; the most conspicuous specimens being of mammoths and a gigantic whale. Then, ascending a winding staircase, one came to father's laboratories, with their characteristic zoological smell and innumerable bottles, microscopes, microtomes, books and papers scattered around. And right at the end there was my father's small, cosy, private study where he used to write and think, a monk's cell looking out over the old roofs and towers of the city.

It was all very romantic, but hardly suitable for scientific research. And my father was glad when the new institute, fitted with all the modern appliances, was finished. It was far away, and we could not easily visit him there, as we did in the old building. Before the removal I used to fetch him in the evening, and sometimes we did not go straight home but entered one of the innumerable beer-gardens on the Promenade where one could sit under old chestnut trees and consumed a good simple supper, scrambled eggs or Vienna sausages (a kind of Frankfurter). These peaceful evenings were among the greatest pleasures of my childhood. They ended when the old institute was demolished.

III. High School

In my youth German schools were mere places for learning, in an abstract and dull manner; there was hardly any attempt at a general education, nor any community life. From 8 a.m. to 1 p.m., five long hours, we were filled with 'knowledge', and two hours weekly in the afternoon were spent on '*Turnen*', the conventional German form of physical exercise. In the summer months one of the '*Turnen*' hours was extended to half an afternoon given over to games, football for the older boys, and a peculiar game, '*Barlauf*', for the younger ones. But the playing fields were in a distant suburb, and it was easy to be released from participation on the excuse that too much time was lost on the journey. We had no personal contact with our teachers, and I, being a bad mixer, had hardly any contact with my classmates. In fact, during all these twelve years at school, I had not a single friend among the boys with whom I shared the classroom day by day. Only during the last two years a boy named Straube, a nice quiet fellow, attached himself to me and was introduced to our house, but I think that my pretty sister and cousins had something to do with it. The explanation for this detachment was to a large degree my own taste and preoccupation. I had quite a number of friends outside the school, my cousins and children of families of my father's friends, and I felt no need for another friendship. After the failure of my approach, already described, to the inseparables, P. and K., I became still more reserved, in particular towards the Christian boys. The Jewish boys, on the other hand, were mostly sons of small shopkeepers, of a not very attractive type; they were orthodox, kept the Jewish laws (on 'Kosher' food for example) and holidays and attended the synagogue. There was a class of Jewish religious instruction at the school, which for the first two or three years was obligatory, later optional. I was not interested in it and gave it up as soon as possible. On Saturdays the Jewish boys did not come to school as they had to go to the temple. As not much teaching occurred on these days anyhow, I asked my father whether I might not stay away too. He decided: 'Well, you need not go to school, but then you have to go to the service in the Synagogue.' This I tried, but only once. It was so dull and strange that I preferred school. This decision produced a kind of barrier between me and the other Jewish boys. However, you may rightly say I was a prig.

The teaching itself was determined by the humanistic tradition. Latin was the central subject; at least one hour's lesson every day, and on some days even two. If I add up those many hours spent in the classroom and with homework and compare them with the resulting knowledge of the Latin language, I cannot help doubting the efficiency of this kind of education. Only a few years after leaving school I was unable to read Latin, and today I can hardly decipher a sentence of Julius Caesar even with the help of a dictionary. The same holds for Greek, with the difference that the Latin literature of Cornelius Nepos, Caesar, Cicero and Horace, as presented to us at school, seemed to me rather dull and uninteresting, while I enjoyed Homer and the Greek dramatists. I am a bad linguist in general, as my memory is weak, yet the enchantment produced by the hexameters of the *Odyssey* and *Iliad* was strong enough to make me one of the best Greek scholars in my form. In 'Prima' I could read Homer fluently. Yet even this was no lasting acquisition. Five years later, after my Ph.D. graduation, I went to Greece for a holiday and took a little volume of Homer with me in order to read it on the classical sites. The first attempt, however, in Corfu, was a hopeless failure; apart from some ten or fifteen verses, which I still knew by heart, I could hardly understand a line. So after I arrived at Athens I bought a German translation and enjoyed it thoroughly.

Is it worth while during the most receptive years to spend so much time on these languages if so little remains? If the result of education is the background which remains when all school knowledge is forgotten, I had my proper share: the classical world is not a strange world; it is like the half-forgotten world of childhood which I am trying to conjure up on these pages, and later when I travelled in Italy and Greece I felt as if I were coming home. Yet is it necessary to spend so much time and labour on grammar and syntax? Would not a more superficial knowledge of the classical tongues, combined with more extensive reading of good translations and of modern historical books be more efficient and save time for other important things? Scope could be found for matters relating to our modern life: science and medicine, the elements of law and sociology, economics, agricultural and industrial production, and general history of all nations.

What we learned of history, apart from the usual representation of Greece and Rome, was centred around the 'Holy Roman Empire of the German Nation'. There were long lists of emperors and battles, but very little of the life of the people and their cultural development. Within this framework there was a much inflated history of Prussia, beginning with endless lists of the Brandenburg Electors; then still within this framework much local Silesian history, with lists of the Dukes of Breslau and Lignitz. That these 'Piasten' princes were Poles was hardly mentioned, as the whole historical instruction was extremely nationalistic. The degree of this colouring depended of course on the personality of the teacher, and there were great differences. We had, as I have said before, some very liberal masters whose nationalism was of that amiable romantic type which found expression in the democrats of 1848 and the Paulskirche national assembly. On the other hand, our headmaster, Director Eckhart, a tiny neat man with a white beard, who taught us modern history in

'Prima', was the typical Prussian civil servant. At the heart of modern history lay the supremacy of Prussian state-craft and politics, as exemplified by the Great Elector, Frederick the Great, the heroes of the wars of liberation against Napoleon and, of course, Bismarck. But to do him justice, all that was his honest conviction and was professed without malice against other peoples and races. He was perhaps rather narrow and strict, but was also just and kind. Anti-semitism was not allowed under his regime.

Modern languages were poorly taught at our school. We had some French, but only learned reading, without being able to speak. There was also an optional course in English which I took for two years; but as it was in the afternoon and not particularly interesting, I gave it up before acquiring a proper knowledge of the language. How little did I foresee at that time the catastrophic events which would make it necessary for me to write in English in order that my grandchildren might understand my words.

Mathematics was, according to the humanistic tradition, a main subject, a part of classical education. I was not bad at it, but not deeply interested; in the lower forms, where the syllabus was dominated by Euclid, I was even bored. We used not the original, as is still done in some British schools, but text-books adapted for school use. I think that in this way the little attraction which this kind of mathematics might have for a boy by reason of its antiquity and classical form is lost, and by and large I cannot see much educational value in the matter. The claim that the power of abstract thinking is trained by the study of Euclid's rigorous proofs seems to me an absurdity, a remnant of a time when the psychology of children was unknown and a child was considered a grown-up in miniature. With rare and altogether unimportant exceptions, no boy under fifteen will ever grasp the idea of a logical structure like geometry, or of rigour in demonstration. At best he will play with theorems and problems as he plays with crossword puzzles. But in general he will simply learn theorems and proofs by heart as he learns the vocabulary of a foreign language. Accustomed to learn classical languages which obviously are of little practical use, he will not object to learning theorems which have no more practical use. But to expect that this learning improves the capacity of thinking in the brain of an average boy seems to me unjustified. I understood the meaning of Euclid's work as an advanced student in Göttingen when David Hilbert introduced us to the general idea of proof theory and his modernized form of an axiomatized Euclidean system. The misuse of Euclid in schools has the same effect as the learning by heart and dissecting of poetry. I hated Euclid almost as much as I hated Schiller, whose dramas were maltreated line by line in endless dull lessons.

However, when in the higher forms analytical geometry was introduced, I became rather interested. Here was apparently a general and powerful method for dealing with geometrical things, reducing figures and shapes to algebraic and even arithmetical relations which appealed to my taste—although at that time I could hardly have said what the attraction of the thing was. Anyhow, I became quite a good mathematician in Prima. This was also due to my having a good teacher, the only one with whom I came into personal contact and of whom I have preserved a grateful memory. This was Dr Maschke, a small, sturdy man with a kindly round face and a big moustache. His way of teaching mathematics was different from that of all my previous teachers and was quite modern. He did not state theorems and give demonstrations, but he told us something about the meaning of the whole theory to which a theorem belonged, about its history and its practical applications. And he was an enthusiast who loved his subject, as well as mathematics and physics. We had two hours weekly for natural sciences, which consisted of a little of everything: botany, zoology, chemistry and physics. Maschke's physics in Prima was the only lesson to which I looked forward and in which I took a lively part. He soon noticed it and invited me to help him prepare the experiments for the following day. So I spent many a happy afternoon with him in the physics classroom connecting and switching electric circuits or adjusting lenses on an optical bench. It happened at that time that Marconi's first experiments with wireless signals were becoming known, and Maschke decided to reproduce them. We built a coherer and were extremely proud and happy when we succeeded in transmitting Morse signals from one room to the other. I was sent to fetch Director Eckhart and saw with pleasure on the face of this extreme exponent of classical education the expression of amusement and unwilling admiration for this triumph of science.

Apart from these first adventures in physics there was hardly anything of interest for me at school. But there was much boredom as well as disagreeable excitement. These were caused by the weekly or fortnightly papers which we had to write in class. I never felt at ease with tasks which had to be finished in a given time, an hour or two. Even when I had spent a lot of time and trouble on learning—which I did not always do—I was never sure that my memory would not fail me in the classroom, or that I would not commit some terrible mistake in grammar or syntax or arithmetic. Sitting there before a blank sheet of paper, trying to concentrate on the task, feeling the minutes pass by, the meaning of a word or the construction of a sentence apparently wholly mysterious (although with a conviction that it was really somewhere in my brain and only yesterday could easily have been reproduced)—that was a kind of torture which I can still vividly recollect and which appeared many times in my dreams years after leaving school. I felt that no effort of mine would ever make me write brilliant papers and bring me to the top of the class where P. and K. resided, only temporarily attacked and even defeated by two Jewish boys, Masur and Dienstfertig. So I gave up trying and was content to be just average. And as my father did not object I was quite happy in this position.

Once or twice a year the whole class, under the guidance of some masters, went on a whole day excursion. In the lower forms this was quite pleasant. We travelled by train, railway or steamer to one of the lovely spots surrounding the city. There was Sybillenort, a castle and park belonging to the King of Saxony, but open to the public; the Oder forests; the Zobten hill rising out of the Silesian plain; the Trebnitz range of wooded hills, and many others. After some two or three hours' walk we settled down in an open air restaurant or beer-garden for a frugal meal; then we played games or sang folk-songs.

But in the higher forms these excursions became less and less attractive. There was a group of boys who were keen on drinking considerable quantities of beer in imitation of the University students, with irritating and sometimes catastrophic consequences. In order to avoid this, the masters who were responsible for us arranged the excursion in such a way that its main feature was a walk of six, eight, or even more hours, so that little time was left for sitting in a beer-garden. Walking in the morning hours through the lovely valleys and forests of the Sudeten hills was most refreshing and exhilarating, but the heat increased, the numbers on the milestones did not seem to increase correspondingly; most of us were not trained for marching long distances, and in the end we were completely tired out. I remember I once arrived home in such a state of exhaustion that my father wrote a letter to the director complaining of this unreasonable treatment. It was certainly no pleasure for many of us. Yet what little personal relation I had with my comrades came from these excursions. Once or twice the contact of a whole day spent together with nature produced the flaring up of a friendship, but in no case did it withstand the routine of everyday school life.

My real life was outside the school, and so were my real friendships. Apart from my cousins Hans Schäfer and Reinhard Jacobi, there was another second cousin, who became my dearest friend for many years, although he was two years younger and went to another school. His name was Franz Mugdan.

My grandmother Kauffmann had three nieces, named Rosenthal, who married three brothers Mugdan. One of these couples, Hugo and Käthe Mugdan, were the parents of my friend Franz. Hugo died young and left his wife with four children and very little money. They lived quite near us, only three minutes' walk from our house, Zimmerstrasse 5, and we became great friends. My sister Käthe and Bertha Mugdan were of almost equal age, and they are still great friends today. The two youngest Mugdans were twins, Albrecht and Susi. My friend Franz was a most gifted and attractive fellow. He was, unlike myself, always at the top of his form at school, and he grasped everything so easily that he was never seen to do homework. He was good at all sports, much stronger and more enduring than I, in spite of being almost two years younger. He became a leader in our wild games in the Kleinburg gardens; for my cousin Hans Schäfer, who was the eldest of the Kauffmann grandchildren, suffered from an infirmity—he had fallen from a horse in Tannhausen when a small boy, and shattered his left arm, which remained rather stiff. Therefore in all games consisting of two opposing parties the chiefs were mostly 'Franz and Max'. Later when we began to be interested in real sport it was again Franz Mugdan who transformed the slow peaceful 'Victorian' tennis into something approaching the real thing, with hard hitting and running.

In the winter months he was my regular playmate at home, and we had a marvellous time. Our parents provided us with well assorted, rather expensive boxes of bricks, 'Anker-Stein-Baukasten', which were produced by a firm in continuous series, each new box supplementing the previous ones, and they came with well printed books containing drawings and cross-sections of awful

33

houses, castles, churches, etc. Our elders considered the construction of these edifices according to the exact plans as something of high educational value; for if one made one little mistake at a lower cross-section and used a wrong brick, one could get into trouble later and be unable to finish the structure.

We were soon fed up with this kind of accuracy test and began to follow our fancy; but the results were, from the aesthetic standpoint, even more hideous than the printed models and provoked ironic remarks from parents, aunts and uncles. Hence I decided to invent something new and wonderful, and I succeeded. It was a way of constructing bridges of a wide span, more than a yard, out of bricks, the longest of which was no more than 3 inches. I used no glue, only the principle of the vault, put into effect with the help of small, blue, wedge-shaped bricks which were really intended for roofs. These bridges were so elegant and surprising that they evoked even an admiring remark from my critical grandfather Kauffmann. I still remember the exciting moment when the last wedge was being fixed into the top of the polygonal arch resting temporarily on a pile of books, the books were removed: would the bridge stand, would our purely intuitive calculation of the forces be correct? I think that this was the first time I had that kind of scientific adventure which consists in bringing a rational construction to the test by experiment, and it gave me the first idea of the meaning of definite rules of nature.

We also used our bricks to build whole towns of miniature size, for our tin soldiers. We had legions of those small figures, about an inch high and quite thin, but cheap and therefore obtainable in considerable numbers. We used to reconstruct the stories told in our history books. There were always two cities, one for each of us, and two armies. The armies were not fighting all the time but were employed in peaceful occupations constructing roads and houses, in the manner of the good rulers described in our books. One great invention made this peaceful life more attractive—street lighting by gas. At this time the use of acetylene for bicycle lamps became popular, and we at once constructed gas factories with the help of an inverted tumbler filled with water and some pieces of calcium carbide. The gas produced was carried by an elaborate system of glass pipes with rubber tube connections along the streets of our miniature towns to be burned in little glass jets. The result was a marvellous illumination, accompanied by small accidental explosions which increased our pleasure. My stepmother objected strongly because of the penetrating smell of the acetylene, but my father decided that we should be allowed to continue—I think he thought it a good training for handling simple apparatus, and I am sure he enjoyed the idea and the spectacle. Yet the thing came to an abrupt end. For one day after having finished playing we were too lazy to take the remnants of the calcium carbide to the dustbin in the court and threw them out of the window. Half an hour later the bell of our flat rang furiously and the lady of the flat below appeared, greatly excited, and complained that a terrible smell was coming in through one of her windows just below our nursery. It appeared that they had a flower-box in front of the window, on which our calcium carbide had fallen, and when the flowers were watered a fair amount of acetylene developed which penetrated their rooms. This incident caused the victory of Mama's opinion about this nasty game over Papa's milder views, and street lighting was prohibited.

As in the real world, the peaceful technical developments of our tin-soldier states and cities were interrupted by wars and violence, which happened regularly about half an hour before bed-time, leaving a scene of devastation. As Mama insisted on our packing all the toys and bricks neatly in the boxes after the game, we found that, the pleasure of the fight apart, the resulting chaos spared us the heart-breaking task of pulling down our neat cities and castles systematically. For these battles, we had little guns with springs from which peas or shot could be fired; the noise of battle being supplied by our own voices. After a while we found this not sufficiently realistic. We succeeded in purchasing some gunpowder, but our first attempt to use it produced such a violent explosion that it was strictly forbidden. Then I invented another method. The barrels of our little guns were replaced by pieces of glass tube, sealed at one end. Into this tube the top of a match was inserted, and a shot ball was placed in front of it. The closed end of the tube was heated with a short piece of candle and—bang!—off went the bullet, with an explosion just loud enough to give a tinge of realism to our battles without unduly attracting the attention of the grown-ups.

My father did not interfere with these militaristic games until one day—I must then have been between thirteen and fourteen years of age—I boasted at supper about the crushing defeat I had inflicted on Franz's army. Then he said: 'Shall I tell you about my own experiences in the war of 1870–71?' My sister and I, of course, were delighted at this suggestion, expecting to hear exciting stories about the battles of Wörth or Sedan and the siege of Paris. However, it came differently. Father told us how, at the outbreak of the war, he was a student of medicine and, carried by the wave of patriotism, tried to join the army as a volunteer, very much against the will of his father, who as a democrat of 1848 was deeply suspicious of the wars inaugurated by Bismarck. He was accepted as a private into the army medical service and had to serve as stretcher-bearer and medical attendant. He did not tell us about battles and glory, but about trains crowded with soldiers who had left their personalities behind and become numbers in a great machine; about field ambulance stations where wounded and maimed men were brought in and operated on; about field hospitals full of suffering, death and horror; he spoke about the fate of the civilians in the battle zone, whose houses were burning, and about the famine in the beleaguered city of Paris. He was a good narrator. He did not use resounding phrases, nor did he complain; he gave us simple facts, a glimpse of the reality of war, without the romantic disguise and the patriotic sentiment with which we were taught history at school and from our books. He did not mention his personal discomfort and sufferings, but he spoke about the loss of his dearest friend, Mitscher, who was an officer in the army and was killed in action. The point he stressed with a suppressed passion was the degradation of the human personality to a cog in a wheel of the war machine, deprived of will and dignity.

I do not think that the impression produced by this narrative—and in others which followed at intervals by my father and Uncle Joseph—was very noticeable. We continued our games, imitating what we thought to be the life and struggle of peoples and states. But I preferred more and more the 'civil' developments, much to the disgust of cousin Franz, who had not heard my father's tales and, being two years younger, had still more of the fighting spirit of the healthy boy. So quarrels ensued, and the whole kind of game petered out. Finally there was a day when, on a walk in the park, I informed my younger comrade of what I had heard from my father, and although my eloquence was certainly not up to the task, impressed this passionate boy to such a degree that he became from that day a fierce hater of everything connected with militarism and war.

Here I shall say a few words about Franz's later fate and tragic end. He was one year ahead of his contemporaries at school, so that he left school not two, but only one year after me. And it seemed a matter of course that he should take up mathematics as I had done. He grasped things with incredible ease and bewildered me very often by knowing propositions and whole theories which I had hardly heard of. For instance, as early as his second semester he took a course on the theory of numbers and overwhelmed me by telling me the beautiful theorems about congruences of which I had no idea. But then, on the other hand, he was astonished when I told him that I had found a proof of one of these theorems and it turned out that it differed from that in his textbook—Dirichlet, or whatever it was. He knew every proposition or demonstration which he had read in a book, and understood it; but he hardly ever found a modification of a proof or a new fact. He was highly gifted but purely receptive while I had, from a rather early stage, the urge to assimilate ideas by reconstructing them in a new shape, and to find new combinations. That difference was already apparent in our games; the initiative was mostly mine. That seemed natural in view of the difference in age, but later it turned out to be more than that. And Franz was clever enough to discover it very soon. After two years of mathematics he suddenly dropped it and joined the medical faculty, much to the bewilderment of his relatives, who knew his splendid reports as a mathematics student. He studied mostly in Freiburg im Breisgau and specialized in mental diseases. When I was studying in Zurich in 1902, he came very often to Switzerland to spend a weekend with me mountain-climbing, and we went on long and strenuous tours during the holidays. There he was in his element, very much superior to me in endurance and strength, but always patient with my poor performance in climbing and in suffering the harshness of the weather. I owe to him much of my physical development, and the energy to overcome my asthmatic handicap. On these delightful though strenuous trips we discussed all the problems of the world in the most free and confidential manner.

He married a fellow student soon after their common graduation and qualification as doctors, and after some years of assistantship in Freiburg he settled in Neckargemünd, near Heidelberg, as the chief of a small sanatorium for mental diseases. During this period I saw him seldom, and we never corresponded. Then the war broke out. He did not serve in the army and therefore stayed at home. Once, in the middle of the war, he appeared at my house in Berlin, a changed man; profoundly embittered and disgusted by the slaughter in France and Russia. His younger brother, Albrecht, one of the twins, had been killed in action. But it was not so much this personal grief, though it had struck him hard enough, but the madness of humanity as a whole which had afflicted this doctor of the mind with a hopeless depression. I never saw him again. During the dreadful influenza epidemics in the last year of the war he worked to the limits of his capacity and did not stop when he caught the infection himself. He still went out to see cases although he had a high temperature himself and one day he was brought back unconscious. He died a few days later, and I am sure that this heroic end was more a suicide of desperation than a sacrifice for the benefit of others.

When the tin soldiers receded into the background, the interest in all kinds of mechanical devices remained. Once, it must have been in my fourteenth or fifteenth year, I was down with scarlet fever, then (and, I think, still today) a rather serious illness involving total isolation and a long period of convalescence. At this time I brought my skill with the fret-saw to a high degree of perfection. I remember that I produced a yellow mailcoach of the type which was still used on the roads in the mountain valleys, and a railway dining car fully equipped with tables, chairs and kitchens. I was very sad when all that had to be destroyed because of possible infection. Later I made a lot of similar things, but became more and more interested in constructing simple machinery and physical apparatus, stimulated by my schoolmaster Baschke. I received much advice from one of my father's assistants, Dr Lachmann, who persuaded me that I ought to have a real lathe for metal-work, and this idea became an ardent desire. Yet a lathe is a rather expensive gift even for a grandfather Kauffmann. So I set to work to get it by producing a proper impression of my inventiveness and skill.

Now grandfather's greatest concern, apart from his business, was the Breslau Symphony Orchestra whose conductor, Raphael Masckowsky, a Pole, was a great friend of his. So I built a model of the orchestra. The musicians sitting on the inclined platform were made of sheet metal, dressed realistically (with my sister's help) in black suits and white shirts stuck on with glue; their arms holding the fiddle bows or wind instruments were movable and connected by strings to a cranked shaft under the platform, as was also the right arm of the conductor holding the baton. If you turned a handle on the back the whole orchestra moved in perfect unison. The faces of the ordinary musicians were painted by my Aunt Gertrud Schäfer, who was well trained and quite gifted in drawing and painting. But the faces of the conductor, the leader and the solo-quartet were real portraits cut out of postcards which could be bought at a music shop. On grandfather's birthday (which by the way coincided with my own, 11 December) the toy orchestra was presented to him; my sister turned the handle while my cousin Helene Schäfer and I played a movement of a simple Haydn symphony à quatre mains. The effect was tremendous. Grandfather was moved to tears, and at Christmas, a fortnight later, there was the desired lathe standing under the tree.

From that time my room looked more like a mechanic's workshop than the study of a classical scholar. But I cannot say that the lathe produced many objects. It arrived too late, for it was a period when school demanded a great effort and almost all my time. I learned quite a lot about handling metal, making shafts and screws, etc., and I built some little instruments for Maschke's physics lessons. A big project, the construction of a little steam-driven railway engine, came very near perfection, but not quite; the torso of this engine decorated my room for many years. When I was already a student at the University and took a course on astronomy I built a model of a telescope in parallactic setting with all its possible errors, inclination of axis, eccentricities, etc.; it was surrounded by a set of circular wires showing the spherical triangles and their deformation when the telescope was moved. My professor, Franz was his name, found it very useful and accepted it for his collection of instruments. Maybe it is still there.

This technical interest absorbed so much of my time that my standard at school began to decline. There was an important examination, 'das Einjährige', at the end of 'lower secunda' (three years before the finals), the passing of which secured the right to serve only one year in the army instead of two years in the infantry or three years in the cavalry and artillery. I passed this all right, yet without any distinction; but afterwards my position slowly deteriorated towards the lower half of the class. My father was worried and decided to get me a tutor to do the homework with me and try to deflect my mind from its too exclusive occupation with mechanics and science to other spheres of the intellect.

The name of this tutor was Hans Lewinsky, and he was the son of a Jewish lawyer's widow who lived in very restricted circumstances not far from us. He was only two years my senior, in the top form (Ober-Prima) of another school, the Johanneum. He was a quiet, earnest but friendly fellow who had an altogether good influence on my character, which was inclined to a kind of narrow-mindedness, through concentration on one single line of thought or occupation. His help improved my position at school considerably and even succeeded in imbuing me with a slight interest in the classical subjects and in German literature by reading history and novels with me. But he was also essentially interested in science, and after having got his leaving certificate (Abitur) he began to study medicine. My father was his teacher in anatomy, and in this way I learned for the first time something about this side of father's activities. Hans Lewinsky was perfectly devoted to him and declared that he was by far the most interesting and stimulating of all his professors.

Lewinsky joined one of the students' societies, the 'Literarisches Verein', which was recruited almost exclusively from Jewish students. At that time the liberal traditions of the German students were already completely lost. There were several types of fraternity or society: the conservative and aristocratic 'Corps', which from the beginning had never admitted Jews, nor for that matter sons of artisans or small shopkeepers of any confession. Then there were the Burschenschaften—once, during the period of the 1848 revolution, in the forefront of the fight for liberty and democracy, but now degenerated and just as reactionary and anti-semitic as the 'Corps'. There were also gymnastic societies and singing societies, a big Catholic students' society, and finally a great number of scientific and literary societies. Some of the latter admitted Jews, while some were practically exclusively Jewish. The custom of duelling was quite general. The 'Corps' and *Burschenschaften* practised it as a kind of sport. They had what were called *Bestimmungs-Mensuren*; the members of two of these societies met at regular intervals in a back room of an inn and one man after another had to fight a man of the other party. Officially this was forbidden by the University authorities, but it went on without any inter- ference as long as no serious accident occurred. This was extremely rare, as only parts of the heads were exposed, the eyes being protected by strong wire goggles and the arteries of the neck by bandages. There was also a 'doctor' present, a senior student of surgery. The weapon was a sharp rapier (foil) which was used according to strict rules: No part of the body was allowed to move other than the right arm wielding the foil. It was rather a stupid and graceless kind of sport. But a lot of blood was shed since the fight did not stop when the first wound was inflicted. It was extremely painful to get a second hit just across a previous wound. But that was not what the beginner feared most. It was the infamy connected with the slightest wincing or moving of the head when the sharp blade flashed down. Well, this was considered education for heroism by the young Teutons. Apart from these sporting duels there were many duels provoked by 'insult', usually in the course of some squabble or brawl after the consumption of beer, or perhaps a little collision on the street, followed by a remark like 'stupid fool'. In these cases all the formalities of real duels were observed, with seconds and attempts at reconciliation. The conditions were often more severe and occasionally they led to severe and even fatal casualties. Most of the students' societies adhered to a ridiculous code of honour which was an imitation of that of the army officer corps. Only a few scientific societies (such as the mathematical one which I joined later) refused to 'give satisfaction'. There was also a bellicose Society of Jewish Students who made a particular job of being excellent fencers and of avenging any anti-semitic remark with blood.

Lewinsky was a peaceful soul, but nevertheless he once got involved in a brawl which led to a challenge and, according to the rules of his literary society, he had to fight. I still remember the excitement at the time; for he confided the whole story to me, the progress of his training and all the mysteries of the 'affair of honour'. One day he appeared heavily bandaged and hardly able to speak, but proud that he had not flinched. My father was very angry as he hated this brutal sport. But the 'mentor' was not dismissed and later exhibited with pride some 'nice' scars on his kindly face. When I became a student he tried to persuade me to join his literary society, but I obstinately refused. I was not much interested in literature, considered the fellows of this society to be snobbish and disliked altogether the customs of German students, their drinking, singing, fencing and their idiotic code of honour; this dislike was even more enhanced by the fact that these Jewish boys were imitating the

39

very people who refused to acknowledge them as their equals.

Lewinsky became a physician in Mainz. I met him only occasionally later and finally lost all contact with him. He succeeded in getting out of Germany and came to England. But there he committed suicide. He was always a little strange, depressed, pessimistic. If I had known he was in England I might have been able to help him; but now it is too late. One more on the Nazi score.

Meanwhile my sister had developed from child to teenager. These creatures have a strong tendency to band together in sentimental friendships, and soon our house was lively with girls, chattering, singing, playing, dancing. A boy of the same age, between fifteen and seventeen, is a nobody when young women dominate the scene. There were our cousins Helene Schäfer, Alice and Claire Kauffmann, and many of their school friends. My stepmother had taken on a Fräulein who was at the same time housekeeper and my sister's governess. This Fräulein Beissenherz was a nice, kindly, fair young woman who had two brothers and innumerable (I think six) sisters of all ages. Two or three of these who were of the appropriate age, also belonged to my sister's set—pretty fair-haired girls who liked to visit us as their home was not only poor but rather depressing. For old Beissenherz, a schoolmaster who had been dismissed because of rather sordid conduct, had become extremely pious and forced on his family a rigorous 'religious' life in which praying and preaching were abundant, but dancing and theatre forbidden. The effect was that both sons got into bad company, and were involved in shady affairs and disappeared; and—as far as I know—only one of the girls married. Yet our own Fräulein Beissenherz was a good and faithful person who was devoted to my stepmother and continued to visit her when the old lady was living alone in Breslau, even after the Nazis had made such intercourse dangerous.

Then there was the Jänicke family. Mother Jänicke, Betty, was a rather stout woman with infinite energy and a still lovely face, who had been one of the (rather numerous) loves of my father before his marriage; he only narrowly escaped marrying her. Dr Jänicke, her husband, was a town councillor of Breslau who later become deputy mayor of the city, a pure Aryan of rather noble blood, related to Silesian aristocracy, the Seidlitz's and others. He was a beautiful man with a beard, very dignified and in complete contrast to his small, stout, black-haired spouse. But they were an exceedingly handsome couple, and their children inherited the noble features of both. There were four of them: a son Wolfgang, of my age, and three daughters. The eldest, Eva, was a real beauty, for many years 'the Beauty' of Breslau society, and the youngest, Käthe, also very lovely. Eleonore, of my sister's age, was no beauty, but graceful and attractive—at least so she appeared to me. For she was my first love, just as Wolfgang Jänicke was my sister's first love. That is why I have described this family in some detail.

My romance began during a holiday in the Riesengebirge, the highest part of the Sudeten range. When my father married for the second time, our summer paradise in Tannhausen came to an end. My stepmother refused to join the Kauffmann family, which had not received her with particular kindness. Her own family, the Lipsteins, owned a summer house at Cranz, a seaside resort near Königsberg in East Prussia. There we went two or three times for the summer vacations. There was, of course, violent objection from grandmother Kauffmann who remembered the débâcle of Norderney (described above) and prophesied the most terrible consequences for my health from the sea air. Yet we went, and I got no asthma. The Baltic climate is much milder than that of the North Sea, and it did me a lot of good. We were very much spoiled and pampered by the new relatives, and we enjoyed it in spite of some homesickness for the woods of the Weistritz valley. A year later, Mama decided to take us to Brückenberg, near Krummhübel, in the Riesengebirge which was the favourite summer resort of the Breslauers. The Jacobis and the Jänickes used to go there every year. But that year was a dreadful one; it rained almost continuously, until the mountain streams became torrents, houses and bridges being torn to pieces. In fact, we were isolated for several days from the outer world and pioneers had to be called from the nearest garrison to build an emergency bridge over the raging Lomnitz river. So we had to spend our days mostly in our little hotel or in the farmhouse where our friends lived. We children did not mind, for there were lots of friends to play with, and amongst them the Jänickes. It was then that Lore and I became great friends. We were children and loved one another as children do. We had great fun and enjoyed the indoor games during the rainy days as much as the walking and climbing on the hills, when at last the sun broke through the clouds. But Mama did not enjoy this holiday at all, rain and noisy children and mother Jänicke all day long. She never repeated the experiment, and though she was most lenient to all our wishes, on this point she was adamant. The following year we went to Cossensass in the Tyrol, which was certainly grander and more beautiful than Krummhübel. My health improved in the mountain air, and we had a good time. Yet we missed the gay company of the previous year.

At home, in Breslau, I did not see much of Lore. There was much visiting between her and my sister, but boys and girls of fourteen or fifteen generally kept apart, and just because I was attracted I was shy to show it. But slowly our friendship developed, and when I entered Prima we used to meet, quite 'accidentally' of course, on our way to music lessons or similar occasions. At that time girls were still subjected to strict rules of behaviour, and it was not easy for them to meet young men. The best opportunity was the ice-rink, as I have mentioned before, and so we looked forward to the Christmas holidays when the ice on the Stadtgraben would be thick enough and we had plenty of time. The Christmas holidays of my last school years are still in my memory, a time of bliss and happiness.

Holidays really began before the school closed, with my birthday on 11 December. There were ample presents and surprises; such marvellous things as the first bicycle, an electric motor for my experiments, plenty of books and sweets. Then we had a party, and Lore came with her elder sister. We also had to visit grandfather Kauffmann, whose birthday, as I have mentioned, coincided with my own—and also by the way, with that of my cousin, Alice Kauffmann. From that day on the preparations for Christmas began, the producing of little surprises for friends, parents and grandparents. We had a

Christmas tree and celebrated this festival in the ordinary German fashion, although we belonged to the Jewish community. I think my father considered it as the old Yule festival of the German people, and he wanted to be a German, though a member of a religious minority. We even sang the lovely Christmas carols, 'O Tannenbaum', and 'O, du selige, o du fröhliche, gnadenbringende Weihnachtszeit', without much thought of their relation to the Christian idea of redemption by Jesus's birth and suffering. We knew the Christian legends quite as well as the Jewish ones, we loved their poetry and were hardly aware of the full meaning of the events which led the Jews to deny the final and greatest fruit of their religious genius.

On 23 December the tree was decorated; we children had to help by covering nuts with thin silver or gold tinfoil, and fixing threads to sweets. But the hanging of these and other decorations upon the tree was done by the parents with Fräulein's help; we were in the adjoining room playing and excitedly discussing the chances of having our wishes fulfilled. The following day we spent the morning carrying to our friends and relatives the baskets containing our parents' and our own gifts, as well as presents for many poor people, former maids of ours and so on.

Early in the afternoon grandmother Born and Aunt Jettchen arrived and very often also Uncle Hermann, father's brother (who had settled in Beuthen, Upper Silesia, as a specialist in gynaecology). There was a big tea with plenty of cake, and then came the solemn moment when the door was opened and we stood gaping at the tree, shining with the light of many candles. Fräulein Beissenherz played 'O Tannenbaum' on the piano and we sang the old carol, and then we were led to the big tables covered with presents. Fortunately when Käthe and I were growing up there was still a baby in the family, which is indispensable for a real Christmas, namely our little brother Wolfgang, about ten years my junior.

At six o'clock grandmother Born, Jettchen and Hermann wrapped themselves in their winter cloaks and marched off to attend the celebration in the Jacobi family. The children there were always cross with us because they had to wait so much longer for their gifts. We also went out, to our Kauffmann grandparents, where we found all our uncles, aunts and cousins, besides numerous friends. That was a magnificent affair, but rather formal and stuffy. For instance, when electric lighting had appeared in Breslau in the 1890s, the wax candles on the tree were often replaced by electric bulbs, though they were a rather ugly substitute. There was a big and solemn dinner where we children had our own little table until Hans Schäfer and I became Primaner and were admitted to the table of the grown-ups. Very often we had prepared little surprises, such as the performance of a play written by Uncle Schäfer, or of a piece of music, like the Children's Symphony by Haydn, or a series of funny pictures and rhymes with allusions to events in the family circle. On the following morning, Christmas Day, we visited all our cousins and friends to admire their trees and presents.

And then we waited for the cold weather, which generally came between Christmas and the New Year, and for the opening of the ice-rink. That was where we had our greatest fun, especially when we reached the age when the other sex became attractive. For on the ice we could meet, boys and girls, without interference or supervision from the grown-ups. I was a good skater and enjoyed that graceful sport. But when I was practising a complicated figure in a corner to the rhythm of the music my eyes were wandering over the rink looking for a slender girl in a fur cap, and when I saw her sport was forgotten and flirtation began.

That was just in the last period of my school-days, and though it distracted my mind, it did also kindle my ambition and had no bad influence on my performance at school or elsewhere. One day my music master Auerbach said to me, after I had played the E-minor Prelude by Chopin: 'Max, you are in love! I can hear it in your playing.' He was, by the way, an expert on this subject. Mothers used to be reluctant to allow their daughters to have lessons with him without a chaperon.

The musical incident seems to me not without significance for an understanding of the kind of chap I really was at that time. The early death of our mother had deprived us of the natural outlet for our feelings, and neither Fräulein Weissenborn nor our stepmother was able to replace her. My sister was in that stage of transition between child and young lady which makes a confidential relationship with a brother somewhat difficult. So a kind of shell had developed around me, and nature had not given me the gift of expressing emotions in words. I would not say that I was deaf and mute in respect to poetry. For there are three or four serious poems written in my later life—all addressed to my wife—which are not bad at all; and, as a student, I was quite popular as a producer of funny rhymes for our festivities. But language was not my way of expressing my innermost feelings. Verses did not come to me as an inspiration, without effort, and I had the (mistaken) idea that all possibilities of language had been exhausted by endless generations of poets. Music is different. Even if you have no gift for composition you can put a morsel of your personality and of your mood into the playing—provided you have some technical skill.

It seems as if on the occasion mentioned this condition was sufficiently fulfilled. But altogether this was a weak point and remained so all my life. My father had a cousin who was a music teacher and rather poor, and he thought it more important to assist her by letting her teach his children, than to provide them with the best instruction available. We called her Tante Römer; she was a kind woman with a lovely face and though a good pianist, not a good teacher. She was not deeply musical, knew very little of the theory of harmony and counterpoint and of the historical background of the art, and she concentrated on drilling us in a superficial brilliance of technique, with the purpose of making a good show at the annual concert of her pupils before an audience of parents, relatives and friends. Her method was quite lost on my sister, who disliked practising scales and études, while I was more patient and sufficiently gifted to please her. I learned rather complicated pieces by Beethoven, Chopin, Weber, Mendelssohn, which I played with an apparent skill, but in quite an amateurish way, slowing down when at the difficult bars, hitting innumerable wrong notes and, worst of all, ignoring the style and the musical expression.

My Kauffmann grandparents, who knew more about music than my father, urged him constantly to provide better lessons. But only after my father's death, during the last half-year of my school-days, was I persuaded by them to look for another teacher. It was not easy to dismiss good Aunt Römer. My new master, Max Auerbach, was a member of a well-known family. One of his ancestors, Berthold Auerbach, had written many novels, mainly about village and peasant life; two of his brothers were professors, one a physicist in Jena, the other a chemist in Berlin. Max Auerbach himself was an excellent pianist and might have become famous if he had been more industrious and less involved in amorous adventure. He was an excellent teacher, and the little knowledge of music I have I owe mainly to him. I had to begin almost from scratch. He set about eradicating my false brilliance and replaced it by a solid technique based on Bach and Mozart. From him I learned the elements of harmony and the style of composition. If I did well he ended the lesson either by playing some four-handed piece, e.g. a transcription of a symphony, or by studying an opera, where he sang all the parts from bass to soprano, with great skill and enthusiasm. In this way he introduced me to Wagner's work, which for a long time fascinated me more than anything else.

I continued these lessons during the first year of my studies in Breslau, and again later when I returned to my home University for a winter semester. Altogether this systematic musical education did not last more than about three years. It gave me some foundations on which I later tried to build unaided.

After this digression, I return to the story of Lore. I think that she was very near to breaking my shell of shyness and reluctance, as her influence on my musical performance shows. But she did not succeed. If there is any nut-cracking to be done in love, the role of the sexes must be the opposite. The male is the natural aggressor. There was no lack of males who tried.

It was a bitter disappointment when I discovered that she preferred another boy to me, one who knew the female heart. I certainly did not. She herself later (not long ago) told me the following story. It had completely vanished from my own memory but nevertheless is likely to have been true. The day of my successful finals at school I met her in the street by the usual 'accident'. I was on my way to grandmother Born to introduce myself as a 'Mulus' (Mule), the term applied to those who were no longer schoolboys and not yet students. I told my girl-friend that grandmother had a parrot, also called Lore (a common name for parrots), adding: 'We call her Lore Born—doesn't it sound nice?' Thus far I went in disclosing my most secret wishes. How little I knew about the hearts of pretty girls as to compare them with parrots! I suppose that my conversation was often on that sort of level, and I do not wonder that she did not relish it. When I found this out it was a bitter experience and it helped to increase the thickness of the shell around me. But Lore and I always remained friends and I did not give up courting her when I met her again, even by letter. But these attempts ended abruptly, many years later. I was then no longer a bashful youngster but a learned doctor of philosophy, a rather elegant and worldly young man who had travelled widely over Europe and made friends with people of many nations, men and women. I was on my way to Greece—a journey already mentioned in these pages—and intended to embark on an Austrian boat at Trieste. When I learned that Mama Jänicke (who had since been widowed) was going for the Easter holidays to Venice with her two younger daughters, I decided to join them. So we travelled together over the Semmering and had some delightful days on the lagoons, the Piazza San Marco and the Lido. I fell in love again, but alas, there were two girls, my old friend Lore and her sister Käthe who had become a very sweet young lady—and I liked them both, and was not quite clear which of them liked me better. After two or three days of three-cornered flirtation I took the train to Trieste according to plan and sailed to the land of the Hellenes, where I roamed about and filled my soul with the beauty of the classical landscape and architecture. One day, sitting under the columns of Poseidon's temple at Cape Sounion and looking out over the incredibly lovely coast of Attica and the islands of the Aegean Sea, I recalled the pleasures and dangers of the journey, and I was content. Love which could waver between two sisters was not deeply rooted. My father's entanglement with their mother and his narrow escape had with some variations been repeated in his son's adventure with the daughter. His instinct was right, though I still could hardly define what the obstacle was. Anyhow, I was free and enjoyed it.

But my friendship with the two girls continued. Käthe, who had an excellent soprano voice, became a professional singer and later married an Aryan musician. The other two sisters, my friend Lore and her lovely sister Eva, both returned to the Old Testament by marriage and had to pay for it. Eva was first unhappily married to an Aryan and then divorced; later she married a Dr Cohn, who changed his name to Corneel. Lore also married a Dr Cohn, an ophthalmologist in Breslau, who was christened under the name of Colden. But this disguise did not protect them from the Nazis and both emigrated to England. Lore's fate was rather sad. While still in Germany she lost her eldest daughter, and a short time later her husband. Her son had to break off his study of architecture and worked in England for many years as an agricultural labourer; later he volunteered as a miner. Her daughter is a nurse in London. Lore herself joined the staff of Kurt Hahn's school, Gordonstoun, in Scotland where she taught French and German.

IV. My Father's Death. Breslau University

I was in my sixteenth year when my father fell ill. The first attack came when he was just arriving at the Jacobi's for a dinner party. Uncle Joseph saw at once that it was very serious and he told me so the following day. Though I did not know what angina pectoris meant I could see from father's face and his changed expression that he was very, very ill. From that day our whole life was under the shadow of his suffering. He recovered slowly and was able to take up part of his work. I am certain he knew his days were numbered and he was anxious to finish some of his research. He had just enjoyed the first acknowledgement and success. For his work on the development of the embryo he had been awarded the Gold Medal of the Senckenberg Institute in Frankfurt am Main, a highly prized distinction. The University of Breslau had appointed him honorary full professor, a title given in rare cases to members of the staff who merited a professorial chair though there was no vacancy. He was deeply immersed in some new research on the Corpus Luteum. I know very little about these biological problems but have learned from medical experts that this last investigation of his was of great importance. It was partly finished after his death by two of his collaborators (Dr Cohn and Dr Fränkel).

He still went to his laboratory once or even twice a day, using a taxi. But he seldom went out for any other purpose. I only remember one occasion: he insisted on attending a luncheon given every year by the director of the Breslau Zoological Gardens in its restaurant in the laying season for ostriches: 'scrambled ostrich eggs' was the main course. When I was in Prima I was allowed to join in these biological repasts. Father never went out in the evenings and we profited from that as he liked to chat with us in his unique way, mixing humour and earnestness, or to read to us aloud—if his breath allowed the exertion. During his last winter he read the whole of *Faust* to us; or rather he recited most of it by heart—a feat not uncommon in that generation. Minkowski, my mathematics teacher in Göttingen, was able to do the same. The last scene of the first part was father's favourite and I shall never forget his rendering of mad Gretchen in prison.

His friends often came to see him. A regular feature was the visit of Professor Karl Weigert, pathological anatomist in Frankfurt am Main, who

46

came every Christmas to stay with his brother, a tiny little man, of no gifts whatever, a clerk in a bookshop in Breslau. Boxing Day was reserved for them; and they came after supper for a glass of wine. Several times they brought another friend from Frankfurt with them, Paul Ehrlich, the founder of chemotherapy and discoverer of salvarsan. Other guests from Breslau were Haydnhain, the physiologist, Neisser, the dermatologist, and of course my Uncle Jacobi. I was allowed to listen to their discussions, which ranged from talking shop to politics and philosophy, but always culminated in most amusing anecdotes and funny stories—a competition in which Weigert was top. He had three versions of each story: in correct German, in Saxon dialect, and in Yiddish, and he produced them, as he said, according to his audience. In our circle he gave all three, which was the real fun. I think that these evenings were the last happy hours my father enjoyed.

He spoke to me about my future. At that time I wished to become an engineer. I made enquiries about the curricula and conditions of this course at different Institutes of Technology and discussed these with father. He then told me he himself had always regretted that his professional study had from the beginning completely absorbed his time to the detriment of all other interests, and he wished me not to make the same mistake. He advised me to spend my first year at the University attending lectures on all subjects which attracted me and not to choose my final study till after this probing. I followed his advice, with the result that I did not become an engineer, and I think this was the right choice.

In the early summer of these years my parents went to Bad Nauheim, a spa for diseases of the heart. During this time we children were put up at different relatives, and I stayed with the Mugdans, much enjoying the permanent company of my friend Franz. Later in August we joined our parents in our childhood paradise, the Weistritz valley.

In 1900 no such hospitality was offered to us, and it was decided that we children, in Fräulein Beissenherz's care, should spend our holiday in Glücksburg near Flensburg, Schleswig-Holstein—a rather strange choice considering the distance and the lack of particular attractions. A schoolfriend of my sister's had a friend in Flensburg; they persuaded their families to spend a summer together, and we were put in their charge. Our departure was fixed for July, but in the small hours of the night before, my father died from a heart attack.

I cannot describe the nightmare of the following days. But after the funeral we went to Glücksburg with Mama. She told me later that the doctors, especially Uncle Joseph, had repeatedly warned her that the end might come any day. She was therefore prepared and considered her main duty the care of us children. I have only a vague recollection of the time in Glücksburg. It seems to be covered by a veil. Yet I remember the view from our house over the lake, with the picturesque old castle in the background, and my participation in the first trip of a new liner built in Flensburg, and an excursion to the battlefield of Düpell.

A short time after our return my grandfather Kauffmann died too. He had been suffering from the common old man's disease of the bladder; but, as he had suffered from mild diabetes my father considered an operation a great risk and dissuaded him from it. But after father's death the old gentleman insisted on being rid of his inconveniences and suffering, and as the great surgeons shared my father's opinion and refused to operate, he went to a second-rank man who undertook the operation. He died a few days later. His wife survived him by a few years.

It was good for me that the period after father's death was my last half year at school, which kept me very busy with preparations for the examinations. During this time I came very much under the influence of one of my father's former assistants, Dr Lachmann, whom I mentioned in connection with my lathe. He was a gifted man of broad interests. My father had liked him very much and was deeply concerned when he showed the first signs of developing tuberculosis of the lung. Some years before the period I am describing now, one of my mother's cousins, Franz Kauffmann, son of Robert, also fell ill, with a very serious disease of the kidney, and was advised to go to Egypt. My father arranged that this Dr Lachmann should be chosen as his companion and medical adviser. After a year or two both came back, much improved. Dr Lachmann married and then emigrated to Los Angeles in California, because of its splendid climate. Yet he did not like life in America, and when he felt perfectly cured he returned to Europe. That was about the time of my father's death. Lachmann became a practitioner in a small Silesian spa, Bad Landeck, where he worked during the summer, while he spent the winter in Breslau, studying at the University. He was then deeply interested in astronomy and mathematics, and it was much as a result of his influence that I took up these subjects. But I owe to him not only the deciding direction of my scientific interest, but also a general widening of my outlook on life. He was a socialist, an active member of the Social Democratic Party, and he was not afraid of proffering views which, in our bourgeois society, were still considered despicable.

My father was, of course, not so narrow-minded and liked discussing these problems with Lachmann and defending his liberalism against the new revolutionary creed. But grandfather Kauffmann, though also a liberal man, would hardly have admitted a socialist into his house or considered his politics worth discussing; and the same held for his brothers. They were self-made businessmen; they believed in the right of the gifted individual to battle his way to a higher social level and to dominate economic life. They treated their employees and workers well and as far as I know none of their half dozen big factories ever had a serious strike. Their ladies patronized the old and the infirm in the villages and workmen's settlements; they visited the families, gave them presents at Christmas and helped in cases of emergency and disease. My grandmother Kauffmann was particularly active in this kind of charity. Months before Christmas she began to prepare parcels for hundreds of families, including not only employees of the works, but also her former house servants, grooms and coachmen. There was never a doubt in her mind that this state of affairs was the right one, the only possible one in a civilized community. Rich and poor, high and low: everybody according to his merits—that is, to his ability to earn money.

For myself, I had seen hundreds of times the army of working men and women stream in and out of the gates of the factories in Tannhausen and Wüste-Giersdorf. I had seen their simple dress, so different from our own elegance, their tired apathetic expressions, and I had taken these things to be natural phenomena, like rain and snow, thunder and lightning. When I began reading newspapers I discovered that these masses, the common people, had their socialist organization. The liberal journalists called it revolutionary and fought it with clever arguments. At school too the socialist movement was mentioned, always with such expressions of contempt and hostility as 'enemies of society', 'subversive forces directed against the state'. I well remember such outbursts from our director Eckhart, and also that I did not quite believe him. I had always a sceptical mind but I did not think it worth while to study the question seriously.

Dr Lachmann was the first person to encourage me to do so. He had read not only Marx and other socialist writers, but also Kant and Hegel. It was from him that I heard these names for the first time. He opened my mind to all the great problems of philosophy and sociology which for him were not academic questions but matters of practical life. Yet I did not at once become an ardent socialist. I also saw the other side, the liberal view of free competition, and I could not help comparing the clever little Dr Lachmann with my grandfather Salomon and his brother Julius, who had risen from humble beginnings to their present greatness by their efficiency and their commanding personalities. I think that socialism at that period lacked one deciding motivation which it has today: the practically infinite productive power based on scientific research. Nowadays a decent life for all is no utopia but a practical possibility. In 1900 this was hardly true and therefore socialism was based to a considerable degree on the motive of envy. Today a rational transformation of society seems possible, allowing room for the gifted personality, within the framework of a general prosperity for the average man. As this situation has developed I have moved in the direction of socialism, and I still hope to see a further development of it in England.

Another interest awakened in me by Dr Lachmann was astronomy and cosmogony. He insisted that I should include a course of astronomy in my first tentative academic year, and as that was not possible without mathematics I had also to take mathematics, although my experience of it at school gave me no relish for the prospect. My further programme for the first two semesters consisted of courses in experimental physics, chemistry, zoology, general philosophy and logic.

But before I could plunge into this wide field of learning I had to finish my classical education by passing the school leaving examination (*Abitur*). It is the only serious examination which I have ever taken in all my life—for the doctor's degree was mainly given on the merit of a thesis, the oral being rather undemanding and easy. If you compare this with the number of competitive examinations which a student in England has to pass after the leaving

49

certificate—one every term and a bigger one at the end of each year, with finals for the first degree and more for each higher degree—you will understand the difficulties which I had, and still have, in performing the duties of a British professor. I was accustomed to a less formal, more individualistic method of grading the merits of students, and I still think that it is by far the better method. Even our 'Abitur' was not as formal as a British exam; it consisted of a paper in each subject—not incoherent questions but an essay, a translation, a mathematical demonstration—and an oral in those subjects if the paper was below a certain standard. For the final results the whole standard of performance in the upper forms was also taken into account. I cannot remember the whole affair being very exciting; the results could almost be predicted from the class marks. Therefore nobody was astonished that I came through a little above the average. My Greek paper was the best of the whole form, to my own great surprise; but my Latin paper was a failure, so that I had to undergo an oral in Latin. P., K. and a few others were exempted from the oral.

It was half a year after father's death (Spring 1901) that Mama, in spite of her mourning, gave a little party in my honour and I danced very happily with Lore and was the hero of the day. Then she took Käthe and me on our first Italian journey. Her old mother and her youngest brother Alfred, who was only a few years older than I and a student of medicine, joined our party. We went via Munich and the Brenner to Verona, Venice, Milan and ended up at Lugano. Käthe and I enjoyed the adventure, as far as it could be called so, in view of the ever-present old ladies. Yet I do not think we were impressed by Italian art and landscape. In this point our education had been deficient. My grandfather Kauffmann loved Italy and visited it several times but he spoke about it in terms which were above our understanding. My father, as far as I know, had never been to Italy and never shown any interest in classical archaeology or the Renaissance. We had no scale for valuing a picture or the façade of a building. We found the Academy in Venice and the Museum in Milan rather boring and amused ourselves by counting the number of haloes of the saints in the pictures and attributing the highest prize to the greatest number. We enjoyed the Lido, the icecreams in the cafés, the gay crowds and military band on the Piazza San Marco, the romance of the gondolas and the serenades on the Grand Canal at night. In Lugano I fell ill with severe bronchitis, and while Käthe and Alfred climbed Monte San Salvatore and rowed on the lake, I was lying in bed, coughing and tormented by asthma. The journey was broken off and I was transported home as soon as my state of health allowed. This was the first of many experiences of a similar kind which spoiled my travels.

A few weeks after our return home I took my daily walk to the University. It was a lovely walk, and I shall never forget it. After passing one or two dull suburban streets and crossing the Stadtgraben and Promenade, I was in the oldest part of the city, full of picturesque buildings and romantic corners. There was a short cut through a narrow lane between high old houses, called 'Sich dich für' (= Sieh Dich vor—be careful, look out) a name suggestive of the dangers lurking in this dark passage in mediaeval times. Now it was quite safe and particularly attractive as there was an Italian sweet-shop in a corner where excellent ice cream and sorbet could be got very cheaply. Then I had to cross the Blücher-Platz with Schinkel's noble bronze monument of the old 'Marshal Vorwärts'; the north side of the 'Ring', formed by a series of stately high gabled Gothic houses; the church of St Elisabeth and several other narrow streets. And then I emerged in front of the University, a former Jesuit College in very rich baroque. It stretched along the Oder river and was pierced by a wide tunnel through which the main north to south thoroughfare of the city passed on over a bridge to the suburb on the right bank of the river. The main University building contained the beautiful baroque *Aula*, the administration and classrooms of the philosophical faculty (corresponding to the British arts and science faculties combined). Chemistry was housed in a modern building attached to the main site; but physics, zoology and botany were about ten minutes' walk away, on one of the islands of the river Oder, the Dominsel. In the interval between lectures a stream of students moved along the bank, over a bridge to the Kreuzinsel, passing the picturesque Kreuzkirche, and over another bridge into the quiet domain of the prince-bishop, full of lovely old buildings, from early Gothic (the cathedral) to late baroque (the bishop's palace). Then out of this sleepy mediaeval world of narrow lanes, palace fronts, old-fashioned gardens, fantastic corners with crucifixes and statues of holy martyrs, turning a corner you stepped into an open space with sober modern buildings dedicated to natural science.

The professor of physics at that time was old Victor Emil Meyer, well-known from his book on kinetic theory of gases, one of the first attempts to make the great ideas of Maxwell and Boltzmann accessible to the student. My first semester was his last; he was ill and hardly able to stand the stress of a one-hour lecture. He may have once been a good teacher but he was then a broken man, and the lectures were extremely dull. No experiment was ever successful. When he performed hydrodynamic experiments the water splashed about so much that the students soon began to appear in raincoats and galoshes and armed with umbrellas. This was quite good fun but I found I could not learn much more than I knew already from school, and I stayed away.

In my second semester he had retired and was replaced by Ernst Neumann, a young mathematician and a member of a family of scientists. His grandfather, Franz Neumann, was one of the founders of modern mineralogy and crystallography; his biography, written by his daughter, is worth reading, as it is not without romance and adventure. (He was the illegitimate child of a titled lady, brought up by her forester; he fought against Napoleon and was wounded at Ligny.) One of his sons, Karl Neumann, was a well-known mathematician, in my time professor in Leipzig. My new professor, Ernst, was a grandchild of the old Franz Neumann by another son. He was, as I said, a mathematician and why he was chosen as a substitute and later as a successor to the physicist Meyer, I do not know; it was perhaps assumed that the brilliant young scholar had inherited all the capabilities of his ancestors. He certainly took great trouble with the course, but being unused to experimental demonstration, this part of the performance was not very successful, and I did

51

not attend regularly. In my fourth and sixth semesters I took Neumann's courses in practical physics and some lectures in thermodynamics and on potential theory; these were very much better. Yet he himself must have felt that physics was not his proper subject, and when the chair of mathematics in Marburg was offered to him, he accepted it. He was always most kind and helpful to me, and many years later when I was professor in Göttingen and editing the physics volume of Gauss' collected papers, I was pleased to renew personal contact with him as one of my collaborators in this task.

My father had recommended me to attend a course by the zoologist Kükenthal on comparative history of development, and so I did. But all the facts and connections which had been so fascinating when explained by my father appeared rather dull and pedantic in a systematic lecture. Yet I attended regularly and acquired some understanding of this fundamental branch of biology, though hardly any single fact has been preserved in my memory. Anyhow, it became quite clear to me that the descriptive sciences were not in my line. There was too much to be memorized, too little causal understanding.

With chemistry I did not fare much better. The professor, Ladenburg, was not a bad lecturer and his demonstrations were impressive. But there was too much of the kind of stuff I could never digest: accumulations of almost incoherent facts which had to be memorized—at least so it appeared to me. My mistake was, of course, that I did not take a practical course in the laboratory where the daily handling of bottles and powders transforms names and formulae into objects with familiar properties. I missed this opportunity solely through lack of time; participation in a chemical laboratory course is a full-time occupation. I have never succeeded in making good this mistake. Fifteen years later I had cause to regret this gap in my scientific education when my research on crystals led me to the calculation of the so-called lattice energy which is connected with a fundamental thermochemical quantity, the heat of formation of certain compounds. I had to rely on the advice of chemical friends and I felt very uncomfortable when I was invited to lecture to chemical societies.

The greatest disappointment of all my courses was that on philosophy. The professor was a Roman Catholic priest (I think his name was Baumann), who spoke mainly on Aristotelian logic and metaphysics. There I heard for the first time the celebrated syllogism: 'All men are mortal; Socrates is a man, therefore Socrates is mortal.' It seemed to me the epitome of triviality. I never felt the need to modify this impression. What the problems of logic really are I have learned by studying mathematics, by practically handling it, and more systematically—many years later—from Hilbert's lectures on the foundations of logic and mathematics. Aristotelian metaphysics I have never learned as I gave up attending these lectures.

The only lectures which I really enjoyed were those on mathematics and astronomy. Of my first three years, I spent two in Breslau, though not without interruption, for I went to Heidelberg and Zurich in the second and third summer semesters. In those four Breslau semesters, I laid a solid foundation for my mathematical education. My teachers were Rosanes and London, both Jews, strangely enough. Rosanes was an elderly, ascetic-looking man, very much the privy councillor in his dignified appearance. He was a pupil of Kronecker and a friend of the celebrated Frobenius. His courses on analytical geometry of the plane and in space were brilliant and his algebra even more so. He introduced us very early into the ideas of group theory and matrix calculus and I owe to him my knowledge of this powerful method which I later successfully applied to physical problems, first to the theory of crystal lattices, then to quantum mechanics. London's subject was analysis; he gave us the introductory course in differential and integral calculus. Later I attended his useful lectures on definite integrals and on analytical mechanics. He was a very good teacher, extremely clear and lucid, though perhaps not very original. His son, Fritz London, was my pupil twenty years later in Göttingen and became well known from his work on the explanation of chemical valency as a quantum effect, which he published in collaboration with Heitler, and from his research on superconductivity.

What was it that fascinated me in those mathematical lectures and completely reversed the indifference, almost distaste, with which I had reacted to mathematics at school? It is a matter difficult to explain to people who are not familiar with mathematics. I had begun at that time, under Dr Lachmann's guidance, to read some philosophy and was deeply interested in the ancient metaphysical problems of space and time, matter and causality, the roots of ethics and the existence of God. Somebody had given me a book on epistemology by Heymans which my cousin Hans Schäfer and I read together and took as a fount of wisdom. It must have been quite a good introductory book for I had never much need to correct the views taken from it. In any case it acquainted us with the paradoxes of infinity which appear not only in the speculations on space and time but on a higher level in considerations on moral law and God, whose attributes include invariably the syllable all-, in connection with mighty, wise, good, etc. We got an inkling of the difficulties springing from the use of this syllable, and we were seriously worried as the whole of metaphysics seemed to be an awful muddle.

Now the notion of infinity also appeared in the mathematical lectures. But instead of being veiled in a mist of paradoxes it was a clear notion—or, to be more correct, it could be formulated in some perfectly clear way according to the case. That was the important discovery: that such words mean nothing unless applied in a definite system of ideas to a definite problem where they can be made significant for that problem. In analytical geometry the infinite points (lines, circles, etc.) were a pure invention made for the purpose of avoiding the introduction of exceptional cases in the formulation of general theorems. Instead of saying: 'two straight lines in a plane intersect at a point, except when they are parallel when they never intersect', one says: 'they always intersect at a point, allowing a set of parallel lines to define an ideal, symbolic point, "infinity".' Then one has, of course, to show that ideal points share the properties ascribed to real points in plane geometry. On the other hand, in analysis the infinite appears in a quite different way. It is not considered as an object, real or symbolic, but as a negative property of sets of objects, namely to be not finite, not countable, and it is subject to analysis with the help of the notion of 'limit': e.g. the set of number $\frac{1}{2}$, $\frac{2}{3}$, $\frac{3}{4}$, $\frac{4}{5}$, $\frac{5}{6}$ . . . has no end, is infinite, but the elements of the set do not grow beyond all bounds, but obviously approach the limit 1. Later, in Göttingen, I was introduced to Cantor's theory of sets, where infinite sets of objects are classified into different types, according to different kinds of infinity (countable and not countable etc.) which were treated like numbers of ordinary objects. All that had, and still has, a great fascination for me. When I first came in contact with philosophy, I felt that the philosophers moving in the realm of infinity without the precautions and experiences of the mathematicians were like ships in a dense fog in a sea full of dangerous rocks, yet blissfully unaware of the dangers. (There are, of course, exceptions—Russell, Whitehead and others—but I came across them only much later.)

Astronomy attracted me in another way. There we have the problems of cosmology related to the infinity of the physical universe. But I soon found out that there was little about these great questions in the elementary lectures of our Professor Franz whose speciality was the moon. What we had to learn was the careful handling of instruments, correct reading of scales, systematic elimination of errors of observation and precise numerical calculations—all the paraphernalia of the measuring scientist. It was a rigorous school of precision and I enjoyed it. It gave one the feeling of standing on solid ground. Yet actually this feeling was not quite justified by the facts. The Breslau Observa-tory was not on very solid ground, but on the top of the high and steep roof of the University building, in a kind of roof pavilion, decorated by fantastic baroque ornaments and statues of saints and angels. Although the main instrument (the meridian circle) was placed on a solid pillar standing on the foundations and rising straight through the whole building, it was not free from vibrations produced by the gales blowing off the Polish steppes. The whole equipment of this observatory was old-fashioned and more romantic than efficient. There were several old instruments from Wallenstein's time, similar to those Kepler might have used. The most modern instrument was the meridian circle mentioned above which once had been used, a hundred years ago, by the great Bessel. It had no electric chronograph, but we had to learn to observe the passage of stars through the threads of the field by counting the beats of a big clock and estimating the tenths of a second. It was a very good school of observation, and it had the additional attraction of an old and romantic craft. I remember many an icy winter's night spent there on the little roof pavilion. There were only three of us studying astronomy, Dr Lachmann, a man named Przebyllock (or some such name) and myself, and we took the observations in rotation, while one was noting down the figures and the third was free to do what he liked. What I liked was to look down on the endless expanse of snow-covered gabled roofs of the old city, surmounted by the massive towers of St Magdalene, St Elisabeth and, on the other side, beyond the pale water of the river, the profile of the cathedral, the Kreuzkirche and many others against the star-spangled sky of a continental winter's night. There on the narrow balcony amongst the stucco saints and ancient telescopes, one felt like an adept of the old guild of explorers of the skies, a scholar like

Faustus, and would not have been surprised if Mephistopheles had appeared from behind the nearest pillar. However, it was only old Professor Franz who came up the steps to look after his three students—he had not had so many for a long time—and he brought with him the soberness of the exact scientist, checking our results and criticizing our endeavours with mild and friendly irony.

As for our results, I suspect they were not very reliable, yet that was not our fault, but that of the observatory on the roof of the University. Professor Franz himself did, therefore, only descriptive work, a thorough study of the moon's surface, which he knew much better than the geography of our own planet. He made strenuous efforts, however, to obtain a modern observatory, but never succeeded. During my student time there were great hopes; for there was the World's Fair at Chicago, where the firm Carl Zeiss of Jena exhibited a set of modern astronomical instruments which, after the show, were bought by the Prussian State for its university observatories. Breslau obtained an excellent meridian instrument and a parallactic telescope. Yet no proper building was provided, and the previous instruments were installed in two wooden huts on a narrow island in the Oder river, just opposite the University building. This island was in fact an artificial one between the river and a lock through which many barges used to pass. The time service for the province of Silesia, which had been done for a hundred years with the old Bessel circle, was transferred to the new Zeiss instrument, but the results were rather unsatisfactory. Dr Lachmann was charged with finding out the reason, and he soon discovered a correlation between the strange deviations of the time observations and the changing level of the water in the lock: the island was not proof against water pressure. Professor Franz's hopes of a more efficient observatory had broken down again. As far as I know, a modern observatory (in Scheitnig) was built only after his death.

There was a time when I was so fascinated by astronomy that I considered the possibility of devoting my life to it. But this fancy did not last. It was shattered by the horror of computation. Franz gave us a lecture on the determination of planetary orbits, connected with a practical course where we had to learn the technique of computing, filling in endless columns of seven decimal logarithms of trigonometric functions according to traditional forms. I knew from school that I was bad at this kind of numerical work, but I tried hard to improve. All was in vain, there was always a mistake somewhere in my figures, and my results differed from those of the others. Dr Lachmann tried to encourage me and, when this did not help, began to tease me. But that made it worse. I do not think that I ever finished an orbit or an ephemerida, and then I gave up not only this calculating business, but the whole idea of becoming an astronomer. If I had known at that time that there was in existence another kind of astronomy which did not have as its ultimate aim the prediction of planetary positions, but studied the physical structure of the universe with all the powerful instruments and concepts of modern physics, my decision might have been different. I came in contact with this new science of astrophysics in Göttingen through Schwarzschild; but then it was too late to

55

deflect my plans of study again, and although astrophysical research has always interested me, I had to be content to follow it from a distance.

Dr Lachmann also attended the mathematical lectures. Each winter he came back from his balneological practice at Bad Landeck to Breslau to indulge in his many interests, foremost among them astronomy and mathematics, but socialist politics and other things as well. In the lecture-room we generally sat together in the front row, and alongside other people who played some part in my later life and may therefore be mentioned here. There was Otto Toeplitz, one year my senior. His father was a Jewish schoolmaster and mathematician who had brought up his only son from birth with the sole object of turning him into a mathematician, and with success. For me all this was a new world. I had hardly heard the name of Gauss, and certainly not those of Euler, Lagrange, Cauchy, Riemann, Weierstrass and others. For Otto they were old acquaintances or even friends. He not only knew the outlines of their scientific work but much of their lives and characters. He quickly took to me and behaved as a self-appointed mentor to a novice. I must say I owe him a great deal; especially the first introduction to the atmosphere of the mathematical sciences, that oldest and most continuous branch of all human knowledge. Then there was Ernst Hellinger, one year my junior (he came, in fact, from the same school as I, but there we had had hardly any contact). He was a little more initiated than I in the mysteries of the profession and in any case a much more conventional type of mathematical student that I was, sober and precise, though not very prolific in ideas.

Then there were some chemists attending the lectures on calculus: one, Rudolf Ladenburg, the son of the professor of chemistry, later became one of my closest friends and still is. Anny Hamburger, a red-haired girl, sister-in-law of Professor London, was one of a pair of almost indistinguishable twins; the other one, Clare Hamburger, was a zoologist and later for many years a lecturer (with the title professor) at Heidelberg. They were neither pretty nor elegant, but very clever, gay and amusing creatures. They escaped the Nazis and safely reached America.

Then there was Wätzmann, a charming, good-looking boy, gifted and modest, who later became professor of physics in Breslau and died very young. Kynast, a clever mathematician, but of a rather dull character; I came into closer contact with him in our third year when we were both studying at Zurich, but I lost sight of him later. Franz, the astronomer's son, a sturdy fellow who, like his father, had the impressive face of an actor, took part in some of our activities, but became a zoologist (if I remember rightly). This was the group to which I belonged. We took our work very seriously and often continued our discussions late into the night.

The rest of the class consisted of rather indifferent types who had no other ambition than to pass the final examination in order to get their teaching certificates. They had taken mathematics because their marks at school were better in this subject than in others; but they would have been just as happy at lectures on history, philology or law. I would definitely not have been, nor would Ladenburg, Wätzmann, Toeplitz or Hellinger.

Yet I was to come into closer contact with this crowd. For Dr Lachmann, feeling responsible for my overall education, disapproved of our 'high-brow' group. He said we were not true students for we missed the enjoyment of youth. He believed we ought to sing and drink and be sentimental—in the style of *The Student Prince* (though that 'classic' had not yet come into existence). I am afraid the good doctor was a little envious of our high-brow circle. A man approaching forty, he had some difficulty in following the lectures, and even more, our discussions; he was distracted by his family and many other interests, and he always had to miss the summer terms, when he practised in Landeck. Now there was a students' society, called the mathematics society, whose members were so impressed by this elderly gentleman attending lectures and working hard for no practical purpose, that they elected him an honorary member. He thoroughly enjoyed taking part in their meetings and urged me daily to join the society. Hans Lewinsky had tried before to persuade me to become a member of his literary society; when he saw that this was hopeless because of my distaste for that kind of activity—and for other reasons, such as the exclusively Jewish membership and the acknowledgement of the students' code of honour and of duels, etc.—he helped Lachmann to bring me into the mathematics society. Together they succeeded. As they both meant very well and were clever enough, I suppose there were good reasons for their efforts. Sophisticated, shy and therefore lonely, I was not a student of the ordinary type. It is not easy to give a description of what such a German students' society was like; one has to have been born in the old pre-war Germany to understand the background and atmosphere of this kind of communal life, which may appear ridiculous and perhaps distasteful to modern eyes. It did not appear so to me, at least at the beginning, otherwise I should not have joined it. The kind of 'scientific' society which I entered was modelled on the rules and customs of the older 'Corps' and *Burschenschaften*; it was a watered down version of them, with a dash of scientific interest and without compulsory fencing.

The meeting-place was a back room of a small beer-hall. The walls were covered with innumerable photos of former members, pictures of groups and single portraits, between which were scattered the symbols of a student's life, tankards of all sizes, shapes and materials, and rapiers, as used in the duels. In one corner there stood a piano, with a defective action and out of tune. The room was never free from the smell of beer and tobacco. There we met once a week, I think it was on Tuesday, at 8 p.m., and settled down around a horseshoe-shaped table. In the centre was the '*Präses*', who had a rapier lying before him.

At a quarter past eight he banged this weapon down on the table, thus giving the signal for the beginning of the 'official' *Kneipe* (*kneipen* is students' slang for drinking). The appointed musician went to the piano, and the old students' song, *Gaudeamus igitur*, resounded. Then, in our society, there followed the part called 'science'. One man had prepared an address on some mathematical, physical or astronomical subject and spoke for about thirty to forty minutes. There were very great differences in the standard of these

57

performances. Most of the fellows reproduced a summary of things they had just learned; only a few gave well considered and prepared lectures on special problems of general interest. In my time the presence of Dr Lachmann very much improved the level of these addresses and of the discussions following them, for he was full of life and humour, and he knew how to develop the most fascinating dispute after the dullest lecture.

About 9 p.m. the scientific part was finished and the *Kneipe* began. Each song was announced by the *Präses*, after a terrific bang of his rapier, with '*Silentium für ein schönes Lied*' (Silence for a beautiful song). They are very lovely, indeed, these old German students' songs, and even if one hears them over and over again, they do not lose their charm. The beer consumed enhanced an almost trance-like exhilaration produced by the feeling of comradeship and the half-military discipline, the so-called 'Komment' (pro-nounced in the French way, *comment*). That is the most strange and grotesque feature of the whole festivity: that every movement, every expression, every drink was regulated by strict rules and formulae in a weird mixture of Latin and German. After each song the *Präses* would present a toast, 'Old fellow so and so', or 'Professor X', etc., and everybody had to empty half a tankard. If somebody said 'Prosit' to you and lifted his tankard, you had to answer and drink half the contents of your glass. This would mount up quite considerably as time went on, and in order to avoid getting soaked too quickly, one would never drink on one's own account. Then there were particular drinking exercises, after which you again contained considerably more beer than before. But the strangest thing was that drinking was not only considered a pleasure, but was also used as a penalty for any of the innumerable offences which you may have committed against the customs or against the respect due to your 'superiors'. If you were asked to sing a line of a song solo, and you could not; if you, as junior or *Fuchs*, were not quick enough in responding to a 'Prosit' of an older fellow (*Bursche*); if you said a word after the *Präses*' '*Silentium*'; if you told a funny story and nobody laughed, etc., etc.; every time the *Präses* condemned you to drink a half or whole litre of beer. And all that was only the introduction. For after an hour or two the 'official' part of the *Kneipe* ended with a mighty song and than the '*Fidelität*' began, a less regulated and often rather wild, sometimes amusing, drinking bout.

There was little opportunity for having a good talk with one's neighbour. Anyhow, the number of subjects was restricted: wine, women and song, as the saying was, and of course, fencing. Although we were a non-fencing society, everybody learned this noble art in order to be capable of self-defence. After a *Kneipe* as I have described it there were very often collisions between the more or less tight fellows of different societies which would end with a challenge. I had lessons from the university fencing-master, and I was soon quite capable of standing up against a fellow who was not much taller than myself. For height was the secret of the art: the taller man had a decided advantage over the shorter. Opponents had to stand stiff as a candle and were allowed to move only the right arm and rapier, and this only in a restricted way, at full stretch, and with all the movement of the rapier from the knuckles as pivot. As sport it was almost useless; it exercised only the mobility of the muscles of the right arm, and it helped to produce the characteristic Prussian stiffness of the body. Yet I remember how we could talk about this silly art, our practice and progress, and on the performances of the participants in the latest duels!

It seems that I must have fitted in quite well in these surroundings, for I was elected *Präses* for one semester, I think after my return from Heidelberg in my second year. But after having finished this office, I was fed up with the 'soaking and merry student life' and ceased to attend the meetings.

The sport which I really loved was riding. I had begun to ride together with sister Käthe, during my last school year, after my father's death, on the advice of my Aunt Luise Kauffmann, Uncle Max's wife. She and her daughter Alice, about a year younger than I, loved this sport; I do not think they were very great experts in handling horses, but both of them looked handsome on horseback and knew it. We had a jolly good time, going through the basics of dressage and 'Quadrilles' in Reich's Tattersall, once or twice a week in wintertime, and cantering over the endless Silesian plain in spring and summer.

Uncle Max was already dead by that time. He had attended the Chicago World's Fair together with his cousin Franz Kauffmann (Robert's son), and both came back very ill. Franz, as I mentioned before, was sent to Egypt with Dr Lachmann. He returned half cured and lived for many years in a precarious state of health, later partly paralysed, but full of mental energy, as his father's successor in the directorship of the Tannhausen factory. Max Kauffmann did not recover but died. His widow, with her two daughters Alice and Claire, continued to live in Kleinburg, and when the girls grew up their house became the centre of a fascinating and gay life.

My Aunt Luise was indeed a woman of extraordinary charm and beauty. She was the daughter of a Berlin banker, Helfft, owner of one of the few private banking houses, like Bleichröder, which survived the period of amalgamation when the big joint-stock banks were founded and devoured in turn all the smaller firms. The Helffts were very rich and owned a splendid house in the Pariser Platz, at the Brandenburger Tor, the noblest square in Berlin. Old Frau Helfft lived there in a princely style, with butler and footman, horses and carriages; she was still beautiful as an old lady, though she suffered from a nervous twitching of the head which we as children found very amusing. Old Helfft himself looked like an aristocrat, with scarcely a trace of his pure Jewish ancestry. The three daughters inherited the lovely features of the parents. The youngest one, Helene, was often photographed as the prominent beauty of Berlin. But she and her brothers were weak characters; their life was a sequence of folly and incompetence. The firm became bankrupt and the whole splendour disappeared. My Aunt Luise was the only one of this breed whose soul was just as perfect as her body. But before I speak about her and what she has done for me, I must mention another family whose fate was strangely similar to that of the Helffts.

That was the family of my grandmother Kauffmann. I have already tried to describe this little brisk lady who ruled over our childhood, but I have not said where she came from. Her family, Joachimsthal, were rich Jewish merchants in Berlin. Some ancestor had helped to provide for the armies of Frederick the Great in the Seven Years' War so efficiently that the King granted him—before the general emancipation of the Jews—the freedom of the city and allowed him to settle in a lovely rococo house on Monbijouplatz opposite the small royal palace of that name. There they still lived in my time; not so proudly and ostentatiously as the Helffts, but in a noble bourgeois style. When later I read a book on the Mendelssohn family I was much reminded of the Joachimsthals' house, although the latter have not produced a musical—nor any other—genius. My great-uncle Georg Joachimsthal, grandmother's brother, was also a most aristocratic-looking Jew, slim, straight, bewhiskered like a colonel—a type which has quite disappeared. His wife, Aunt Lizzy, was English, once a great beauty, as could be seen not only from her many portraits hanging on the walls of the house, but also from her noble features as an old lady. Her son and two daughters combined the good looks of the parents; the younger girl, Ella, in particular, was very lovely. But none of them was lucky. The firm became bankrupt during the old man's time; the house was sold, but they continued to live in a small flat in the attic. For some time I was in contact with the children, but later lost touch with them, and I am afraid they all may have perished together with the rest of the Jewish population of Germany.

After this digression into family history I have to return to my tale about Aunt Luise. She had continued the tradition of the Kauffmanns as patrons of music and musicians. When the grandparents were dead and the house in Tauentzienplatz deserted, the Kleinburg villa was open to all artists who played at the symphony concerts. The conductor, Raphael Maschkowsky, died about the same time as my grandparents; his successor was Dohrn, who with his charming wife became great friends of Aunt Luise. He was also a brilliant pianist and played as such not only with his own orchestra, but also in other cities. I often listened to his playing in Aunt Luise's large music-room, which contained two Bechstein grand pianos. There I also heard many other great artists: Sarasate, Caregno, Therese Behr (whose brother was second conductor of our orchestra, and who later married Artur Schnabel), Busoni and many others. In later years it was there that I met Schnabel and Edwin Fischer for the first time; they were of my generation, and when they flirted with the two daughters of the house they had to put up with the inseparable cousin Max as well. My friendship with Schnabel has lasted in spite of long intervals of separation. We met again in 1933, when we stayed in Selva, Val Gardena, after our expulsion from Göttingen, and later still Schnabel was our guest in Edinburgh. I met Fischer again too after our emigration, when he gave a recital in Cambridge.

These experiences of course stimulated my own musical activities. I began to have a regular trio evening every fortnight. My partners were a curiously assorted pair. The violinist was a girl of about my own age, Ilse Späth, the daughter of a pastor and well-known preacher. She was a typical pastor's daughter in appearance, small and buxom, with a pretty doll's face, always friendly and cheerful, and without any interest or gift of conversation outside music. However, when she played she was quite a different creature, full of temperament and expression, but also conscientious and accurate; thoroughly musical. I do not remember where I first met her, but we played for years together, during my studies in Breslau and later when I came home for my holidays, and we became great friends. Perhaps she was a little in love with me for some time. On my part I rigorously suppressed any feeling of this kind which might have arisen, because I thought her too good for a flirtation which would spoil our comradeship in music, while a serious engagement with a girl from a pious evangelical house ran counter to all my instincts and convictions. I was very kindly received in the pastor's house but found the atmosphere strange. At that time such platonic friendships on the basis of a common interest or activity were still rare. There were girl students, but they were mostly elderly and unattractive women. I was teased a great deal about this girl-friend, but that did not affect me, and our friendship and collaboration lasted until her engagement and marriage to a pastor from Vienna. After the grand wedding party, I never saw her again, but my brother Wolfgang visited her during his stay in Vienna and told me that she seemed to be a very happy wife and the mother of numerous children.

The cellist in our trio was an old professional member of the symphony orchestra whom Mama engaged for a modest sum. I have forgotten his name. He gave us tips and instructions, and he obviously liked these evenings when, apart from the money, he got a splendid supper and could talk about his son, of whom he was extremely proud, for this young fellow had worked himself out of the rank-and-file of orchestral players and was at that time conductor of the Bremen Symphony Orchestra. So we happily played Haydn, Mozart, Beethoven and even ventured on Brahms, with no audience except Mama, Käthe, Wolfgang and occasionally a close friend.

But Aunt Luise heard of my trio and one day—it was perhaps in my third year as a student—she asked me to invite her to one of our evenings. Rather reluctantly I agreed and we practised two trios with care and industry. Mama invited some other aunts and cousins and prepared a fine dinner for the party. But on the afternoon of the performance the old cellist sent me word that he was ill and could not come. I informed Aunt Luise at once that the plan had to be abandoned. She however answered that there were other cellists in the town and she would supply a good one. When she arrived in her carriage, who should step out with her, carrying a cello, but Herr Melzer, the cellist of the Breslau Quartet, a local celebrity not only on account of his excellent playing, but also because of the roughness of his manners; a man much feared by his pupils and even his colleagues. Well, the whole evening was a total disaster. He began quite cheerfully, but after only a page or two he interrupted our playing and declared our conception of a passage wrong. We began again, but his tender musical conscience did not allow him to play his part and let us do whatever we could. He interrupted again and again, criticized us, as if we were professionals, and carped in particular on poor Ilse's technique until the girl's eyes filled with tears. Aunt Luise tried to interfere, but in vain. I wanted to shorten the programme but Melzer would not allow it, we had to go through all the movements, with numerous repetitions. It was hardly any consolation for us that at dinner, when good food and wine had mellowed him, he declared us quite promising amateurs and said he would like to play with us again. We certainly did not relish this idea and decided never again to play before an invited audience.

I still think that the main condition for amateur music to be enjoyed by the players is the exclusion of any audience apart from members of the family. But in later years I often had to yield to other circumstances; for instance, my wife's wish of combining music with hospitality to interested friends.

It was in the winter following my father's death that Aunt Luise arranged weekly dancing-lessons in her house for her two daughters and their cousins and friends. At that time I hated this noble art, or at least pretended to. There were reasons enough for avoiding participation; our state of mourning, and my preparing for the school-leaving examination. But my sister was allowed to take part as these evenings could be considered contributions to education. So I had rather a fight to remain free from it. I think that my real motive was the fact that Lore was not invited. But I also disliked the priggishness of some of the boys who were my cousins' beaux. They liked to sit together in Alice's boudoir with effectively dimmed lights and read aloud, with plenty of feeling, the poetry of Hölderlin, Nietzsche, Rilke and other modern writers. Once I could not avoid listening; but when Richard Kroner spoke the lines: 'Das ist der Herbst, der bricht mir noch das Herz . . .', and pronounced the 'Herbst' and 'Herz' with his lisp, I had to run out of the room, much to the anger of my good cousins. But the same Richard Kroner and Alice Kauffmann became engaged during the first session of dancing-lessons, and married after a long period of waiting. They are a happy couple and I have learned to recognize the value of Nietzsche's poems for match-making.

These dancing-lessons were continued the following winter, when it was much more difficult for me to refuse. Aunt Luise told me baldly that I had no manners and ought to learn how to behave in society. My way out of this dilemma was odd enough! I booked a course in a public dancing-school, run by a Herr Reiff, and attended it for two or three months fairly regularly. I think I learned there the manners of the ballroom and the elements of dancing quite well as I took the whole affair as a duty and not as a pleasure. There were a lot of pretty girls, and I danced and talked with them, but without the slightest personal interest. I must have been, at that time, a most absurd fellow. When my sister asked me after one of these evenings: 'Whom did you dance with last night?' My answer was, 'With the green one', or 'the blue one', meaning the coloured frocks, as the names of the girls were, and always remained, unknown to me.

A year or two later, at home for a holiday, I was present at a house-ball in Aunt Luise's villa and danced with my cousins and their guests. It was a colossal success. I was suddenly discovered to be a perfect dancer, accomplished in the waltz, polka, quadrille, française and the other dances fashionable at that time. But that I had not acquired at Herr Reiff's course; it was the result of my travels and of education by life in a wider world. How this came to pass will be told later.

V. Heidelberg and Zurich. The Neisser House

As the end of my first university year approached I had to face a definite decision about my future: whether I should follow my original plan of becoming an engineer, or continue my scientific studies. There was a third possibility, but it was given very little consideration, namely my becoming a businessman by joining the Kauffmann firm. I had never shown much interest in the factories, much to the disappointment of my grandfather, who naturally wished one of his grandsons—there were only Hans Schäfer and myself—to become a member of his staff and later his successor as chief of the firm. He had no high opinion of either of us; once when he had listened to a high-brow talk between us at his table, on some philosophical problems which were certainly above the heads of two *Primaner*, he remarked: 'You fellows are too learned and too stupid to become good businessmen; you are only good enough for studying.' Luckily my father was not present; but I told him, and I noticed that he was hurt. Then he said grandfather had been joking, and that nobody should be called stupid who was interested in theoretical questions. Cousin Hans studied chemistry, graduated with a Dr Phil. in Breslau and, after some years of training as a dyeing expert, entered the firm, whose chief he was later to become.

My own decision was much influenced by Dr Lachmann—my father and grandfather being dead by that time. I had, according to my father's wish, tried many things, with the definite result that the exact sciences, mathematics, physics and astronomy, interested me far more than anything else. But what career lay open to a man with these subjects? A schoolmaster or a university teacher. I loathed the one, and I was afraid that my gifts were not great enough for the other. Engineering still seemed to me the nearest approach to science from the practical side; for I thought that I had to earn my living, and I intended to do this thoroughly, i.e. to become a great engineer like Werner von Siemens or Abbe, the scientific founder of the Zeiss works. Now Lachmann explained to me that I was a young man of means and that, in our bourgeois society, a wealthy man should use his brains not to accumulate more wealth but to develop the arts or sciences, or whatever cultural occupation appealed to him. He, as a socialist, resented this state of affairs; but until the communist state was in being one had to take things as they were and make the best of them. My duty was obviously to cultivate my scientific interests and to devote my life to learning and research. He thought my gifts sufficient to become an academic teacher like my father but, if I failed, there was still my inherited money on which to live and enjoy my studies. This sermon, repeated on several occasions, was in the end effective. Although I was not at all confident of my talent, this plan coincided perfectly with my innermost wishes.

German students used not to stick to one university, but to change frequently from one to the other. There was much more 'academic freedom' in Germany than in England; no fixed syllabus, no classes, no examinations apart from the finals (doctor's degree and professional certificates), free choice of teacher, and therefore healthy competition among the professors not only within the same but also between different universities. Students flocked around a celebrated teacher and came long distances to listen to a special course of lectures. The different German states competed with one another to attract students, by having the best and most celebrated scholars on the staff of their universities. Some of these centres of learning had of course other attractions. Munich was favoured as a 'winter university' because of its theatres, concerts, the carnival and skiing in the mountains. Heidelberg and Freiburg were the favourite places for a gay summer semester in lovely and romantic surroundings.

My cousin Hans had spent one summer term in Heidelberg and decided to go there again the following summer (1902). He easily persuaded me to go with him, for I wished to see the world, but was still somewhat shy and therefore glad to make my first excursion under the guidance of an 'experienced' friend and cousin.

I took a room in a big house on the steep road called Neue Schlossstrasse, leading up to the castle. There was a terraced garden in front of it, and to reach the entrance one had to pass through a stone gate into a subterranean vestibule from which stairs went up to the first terrace of the garden. In this vault, opposite the entrance, was a big picture of an armoured knight, which is indelibly associated in my mind with the idea of Heidelberg. When I visited Heidelberg in 1937, by then under Nazi rule, together with my friend Ladenburg and my daughter Gritli, I found everything sadly changed, gaiety and romance almost gone; but the knight in armour was still there in the entrance vault of my old digs.

I took my studies in rather a leisurely fashion, as becomes a summer student at Heidelberg. With physics I had no better luck than in Breslau; I tried the lectures of old Quincke, but found them just as dull and inefficient as old Meyer's. Mathematics was better. There was Königsberger, a dwarfish, incredibly ugly Jew, but a great 'privy councillor' in the ranks of the Alma Mater, best known by his biographies of Helmholtz, the physicist, and of Jacobi, the mathematician. He gave a course on differential geometry. It was an old-fashioned kind of mathematics, where things occurred which I had learned from my Professor London in Breslau to consider as utterly forbidden and contemptible; namely the use of 'infinitesimally small quantities', not only as first differentials, but also as second and higher ones. But I soon discovered that one could easily justify each step and make it rigorous by introducing independent parameters, one for a line, two for a surface, etc., and writing their differentials in the denominator. After having soothed my conscience I had great pleasure from this course as Königsberger was a brilliant speaker, full of life and enthusiasm. He came into the long lecture-room from the back and began speaking as he entered, so that by the time he reached the blackboard he had already recapitulated the whole of his last lecture and was in the middle of new developments which were leading dramatically to a climax by the end of the hour. One climax of the whole course was Gauss' celebrated theorem about the expression of the curvature of a surface in terms of the coefficients of the line elements; another Weierstrass' parametric representation of all minimum surfaces by arbitrary functions. I carefully worked out this lecture and have my notebook still; when I read it now I don't think it as brilliant as it appeared to me at the time. Endless calculations are performed where simple reasoning leads to the same results, and Reimann's great work is not mentioned at all.

Another mathematical lecture was by Landsberg on determinants, a clear and straightforward introduction which gave me the foundation for an understanding of Rosanes' algebra of matrices which I attended the following term in Breslau (as already mentioned). I have a good manuscript of this lecture too.

The habit of working out lectures, that is, taking short notes in the classroom and writing a coherent and accurate account at home, if possible the same day, I had acquired in Breslau from Toeplitz. I found it extremely instructive and useful. I persisted with it throughout my time as a student, and I still have the whole collection, which constitutes the equivalent of a good series of text-books. They are all tidily written, without visible corrections, and many of them still serve well as a reminder of a since forgotten idea or trick. Later on when I had gained some experience in lecturing, I always selected a student who could work out my lecture. At the end of term his manuscript was duplicated; so I also have readable manuscripts of all my own lectures. This custom was a tradition in Göttingen where copies of all the important lectures were collected in the mathematical reading-room. A condition for doing this of course is an attitude to lecturing which regards each course as a unique production like a work of art. The lecturer builds up the pattern of his subject, while his personality gives the work character. Felix Klein's lectures, published as autographed manuscripts, are the most brilliant examples of what I mean.

In Heidelberg these mathematical lectures, altogether six hours a week, left me time for other interests. The chair of philosophy was at that time held by Kuno Fischer, whose lectures attracted enormous crowds. The audience of hundreds of students was augmented by other visitors; for even tourists included attendance of one of Fischer's lectures in their sightseeing programme. These courses were therefore held in the 'Aula', the ceremonial hall of the University which had not only a huge floor area, but also a gallery with boxes all round. Even so the demand was so great that one had to book a seat beforehand and stick one's visiting card on it.

Fischer was a sturdy little man with a clean-shaven, very expressive, bulldog's face and the resounding voice of a trained actor. He was one of the great celebrities of the University and had the title 'Working Privy Councillor' on account of which he had to be addressed as 'Your Excellency'. There were many stories about his conceit, mostly apocryphal, I think. It was said that when a young colleague of his spoke to him and began every sentence with 'Your Excellency', he interrupted him with the words: 'My dear colleague, not so much of "Your Excellency", please; now and then is quite enough.' His most popular lecture was on Goethe's *Faust*. Hans Schäfer had attended it the previous year but it was not repeated in my time. Instead he gave a course on the history of modern philosophy, beginning with Bacon, Descartes and Spinoza and ending with Kant and Fichte. It was quite fascinating and gave us plenty to argue about. After a while I found a hitch in it; he presented each of the consecutive systems as if it were the summit of all wisdom, and this conviction was conveyed to the audience with such force that everybody submitted to it—until the trick was repeated once too often. When I found it out my old scepticism regarding philosophical systems returned and has remained ever since. These lectures are published and I have a few volumes, amongst them that on Kant.

A young Prince of Cumberland (a relative of the Duke of Brunswick) used to sit in the front row of Fischer's audience with his governor or tutor. One day when these two gentlemen entered the hall, they found their customary seats occupied by two distinguished-looking elderly men, apparently tourists or stray visitors. When Fischer took his seat behind his desk he saw at once the change before him; sitting in the gallery just above, I could see everything that followed. He was obviously shocked and began to lecture with some signs of nervousness. I suppose that his royalist feelings were rather strong. But worse was to come. After one of Fischer's great resounding sentences, when he made an artful dramatic pause, the older of the two newcomers, a dignified gentleman with long white hair, began to applaud. But as the audience did not follow he desisted. Instead, he gave visible signs of enthusiastic admiration by nodding his head and silently applauding with his hands. Suddenly Fischer interrupted his lecture and said in a sharp, ice-cold voice: 'Who is this man? Is he mad? If he cannot listen properly I shall have him removed.' After a pause full of embarrassing tension he continued in a subdued voice, without any of his usual vigour, and came to an end. When he left the hall the old gentleman followed and gave his card to one of the attendants, whereupon he and his companion were at once taken into Fischer's private room. Hundreds of students crowded around the door waiting for the outcome and the explanation of this mysterious happening. After a while Fischer and the stranger appeared arm in arm, both smiling, and were together carried away in His Excellency's carriage. We learned from the attendant that these two strangers, both celebrated scholars, were emissaries from the Académie Française and had brought Fischer some high decoration or honour from that distinguished society. Luckily their knowledge of the German language had been too poor to understand Fischer's offensive behaviour.

Looking back on this Heidelberg semester it occurs to me that neither the professors nor the romantic atmosphere of the old university town were as important in my life as my friendship with James Franck. I made his acquaintance on the very first day, introduced to him by a law student, Fritz Alexander, who was a nephew of my stepmother's youngest sister Rosa. Both came from Hamburg, from the same school. To cousin Hans and to me it was very soon clear that Franck was our man. He had a peculiarly personal charm; he was clever and the best comrade in all our activities. His father was a wealthy businessman who wanted his son to study law and economics with the idea that he should later join his firm. But James's interests were quite different. He wanted to become a scientist. At that time it was geology which fascinated him. As an obedient son he tried for two semesters to attend law lectures, but was quite disgusted with them. When we met, he had definitely decided to give them up and to change to geology and chemistry. He still entered his name for a course in economics but never attended it. Instead he took courses in his beloved sciences. While he was in the middle of these he reported to his father, who reacted by sending excited letters, then by appearing in Heidelberg himself, accompanied by his extremely elegant wife. Cousin Hans and I decided to help our friend with all our energy. It was no easy job. Both parents considered scientists as beggars, people of no consequence, of the rank of lower clerks. At that time this seems to have been the attitude of the Hamburg merchants in general and particularly of the Jewish ones. It was an amusing battle with prejudice and narrow-mindedness; but in the end we were victorious. It was partly thanks to James's stubbornness, who declared he would rather earn his own living than accept his father's allowance if that depended on his studying law; partly because Hans and I, descendants of great merchants, had found no opposition in our families when we decided to study science.

After that episode we three had a good time, not working too hard and enjoying the life of free students. We had of course a wider circle of acquaintances—I remember a nice Dutchman—but we three were mostly together: at our meals in the old beer restaurants, 'Perkeo' and 'Roter Hahn' and 'Ratskeller', and during long walks through the lovely hills of the Odenwald. Often we took a boat and rowed up the Neckar to the 'Stiftsmühle', a beer-garden near a famous old convent, from which there was a lovely view over the river to the wooded hills, the Königsstuhl. There were little rocky islands in the river; we liked to land on these and to play cards, the old students' game of Skat, drinking the beer which one of us had fetched from the 'Stiftsmühle'. On our excursions Franck would take a hammer in his rucksack with which he chipped specimens of minerals from the rocks. Once we persuaded him to allow *us* to put specimens in his rucksack in order to avoid the frequent halts for taking off this bag. We filled it not only with geological specimens, but with all kinds of rubbish—beer bottles, old shoes, pieces of wood—hoping Franck would present it all the next day to his geological professor. But he discovered the fraud in time and threw the rubbish at our heads. He had a marvellous laugh, and took all practical jokes with good humour. We became very good friends. But I had no suspicion that he would play a decisive part in my scientific development and that we would work together in the same department for many years.

It was during this semester in Heidelberg that I made my second trip to Italy, during the Whitsun holidays, in company with cousin Hans. But it was under the same evil star as my first. In Milan I developed asthmatic bronchitis. Hans, whose knowledge of the Italian language was very limited, had difficulty in procuring a doctor and in nursing me in the primitive hotel where we were staying. So I was a nuisance to him too and spoilt his holiday. Yet in his own *Recollections*, a copy of which I possess, he remembers this visit to Italy with great satisfaction, good soul that he is.

In the winter semester of that year (1902) I returned to Breslau. It is hardly possible for me now to put all the little events of those years spent at my home university in proper order, but one thing I have to record which happened that winter. It was my connection with the Neisser house.

My grandfather's brother Julius Kauffmann had two sons and a daughter. One son, Fritz, died early; the other, Georg, later became commercial counsellor like his father, and head of the firm (followed still later by my cousin Hans Schäfer). The daughter, Antonia, called Toni, married a young physician, the dermatologist Albert Neisser, a contemporary and friend of my father. At the time I speak of Neisser was already famous for his discovery of the gonococcus, and was head of the dermatological department of the University. He was a most remarkable man; he looked like Blue-beard, with fierce burning eyes under black brows, given to energetic gestures, full of vitality, a splendid piano-player, a lover of poetry and art, disposed to wild tempers, but at heart infinitely kind and a faithful friend. He was no easy companion, but Toni was a match for him. She was not beautiful, but had an expressive face and lovely eyes. All the intelligence of the Kauffmanns was concentrated in her, though it was not focused on making money, but on using her money for the enrichment of life through art, music and poetry. They had no children, and in addition to her inherited fortune, had a considerable income, so there was ample scope for her activities. She once told me that she ought to have been born in Florence under Lorenzo de' Medici. Having missed that opportunity she tried to live as near to that style of life as she could in a place like Breslau in our times. She patronized the Breslau School of Arts and discovered there two gifted brothers, Fritz and Erich Erler, who came from peasant stock and had fought their way to this school against great odds. Toni took them under her wing, and they became almost like sons to her. She financed them at Munich so that they might perfect their art there in the studios of famous painters, and the results were most satisfactory. Fritz Erler became one of the leaders of the (then) modern school of painting, the *Jugendstil* (art nouveau). Erich became widely known, under the name of Erler-Samedan, as a painter of alpine landscapes; he spent his summers and many winters at the little town of Samedan in the Engadine.

The Neissers built a big villa on the fringe of the Scheitnig Park. Whether it would be considered beautiful today, I rather doubt. It had a tower, a big central hall with a grand staircase and gallery on which the doors of the rooms of the upper storey opened. It was surrounded by a big and lovely garden, full of old trees, well-kept lawns and flower-beds. There were arboured walks covered with ramblers, along a tributary of the Oder, and fountains around which peacocks strutted. But its main feature was its interior decoration, which had been supervised by the Erlers and other artists. In particular the music-room was devised in a unique manner by Fritz Erler, who painted the walls with frescoes, entitled 'Adagio, Andante, Allegro, Furioso', etc. These were reproduced at that time in Germany and also in England. This room had its enchanting qualities, however much its bizarre style might be despised today. It was an unforgettable experience to listen to chamber music there, for instance the Klingler Quartet, or the Bohemian String Quartet.

Having conquered one of the venereal diseases, Albert Neisser attacked the more dangerous one, syphilis. He needed anthropomorphic apes for his experiments, and so he went to Sumatra to study the disease by using orang-outangs and gorillas. While he worked, his wife Toni employed her boundless energy and mental capacities in studying and enjoying oriental art. They brought home a big and interesting collection of Malayan and Indian *objets d'art* and used them to decorate some rooms of the house, which in this way became still more attractive and extraordinary.

On my return I found my sister Käthe quite at home in the Neisser villa. Albert Neisser was a restless fellow. After a hard day's work he did not want a quiet evening but preferred stimulating entertainment, a concert, a theatre or a string quartet in his music-room, and always a convivial dinner party. The guests were young doctors, artists, painters, poets, musicians, who enjoyed a dance after dinner, and Toni had also to provide for that. Hence there were always a few young ladies staying at the Neisser house who had to be pretty and amusing in order to give the Boccaccio touch to the 'Florentine' life.

At that time there were two such girls present. One was Ingrid Thaulow, a nordic beauty of the Brunhilde type, the daughter of a well-known Norwegian painter, some of whose works were hanging on the walls. The other was Lilly Weber, a Swiss girl, who became a great friend of my sister and myself. My sister was called in by Toni to entertain these two girls because they were of her own age, and I sailed in the wake of my sister. But I soon felt that both the Neissers took a personal liking to me. They knew that, after my father's death, our home was rather dull and that we lacked a guiding older friend. Toni tried to be a mother to us, as far as she was capable of motherly feelings, and if I ever had a confidante it was in these years. I owe to her many of my ideas on life; what I learned was in my father's tradition of a positive and active humanism, but there was a definite accent on art and beauty, which had been rather lacking in my parental home. Later, when I came home from Göttingen for the holidays, I often stayed part of the time in the Neisser house, as a third and more intimate son; for the Erlers soon began to gravitate around Munich, and they both married pretty Munich girls. Erich's wife, Olga, was a

70

particular pet of Toni's, as she admirably fitted into the style of the house, in her way of looking, laughing and dressing. Once when we were both staying there for a few weeks we became good pals; for my part I felt rather flattered by this friendship with a woman older than myself who looked like a figure out of a Titian painting. Many years later when I was a married man, Hedi and I had once to wait for a train in Chur. Sitting in the restaurant at the next table was a stout and vulgar-looking lady. It proved to be Olga Erler, and Erich was also there. We exchanged a few words. But I was disappointed, and Hedi, not sharing my memories, strengthened this feeling by some uncomplimentary remarks. After the encounter I never saw the Erlers again. They did not show much gratitude to the Jewish house which helped them on their feet, and I have heard that in the end they turned into perfect Nazis.

But now I have to come back to that Swiss girl, Lilly Weber, and tell her story. It begins about twenty years earlier. One day a party of Swiss mountaineers, after climbing one of the giants of the Engadine, descended leisurely and without a rope the last harmless slope of the Morteratsch glacier. Here, in view of the restaurant, one of them, Weber, stumbled, fell and slid down the steep foot of the glacier. Severely injured, he was brought to his hotel, and the celebrated surgeon of Samedan, Dr Bernhardt, was called in. Neither his skill nor the devoted nursing of Frau Weber could save the patient's life. But a friendship developed between these two, and some years later Frau Weber became Frau Bernhardt. She was of good Swiss stock; her father, named Imhoff, lived in Winterthur and was a great bibliophile and numismatist. Later, when I studied in Zurich, I spent delightful afternoons in the pleasant house of the old Imhoffs, admiring the rare and lovely books, Roman and Greek coins, listening to the old gentleman's entertaining explanations and enjoying the patisseries at the old ladies' tea-table.

The connection of the Bernhardts with our family was of long standing. My grandfather's youngest brother, Dr Adolf Kauffmann, used to spend his summer vacations in the Engadine, long before St Moritz and Pontresina became fashionable. He used to stay in Samedan and here he met young Bernhardt, I think he was the son of the local chemist. He patronized the young student of medicine and invited him to Breslau; here Bernhardt met Neisser, and a great friendship developed, which was later shared by the wives. Frau Weber-Bernhardt was a grand lady of a similar type to Toni Neisser, deeply interested in art and music. Dr Bernhardt was a very successful physician; he introduced the treatment of tuberculosis of the bones by exposing the body to the rays of the sun. He was among the first to have his patients lying in the open air in mid-winter. When I first visited them, they still lived in Samedan; but soon afterwards they built a stately house in St Moritz, beautifully situated on the slope opposite the snowy range of the Bernina group. This house contained an excellent collection of paintings, among them many Segantinis and Hodlers. Frau Bernhardt had three daughters by her first husband, Lilly, Helene and Ella Weber. After her second marriage three more daughters were born, and finally a boy; he was younger than Lilly's eldest son and was referred to by her as the baby-uncle.

It was the eldest of the Weber girls, Lilly, whom my sister and I met at the Neisser house. In the years following, the other two also appeared, but though they were nice and pretty girls, we did not take to them in the same way as to Lilly. She was a most charming creature; not very pretty, but graceful, full of life and a harmless coquette. She flirted with all attractive men, but in such a way that none had a right to feel preferred. For with all her liveliness and temperament, she was very practical and matter of fact. She told us frankly that she would marry only a rich man—'but he must be handsome', she would add with her charming smile. And so she did, for she married Dr Hans Sulzer, a member of the great industrial family with big factories not only in Winterthur but also in Ludwigshafen in Germany.

Through Lilly, my sister and I heard a lot about Switzerland and Graubünden in particular. We were impressed by the fact that she knew members of the old noble families of the Engadine, the von Salis and von Planta, whose ancestors appear in the historical novels of Conrad Ferdinand Meyer. One of Lilly's sisters later married into the von Salis family. She invited us to visit her in Samedan, and we promised to go.

I was the first to fulfil this promise. At the end of the winter semester I again decided to spend the summer at another university in order to widen my views on science and life. It was only natural that I should consider Zurich, where, in addition to the Cantonal University, there was the *Eidgenössische Technische Hochschule*, an important school of science and engineering. My friend Toeplitz approved my choice of Zurich as a mathematician of great renown, Hurwitz, lived there. And so I went to Switzerland.

Again beginning my account of this semester with my studies, I have to confess that I went only to two mathematical courses, one (four hours a week) by Hurwitz on eliptic functions, the other (two hours) by Burkhardt on Fourier analysis.

Hurwitz was a tiny man with the emaciated face of an ascetic in which burned two unnaturally large eyes. He was ailing and very frail. But his lectures were brilliant, perhaps the most perfect I have ever heard. The course was the continuation of another, on analytic functions, which I had not attended; I therefore had some difficulty in following and had to work hard, reading many books. Once when I had missed a point in a lecture I went to Hurwitz afterwards and asked for a private explanation. He invited me and another student from Breslau, Kynast, whom I have already mentioned, to his house and gave us a series of private lectures on some chapters of the theory of functions of complex variables, in particular on Mittag-Loeffler's theorem, which I still consider as one of the most impressive experiences of my student life. I carefully worked out the whole course, including these private appendices, and my notebook was used by Courant, when he, many years later and after Hurwitz' death, published his well-known book on analytical function, the so-called Courant–Hurwitz.

Burkhardt was a strange man, apparently always absorbed in his mathematical ideas and hardly aware of his surroundings. He never looked one straight in the face and seemed to be afraid of the students. His course was clear and interesting. It gave me a solid foundation for my knowledge of orthogonal functions which was later most useful in physical applications. I also tried to attend a course given by the astronomer Wolfer but found it so dull that I soon gave up.

I lived in a simple boarding-house in Zürichberg-Strasse. It was very cheap indeed. I could have afforded better quarters, but I decided to live as simply as any other student and to use my money during the holidays for travelling. The boarders who met during the meals were a strange international crowd. There were two Bavarians, one the rather degenerate son of a Munich painter, Echtler, the other (whose name I have forgotten) a sturdy son of the highlands, who was very friendly with all the waitresses of the little wine and beer restaurants which we frequented. When he boasted of a new conquest Echtler, who lacked the robust manliness of his friend, could never conceal his envy and jealousy. Then there was an old Jewish doctor, reminiscent of Dr Lachmann, with his wife; he was highly intellectual and very frail, almost entirely soul and brain, without body. I believe he had been tortured in a Russian pogrom and fled to Switzerland. Then there were always a number of real Russians, men and women, all revolutionaries and more or less looking as Russian nihilists were supposed to look. One of them, Maitoff, a blond giant, became quite attached to me, with consequences which I shall report presently. Finally there was a young student from Luxembourg, Funk, a quiet fellow with whom I had hardly any connection until the end of the term, when we suddenly developed an ardent sentimental friendship, revealing to one another the deepest recesses of our souls, rather like school-girls. Nothing came of it. After we had separated at the end of the term, we exchanged one or two letters, and then he disappeared from my life.

I had, of course, met non-German people before, but on a quite different basis. My cousins had had French and English governesses and even an Italian one, Gemma Prunetti Lotti (a beautiful girl, who later married the poet Otto Julius Bierbaum and became widely known from his poems in her praise). My father occasionally had foreign visitors, Englishmen and Americans. They came and went and had little influence on our life and our scale of values. But here in Zurich it was different. Looking back, I am aware how strong the nationalistic tendency in our German schools must have been; I remember the shock which I felt on realizing that these people in my boarding-house considered their own countries and their own ways of living as in no way inferior to 'ours'. Some of my schoolmasters had represented Switzerland and Luxembourg as little appendages of the proud Reich separated only by the accidents of history and with no right to a proper existence of their own. Now I had a practical lesson of the real situation and in particular of the national pride of the free Swiss. Another good lesson I had from the Russians. I had written a picture postcard from a weekend excursion and addressed it to Herrn P. Maithoff, spelling the Russian name as if it were a composite of the German syllable 'Hoff'. On Monday evening at dinner Maitoff showed this card around and produced an outburst of laughter at my expense by asking me: 'Hallo, what does the word "Mait" mean in your lovely language? If I am

73

to be germanized, do it thoroughly.' I do not know why this little scene impressed me so much that I can remember it; for there were so many Baltic people with real German names among the Russians. Anyhow, it made me think about national pride and its justification, and I had long discussions on the matter with the old Jewish doctor who had no particular attachment to any nation.

I had letters of introduction to families, mostly of professors at the University and the Polytechnicum. But I remember very little of my visits to these houses. I have only a vague recollection of a large party, a picnic in the forests of the Zürichberg. It culminated in the recital, on a lovely meadow, of a new translation of Dante's *Divina Comedia* by the translator himself, a one-time colonel and celebrated man of letters, whose name I have forgotten. The whole academic society was on a very high-brow level, with many recitals of verses and music, but lacking in gaiety. Fun was provided by the people of my boarding-house. After one of them had passed an examination we had a picnic in the woods, where we consumed so much alcohol, in the form of a 'peach punch', that on the way home one of the Russians fell into the deep well at Fluntern, two others were arrested by the police, and as regards myself—I had better omit any account, which would be difficult anyway as I remember nothing of how I got home. Next morning I found myself on my bed, still fully dressed, with a dreadful hangover.

But there was one family whom I want to describe as I shall always be grateful for their kindness and hospitality. They were to me a new type of the genus Homo Teutonicus. Dr Fick was a friend of my father's whom he had often mentioned and described to me. He was a physician (I think an ophthalmologist) and lived in a modest little house in Fluntern, high up on the Zürichberg. He was a tall man with marvellously noble features, long fair hair and a thick drooping moustache. His wife was from the well-known family of Wislicenus which produced many scholars and artists, a beautiful fair woman of the Kriemhild type. I have to quote such figures from the German sagas, as the whole life of this family was intended to be a revival of these Teutonic legends. There were two boys, Roderich and Roland, about sixteen and twelve years of age, and several girls, the eldest, Gisela, almost a young lady, the others Gudrun, Hildburg down to little Waltraut, all fair creatures, healthy, lovely, natural in behaviour. Frau Fick did not wear modish frocks, but garments of her own design, and the girls were similarly clad, though very simply. Life was very simple in the little Zürichberg house. Once when I put some jam on my bread, which was already buttered, there was a great uproar among the youngsters which I did not quite understand, until Frau Fick explained to me later in private that according to her principles of education this was the height of prodigality and strictly forbidden in a Teutonic family. After this of course I conformed to this rule, as to many others which I learned in due course. For I liked the whole family, parents and children, and I spent many afternoons among them; there was a marvellous gaiety, a healthy enjoyment of life and a warm hospitality to the stranger who was so different in every respect.

Dr Fick called himself 'all-German'; but his pan-Germanism was very different from what it was soon to become in the Reich. Once I asked him quite innocently whether he was a Swiss subject. He reacted almost as if I had offended him and snapped back: 'I am neither Swiss nor a subject, but a free citizen of the German Reich.' He was super-nationalistic in a mystical way; for him the heroes of the *Nibelungenlied* were real and noble, personifications of the genuine spirit of the Teutonic tribes, which ought to be revived. But he was a democrat and hated the Prussian militarists and industrialists. Although his attitude came very near to the racial obsession developed by the Nazis, he was not anti-semitic. In any case he spoke about my father with the greatest admiration and love; he liked to tell me stories of their time together as students and repeatedly said: 'Oh, he was a better man and a greater scholar than simple people like myself and most of our friends.' He was a lovable man, old Dr Fick; yet there was something strange, even weird in his ways, in the fantastic ideas on which the life of the whole family was built. I felt this vaguely, though at that time I would hardly have been able to express it. But I had enough opportunity to compare this Teuton with Swiss people, who were of no less pure Allemannian breed, yet so much more natural. The Swiss are the only branch of the big family of German tribes who have preserved a continuity of historical development. The Thirty Years' War, which almost destroyed the German civilization and prepared its transformation into Prussianism, did not sweep over Switzerland, which continued her old democratic way of life.

Although Zurich is a big city with a considerable amount of industry, it is thoroughly Swiss. There are no slums, hardly any poverty, but a bourgeois life even among the socialistic masses. There is the lake, surrounded by lovely hills, full of pleasant country houses, large and small. In the distance you see a chain of snow-clad mountains. On your right you have the long Albis chain, culminating in the Ütli mountain and covered by mighty forests where the ruins of many an old castle can be found. It is lovely to stroll in these hills and woods, in particular if you have read Gottfried Keller. What Scheffel is for Heidelberg, Keller is for Zurich, with the considerable difference that Keller is a far greater poet. His art makes the poor remains of the castle Menagg on the Albis slope come to life again. There is also no lack of hospitable gardens and cellars where you can get a good Swiss wine very cheaply. All that explains why I found six hours of mathematical lectures per week quite enough. The weekends were very often spent on long excursions. Lilly Weber stayed part of the summer with her Imhoff grandparents in Winterthur, and I visited them, as I have already said. Very often I met cousin Franz Mugdan (who was a medical student at Freiburg) at some railway station near the higher mountains and went with him on a long tour. When I described his life earlier I said how much I owed to him for compelling me to overcome any weakness in my constitution, in particular my asthma. We climbed to the top of the Säntis, the Speer and other mountains near the Walensee in Glarus, also the Rigi and Pilatus. The latter excursions may seem rather trivial, as there are railways up these two mountains. But we did not use them, though I must confess that we did not reach the top of Pilatus, because we ran into a dreadful thunderstorm

75

and had to take refuge in a hay-barn, where we stayed all afternoon and the whole night, and we were glad to get safely down the next morning. The longest walk was that from Glarus over the Klausen Pass to the Gotthard road and for me it was a test of energy; for I had a bad attack of asthma during the night in the stuffy room of a village inn and could hardly breathe in the morning. But after having slowly climbed a few hundred metres it became better, and I did it.

The semester in Zurich ended by the middle of July. I intended to spend a fortnight in the Engadine with the Bernhardts and then to join cousin Franz and his sister Bertha for a long hiking tour through the Bernese Oberland. So I asked for money from home and received it at breakfast, 250 Swiss francs, a sum so extraordinarily high for this modest place that it caused a stir among the other guests, in particular the poor Russians. The same evening there was a Russian dance to which I was invited. It was a strange affair. Many girls and some men in Russian costumes; wild Russian dances, and a lot of alcohol; a performance of a revolutionary play in which some grand-duke behaved abominably to his peasants and was murdered in the end; a lot of kissing and fraternizing. Late at night Maitoff took me into a corner, told me what a marvellous chap I was—for a German, at least—and produced some fantastic and sentimental story that he must say goodbye to me, as he had just got a telegram ordering him home on an important secret mission, that he had to leave the same night, but had no money. Could I give him some? Well, I could hardly say no as they had all seen my wealth arriving in the morning, and so I said: yes, I could; how much? The answer was: 'Oh, 200 fr. will do; you know it is frightfully far to Moscow.' I was a little worried but he said that he would stop in Berlin on his way, where he had friends, and he would return my money from there. After he had promised this and confirmed it by violent handshaking, I gave him the money still under the emotional spell of an evening spent among those passionate and excited foreigners, and under the influence of wine and vodka.

Then I waited for my money; but nothing came. I asked the other Russians for Maitoff's address but they declared they did not know: 'You see, he is on a secret mission.' The day of my intended departure approached, and I had not enough left to pay my bill. So I had to go to Frau Fick and confess my stupidity. They all laughed heartily, and as they knew me to be a young man of means they lent me a sum sufficient for my journey.

When a few weeks later I read in the newspapers that some grand-duke had been assassinated, I did not relish the idea that my money might have had some part in a murder. I tried to find out whether Maitoff was involved in the incident, but in vain. And I have never heard anything of him, nor of the whole lot of Russians again.

I have done the journey to Samedan very often since, but every time I have felt a deep and growing elation when the little train from Chur climbed the winding track, higher and higher up to the Albula tunnel, and then after ten minutes in darkness, emerged into the glare of the valley over which tower the white giants of the Bernina group. Yet this first time it was rather disappointing. The weather was bad, and in the train there was a family from Breslau with a fat daughter, who could not understand why I was going to Samedan and not with them to the fashionable Pontresina, which I was unwilling to explain. Then when the train emerged from the tunnel there was no Engadine but only a thick whitish fog. And at Samedan the good hotels were extremely expensive, and only after some searching did I find a simple clean inn where I got a cheap tiny room.

When I called on the Bernhardts, there were Lilly and her five sisters, all gay and lovely, and the dignified doctor and his wife. I think they all took me for a little savage, a northern bear who had to be taught the higher accomplishments of civilization. Whenever I came back from a long day's tramp, they did not give me much rest, but took me to the tennis court or gave me dancing lessons. Once or twice they arranged excursions. One of these is fixed in my memory, on the Muottas Muragl, the hill which forms the corner where the Pontresina valley branches off from the Engadine. There I had for the first time a full view of the Bernina Alps in all their glory, and that in a most pleasant and entertaining party of young folk with old names—Salis and Planta were there—natives of this beautiful country who spoke Switzer-Dutsch or Romansh among themselves. But for most of my walks I went alone. Once, I remember, I marched all the long way up to Maloja, and after having consumed quite a lot of a good Veltliner wine I felt drowsy and stretched my limbs on a lovely meadow overlooking the Lake of Sils. There I fell asleep and when I awoke the shadows were already lengthening, and the last postchaise down the valley had gone, and I had to walk back the whole long way to Samedan where I arrived after nightfall, tired but happy. At that time the Engadine was not as crowded as it is now, only a few horse-drawn postchaises provided transport. It was not easy to leave this paradise, but I had promised to meet the Mugdans for a longish hike.

There would be no point in reporting that trip if it had not been the only time in my life when I did something approaching real mountaineering, the only time when I reached the 4000 metre (13 000 foot) level, to which only a small number of alpine summits rise. We started from Lucerne and went over the Kleine and Grosse Scheidegg, the celebrated track at the foot of Mönch, Eiger and Jungfrau, to Lauterbrunnen. There we met Franz's and Bertha's mother, and with a guide, we crossed the mountains, over the Tschingel pass to Kandersteg. This was the longest and most fatiguing day's walking I have ever done, 14 hours or so, and more than half of it over ice, always on the rope. But it was grand. In Kandersteg we spent the evening drinking beer and playing Skat, instead of resting; with the result that my heart failed when going to bed. A doctor was called in, but after a rest of two days he allowed me to go on. So we crossed the Gemmi, most of the time in a dreadful blizzard, and reached the Rhône valley at Leuk. Then we went to Zermatt, and after a few smaller trips Franz and I, with a guide, climbed the Breithorn, a colossal ice-covered dome between Monte Rosa and the Matterhorn, about 4100 metres high. The hardest part of this tour was the endless steep track back to the Theodul Hütte, where we spent the night. I was always afraid of these nights in alpine huts, not because of discomfort, but because of the smoke which easily produced an asthmatic attack. But here on the Theodul pass, between the Matterhorn and Breithorn, were only serious mountaineers who after a hearty meal watched the marvellous sunset over the glaciers and then went straight to bed; for the next day began for all of us at 2 or 3 a.m. to avoid the thawing of the surface snow by the morning sun. The ascent from the hut to the summit was no hardship and was an unforgettable experience. The track was mostly over ice, and often so steep that the guide had to cut steps with his ice-axe and while we were waiting for the next step we had time to enjoy the view of the white mountains around us, first illuminated by the stars, then by the reddish light of dawn. When we reached the summit, the sun was just above the horizon. The way down was easy and fast, since we could slide down many slopes on our seats. We added an extra climb up the so-called Kleine Matterhorn, a rock of similar shape to the 'great' one, but much less formidable.

In later years I never repeated such extended tours. I was content to climb the smaller summits of the Engadine, Piz Languard, Piz Corvatsch, etc. Franz Mugdan, however, became a real mountaineer, while I resigned myself to less ambitious sports, my last one being golf, in loyalty to my new home country, Scotland.

VI. Student in Göttingen

The following winter (1903/4) was the last one I spent as a student in Breslau. The reasons for leaving were many. First of all I found the lectures at Breslau a little stale after those of Hurwitz. Then life at home was not very attractive. My stepmother did what she could for us, and there were many parties given for my sister, yet I disliked her set, or at least her numerous admirers. Cousin Franz was away at Freiburg and I had no other intimate friends. My mathematical friends, though decent fellows and clever enough, meant nothing to me beyond science. My only refuge was the Neisser house, where I spent many evenings and weekends. There I found plenty of amusing and interesting people and I could write a long list of celebrities whom I met at Toni's dinner table, such as Gerhard Hauptmann the poet, and Richard Strauss the composer, to mention only the most exalted names. But I was a young student and had to be content to listen to the great ones from a distance, without coming into real contact with them—apart from a few musicians already mentioned, Artur Schnabel and Edwin Fischer, and one man from a very different sphere, Arnold Berliner, later widely known as the editor of the popular scientific journal *Die Naturwissenschaften*. At that time he was on Rathenau's staff of the AEG (*Allgemeine Elektrizitätsgesellschaft* = General Electric), director of the electric lamp factory, and an expert on gramophones and other technical things. He was a cousin of Neisser, and like him was full of life and energy, devoted to art, music and literature. As soon as he found out that I was studying physics, he became interested and involved me in fascinating discussions on scientific and philosophical problems. I was very diffident about my own talent, and the fact that Berliner took me seriously gave me enormous encouragement. He even went a little further in this respect. Once he offered me a wager that I would be a professor in less than ten years' time, and he did this not in private conversation, but at Toni's dinner table, to my profound embarrassment. However, he won the wager, though by a narrow margin, and he was paid, according to the conditions, in bottles of wine.

The Neisser house was pleasant enough; but they were preparing a second journey to the Dutch East Indies, in order to continue the work on syphilis, and I was loathe to spend another year in Breslau without this refuge.

Toeplitz declared that if I wanted lectures of Hurwitz's standard there was only one place in Germany—Göttingen. He himself intended to go there, after having got his doctor's degree in Breslau. I knew nothing of Göttingen; I was not even quite clear where it was, it sounded similar to Tübingen, which I knew was near Stuttgart in South Germany, and I was quite astonished to discover that Göttingen was in the North, in the Prussian province of Hanover. My heart's desire was to go to Munich, not because of a particular teacher, but in order to continue mountain climbing in the summer and to enjoy music and carnival in the winter. If I had known that Sommerfeld was there and what kind of man he was I would have gone to Munich. But I did not and Toeplitz did not tell me; he was such a 'pure' mathematician that he may be excused this omission. I, on the other hand, already had an inkling of what mathematical physics really was and that it interested me more than pure mathematics. I had attended a lecture on Maxwell's theory of electro-magnetism, given by a young lecturer at Breslau, Clemens Schäfer; it was a bad lecture, as lectures by young and inexperienced men are bound to be, but the matter was fascinating in spite of the poor and foggy representation. It was perhaps just as well that I did not go to Munich; I might have won my wager with Berliner a little earlier and acquired a better technique in my subject, but very likely I would have lost my independence of thought and what originality I may have in tackling the problems of theoretical physics.

Apart from Toeplitz's advice there was another incentive for me to go to Göttingen, namely that my stepmother gave me a letter of introduction to a professor of mathematics, Hermann Minkowski, whom she had met years before at dancing lessons and balls at Königsberg. There were three brothers Minkowski, all very gifted. One was a successful merchant and financed the studies of his brothers; the second became a physician, later well known for his fundamental investigations on diabetes; the third brother, Hermann, was already one of the leading mathematicians when I met him. He and David Hilbert were the Castor and Pollux of the mathematical world, intimate friends from their school-days at Königsberg and inseparable at the University. The dominant figure in mathematics in Germany at that time was Felix Klein, who held another chair in Göttingen. He had already passed the peak of productivity but had an immense influence on all things concerned, however remotely, with mathematics. He persuaded his faculty (1895) to nominate young Hilbert to a second professorship, to which the Prussian Ministry of Education agreed. From that time on Hilbert was the star of mathematical Göttingen. But he missed his friend Minkowski and worked incessantly to get him called to Göttingen. In 1902 he succeeded; a third professorship was offered to Minkowski and the dioscuri were re-united. (There was a fourth chair of applied mathematics, held by Carl Runge, well known also as a spectroscopist.) In this way the young Jewish scholar Minkowski broke the anti-semitic barrier at a Prussian university, which men like my father had been unable to overcome. The philosophical faculty in Göttingen was in many respects more advanced and liberal than many other faculties in Germany. A short time later another Jewish professor was appointed on Klein's suggestion, the astronomer Schwarzschild.

I arrived in Göttingen together with Fritz Alexander, to whom I owed my acquaintance with James Franck in Heidelberg, and we took rooms in the same house, Walkemühlenweg 24, which belonged to a teacher at an elementary school, named Heidelberg. But that coincidence of names did not mean a return to my happy-go-lucky life of the summer before. My private existence during the first year in Göttingen was not particularly pleasant and has left few impressions in my mind. The little town and the country surrounding it, which in later years I learned to love almost more than my homeland, did not attract me at the beginning. I compared the town with Heidelberg, the hills and woods with the Odenwald and the Sudeten forests; everything seemed to be small and tame. There is one incident which I remember. Alexander and I were returning from a cycling tour to the lovely Bremker Tal, when on the lonely road two tramps jumped out of a ditch and attacked us—whether in earnest, for robbery, or in fun I do not know. They tried to push their sticks into the spokes to make us fall off, but we succeeded in getting away unscathed apart from a painful blow I got on my back. This infuriated me so much that the very next morning I went and bought a revolver and ammunition (which one could get without a police permit). Then we put some strong boards against our garden wall, drew circles on them with chalk and used them as targets for revolver practice. You can imagine the attention which was given to our noisy sport by the neighbours. The next day when I came home from my lectures I found Herr Heidelberg waiting for me in great excitement: 'A policeman has been here, because of your shooting—it is forbidden to use firearms in private gardens, it will cost you at least a hundred marks!' That gave me quite a shock, for a hundred marks was about my monthly budget and a little more than that sport was worth. But my good landlord continued with a sly grin: 'I persuaded him to drop the case; I simply asked him: Aren't your children in my class in the *Bürgerschule*?' I report this story as one of the rare cases where civil authority prevailed over the mighty Prussian police.

My scientific life was inspiring and fascinating right from the beginning, just as Toeplitz had predicted. I concentrated on mathematics and physics, represented by Hilbert and Voigt. Hilbert's research work is clearly divided into several distinct periods. After his first great success in the 'Theory of Numbers' he solved the central problem of the theory of invariants. Then he attacked the foundations of geometry. When I came to Göttingen everybody was eagerly studying his book *Foundations of Geometry*. Though it had precursors, such as Pasch's writings, it can be considered as the first great step beyond Euclid. It rests on the same idea: to derive the whole of geometry from a finite set of axioms by pure logic, but it formulates the logical foundations, the different possible systems of axioms, much more rigorously and completely than the great Greek did 2000 years earlier. It culminates in the general idea of the axiomatic method as applicable not only to any branch of mathematics but also to physics, even to any system in the wide domain of human abstract thinking. After having finished this great programme, he attacked a new field, opened by the Danish mathematician Fredholm, the 'Theory of Integral Equations', which in Hilbert's hands was transformed from a special branch of analysis into a fundamental investigation of the general features of linear relations between infinite variables, containing as special cases not only Fredholm's equations but also the theory of functions and all the innumerable linear differential equations which occur in different parts of mathematics and mathematical physics. This work later became the foundation of modern quantum mechanics, where even physicists use the expression 'Hilbert-Space' for the extension of an infinite number of dimensions by which the states of physical systems and their changes are described.

When Hilbert was working on such a new subject, he used to announce a lecture series with a harmless title, such as 'Differential and Integral Equations' or 'Calculus of Variations', but developed his theme within the framework, sometimes with revolutionary consequences. I joined one such course in my first semester in Göttingen. I have already mentioned the system whereby one particular student was charged with working out the lecture very carefully and with producing several typewritten copies, one of which was made accessible to the students in the Mathematical Reading Room. In my first lecture, Hilbert asked students to come forward with a sample of their notes. About half a dozen offered him their manuscripts, amongst them myself. At the beginning of the second lecture he said: 'There is one manuscript by far excelling all the others, and I beg Mr Born to take over the duty of working out my lecture for me and the Reading Room.' I was most embarrassed as I had to step forward to take back my manuscript under the eyes of the vast audience, but I was also very happy—and with good reason. For now I had the closest contact with the great mathematician; every time Hilbert had read my summary of a lecture there was a long discussion with him about mistakes and possible improvements, and I had the opportunity of getting a glimpse into the working of one of the most powerful brains of that period. But more than that; Hilbert took a liking to me right from our first meeting and soon treated me as a trusted young friend.

This personal relationship was helped by an incident connected with my stepmother's letter to Minkowski. After having duly delivered it, I received an invitation to dinner (German style, at 1.30 p.m.) from Frau Minkowski for the following Sunday. I was received most cordially and enjoyed an excellent meal; there was only one other guest who later was to be my great friend and of considerable influence on my scientific development. He had a very un-German face, fine and noble features, a long nose and high forehead, and a faint foreign accent. He was introduced as Dr Constantin Carathéodory; I had already heard of this brilliant young Greek who was about to be admitted as lecturer in mathematics. He amused us by telling stories about his work as railway engineer in Egypt. Later I learned that his father, although a Greek, was Turkish Ambassador in Brussels and a leading diplomat; he appears in pictures of the Congress of Berlin along with Bismarck and Disraeli, as the representative of Turkey.

After dinner I wanted to leave, but Minkowski asked me whether I would not like to join in a little excursion to a ruined castle in the neighbourhood, the 'Plesse'; Carathéodory would be in the party and also Hilbert and his wife.

This gave me a thrill and I accepted eagerly. This first walk through the fields and woods to the 'Plesse' will always be one of my cherished memories, although I repeated it later many a time, also in pleasant company with dear friends and with my family. When we met the Hilberts at their garden gate, he at once pounced on me: 'Hallo, there you are—that is nice. I must tell you that I have found a shorter way of proving that theorem of the last lecture'—and so on. The Minkowski's were quite surprised that no introduction of their young guest was necessary.

After having talked shop for a while, Hilbert began to become 'human' in his peculiarly droll way, discussing with Minkowski the events of the week, academic and political, and commenting on them. I was familiar, from my father's circle, with an open and witty discussion, but that between Hilbert and Minkowski was of a different kind. Medical men, even though scholars and scientists, are simple and human and straightforward, compared with mathematicians, whose brains work in spheres of higher abstraction. I was not bred in an atmosphere of tradition and accepted values; but I had never heard that kind of critical and ironical dissection of ideas and conceptions which Hilbert produced when properly stimulated. And Minkowski obviously enjoyed stimulating him, until he went in his rage far beyond his own convictions—then Frau Hilbert stopped him with a sharp 'But David!', whereupon he often went to the opposite extreme, praising some ridiculous thing he had just condemned. There are many stories about Hilbert preserved by his pupils and friends, and I hope they will one day be collected; I remember a few of them but find them untranslatable.

I have already said that Hilbert's regular courses were always leading into new country. One of these was on advanced mechanics, built on the Hamilton–Jacobian methods and the idea of canonical transformations. What I learned there was later of the greatest help in the development of the mechanics of the atom, in the period (1920–25) which preceded the birth of quantum mechanics. One of Hilbert's most remarkable lectures was that on the 'Kinetic Theory of Gases' where he gave the first systematic, and to some degree rigorous, solution to Boltzmann's fundamental equation, expressing the effect of the collision of molecules; Hilbert showed that it could be reduced to an integral equation with a symmetrical nucleus. This lecture was also attended by a young Norwegian, Enskog, who later worked out and published this theory in detail. In England it was taken up by Chapman who published a thick volume on it. I have the impression that in simplicity of conception and clarity of representation none of these later works can compare with Hilbert's lectures, a copy of which I possess.

The most attractive of Hilbert's courses were some shorter ones (one or two hours a week). One, on the logical foundations of mathematics, I have already mentioned. It gave a systematic survey of the axiomatic method in all branches of mathematics as well as physics, and it culminated in a short presentation of Cantor's theory of infinite numbers and of Boole's logical calculus, which was later Hilbert's main tool in his profound researches into the foundation of mathematics.

Perhaps the most fascinating of all his lectures was one on the 'Quadrature of the Circle', a title which covered only a small part of the wide subject, namely a semi-historical survey of all the 'insoluble' problems of mathematics and a systematic investigation of what solubility really means. It began with the proof that $\sqrt{2}$ cannot be written as a quotient of two integers (or as a 'rational' number) and climbed up to the heights of the proof that the ratio of the circumference of a circle to its diameter, π, cannot be constructed with compass and ruler, as it is 'transcendental' (i.e. as there exists no algebraic equation with integral coefficients, such as $5x^7 + 6x^4 - 3x + 2 = 0$ which has π as a root). The proof given was not that of Lindemann, who first solved the problem, but a much shorter one invented by Hilbert himself. My presentation of this proof pleased Hilbert exceptionally well, and he suggested to me—it was in my fourth year of study—that I should do some research in this line for a doctor's thesis. The particular problem he gave me was the proof of the transcendental character of the roots of the Bessel functions. (The number π can be defined as a root of the function $\sin x$, and $\sin \pi = 0$; the Bessel functions $J(x)$ are also wave-like functions, but with decaying amplitude:

and they have roots where $J(x) = 0$.) He also indicated a method, and I began to study these problems. Many a day and many a night I brooded over them, but without the slightest success. It was then, after some months of intense work, that a deep feeling of frustration came over me. I went straight to Hilbert and told him that my fluent and clear writing of mathematical demonstrations was a deception, that I was certain I was no true mathematician and that I would never again attempt to solve a problem from the domain of numbers or any other branch of pure mathematics. He laughed and said the same had happened to him when he was young; I should not be depressed.

And in fact I was not. I saw there were so many things in science where one can apply the average faculties of a normal brain that I decided to wait for my opportunity. It came only a short time later; but before I speak of it I shall mention some of the other lectures and other teachers.

I took only a few of Minkowski's courses, one on 'Geometry of Lines and Spheres' as developed by Plücker, and another on 'Analysis Situs' (today generally called topology). This latter course is memorable because of an incident which was quite characteristic of Minkowski's ways. He began by saying that a good method of introducing the concept of this subject was the study of the famous four-colour problem of the map-makers. They had found out by experience that if you want to colour a map in such a way that any two neighbouring countries are marked by different colours you never need more than four colours, however complicated the map may be. Mathematicians had tried to prove this statement, which is typical for analysis situs, namely independent of the exact shape of the boundaries as long as their number and arrangement is preserved. Then he continued: 'No proof has come forward so far, in spite of innumerable attempts. But as far as I can see only third-class mathematicians have attacked this problem. When preparing this lecture I found a simple proof which I shall present to you.' And he began to explain the fundamental ideas of analysis situs, using the map problem as example, while we listened spell-bound, expecting to witness the revelation of the solution of one of the famous ancient problems. Now this preparation for the demonstration went on for a week, continued into the second, the third and the fourth. It was already summer, just before the short Whitsun vacation, and one morning when we assembled in the lecture-room a heavy thunderstorm was raging. Just as Minkowski entered the door a colossal crash of thunder roared above us. He moved slowly to the desk and said with his soft, deep, vibrating voice, a sad grin on his face: 'Heaven's wrath is upon me for my conceit. My proof is also wrong.' It sounded so sincere that nobody laughed, until he himself began to chuckle. Then he warned us not to waste our time by attempting a solution and continued his lecture according to plan. For the map problem had perfectly served its purpose of introduction.

Several times I attempted to listen to a lecture course given by Felix Klein; but I always gave up in the middle. He was the most brilliant lecturer; in fact too brilliant for me. One needed to be a really pure mathematician to enjoy them, even when he was dealing with applications to physics and technology, which he liked to do; for these things in his treatment lost their own nature and were transformed into play-grounds for mathematical conjuring. At that period of my life I was rather inclined to mathematical rigour—epsilontics as we used to call it in our jargon, because every demonstration began with words like that: 'If a set of numbers a, a_2, . . . is given such that $a_n < \varepsilon$ if $n > N$ and if, etc. etc.' This ε (epsilon) hardly ever occurred in Klein's lectures; when he came to such a critical point he used to say: 'I'll give you an idea of the matter; the rigorous proof can be read in one of the following books (or papers)', and he quoted the literature. He would even add a little contemptuous remark about the people who needed a formal logical proof for such an obvious fact. Now Hilbert's proofs in his lectures were also often rather sketchy; but I never had any difficulty in reconstructing them according to his clear indications, while in Klein's case I found myself very often hopelessly lost and compelled to read a great many dull papers. In their general structure Hilbert's and Klein's lectures were very different: Hilbert was like a mountain guide who leads the straightest and best way to the summit; Klein was more like a prince who wants to show his admirers the greatness of his territory, guiding them in endless winding paths through apparently impenetrable ground and halting at each little hilltop to give a survey over the area covered. I got impatient with this method; I wanted to reach the summit quickly. Therefore I preferred Hilbert. Today my taste has changed. If I ever have time to read mathematics for mere pleasure I take one of Klein's books, for instance his *Development of Mathematics in the Nineteenth Century* or his *Non-Euclidian Geometry* from my shelf, and I enjoy it like a work of art, a great novel or biography. Indeed, some of these books by Klein are an integral part of the civilization in which I have grown up and which is now disintegrating.

There were many other lecturers in mathematics whose courses I attended; I will mention only Zermelo, who gave a very interesting course on Cantor's theory of sets and the logical paradoxes connected with it. He was at that time also interested in the foundations of statistical mechanics and believed he had found a logical contradiction. He had given a simple proof that every dynamic system was almost (or quasi) periodic; that is, it would return very nearly to the initial configuration if one only waited long enough. But statistical thermodynamics led to the conclusion that a system of sufficient complexity, like a gas consisting of innumerable atoms, behaved in an irreversible way, approaching from any initial configuration a final state of statistical equilibrium. These two statements seemed to be contradictory, and Zermelo believed he had demonstrated that the statistical ideas of Boltzmann and Maxwell were futile. This of course infuriated the physicists, and we had lively discussions about it. I remember that I was rather inclined to side with the physicists; I had no decisive argument, but I preferred common sense to formal logic—though I was still somewhat ashamed of it. This preference has slowly grown with the years. Zermelo's paradox was later cleared up by Paul and Tatjana Ehrenfest, in some papers and in their brilliant article in the *Mathematical Encyclopaedia*, by making a sharp distinction between rigorous mathematical theorems which are of little interest (such as Zermelo's periodicity, since the period even for a cubic inch of a gas is of cosmological length) and probability predictions which are applicable to real physical measurements.

This brings me to the courses in physics which I attended. My first attempt was a complete failure. The lecturer Johannes Stark had announced a course on radioactivity, and as there were many more or less sensational reports on this subject in newspapers and periodicals I took the opportunity to learn something definite about it from an expert. In later years I read papers by the Curies, by Rutherford, Soddy and others, and I know that hardly any other part of physics can be built up in such a convincing way from simple experiments by inductive reasoning. Although I was rather spoiled by the deductive methods of mathematics, I was very willing to learn from experience and to generalize the laws thus obtained. But what Stark did seemed to me pure dogmatics. He did not demonstrate experiments but described them in rather obscure terms, and the results were formulated like articles of faith without any attempt at explanation. After a few lectures I decided that if this was modern physics, I would rather have nothing to do with it.

Luckily I had other and better teachers. Stark later crossed my path more than once. He was a genius at handling apparatus and playing with it. But he never understood physics. At that time he was lecturer in Professor Voigt's department, on whose advice he undertook experiments on the light emitted by so-called canal rays, consisting of positive particles produced in evacuated tubes with perforated cathodes. He made two important discoveries; first he proved the existence of the Doppler effect in the spectrum of fast-moving hydrogen atoms, and later he showed the splitting of the hydrogen lines in an electric field, a phenomenon called the Stark effect. For this he was awarded the Nobel Prize and became a great figure in German physics, although he never produced another result of importance. He was a cantankerous man who got into trouble with every faculty of which he was a member. And as he needed a scapegoat for his bad temper he began to hate the theoretical physicists because he did not understand them, and the Jews because among the theoretical physicists there were a considerable number of them, especially one, who after the great war of 1914–18 won world fame—Albert Einstein. Stark was the centre of opposition to the theory of relativity, and later against quantum theory (although Planck, its originator, was a pure 'Aryan'). He joined forces with Philip Lenard, organized anti-semitic demonstrations against 'Jewish physics' and wrote pamphlets and books in the same vein. He became, of course, an ardent Nazi, and as soon as Hitler came to power Stark was appointed to the important post of President of the National Physical-Technical Institute.

There were two chairs of physics in Göttingen, an experimental one, held by Riecke, and a theoretical one, held by Woldemar Voigt. After my experience with experimental courses as given by Meyer in Breslau and Quincke in Heidelberg, I made no third attempt and never came into contact with Riecke. Voigt, however, had considerable influence on my scientific development. He gave a series of lectures covering the whole of theoretical physics; but he also had a flourishing experimental research school where investigations on optics, magnetism, and crystal physics were carried out. I attended his lectures on optics. At the beginning I did not much like them; it was all very clear, but rather dull and tedious because of endless calculations. One of the desks had an inscription carved in it which expressed the feelings of a bored student very impressively with the help of a quotation from Schiller's *Wilhelm Tell*:

Gleich schlägt es zwölf—dann kann ich mich verschnaufen.
Mach Deine Rechnung mit dem Himmel, Voigt,
Denn Deine Zeit ist abgelaufen.

(Twelve o'clock will soon be striking, then I can have a breather,
Settle thine accounts with Heaven, Voigt,
For thy time has come.)

But nevertheless the lecture had its merits, and I was saved from dropping it by a classmate who later became a great physicist and a friend of mine, Max von Laue. How I made his acquaintance I have forgotten. I was quite well known as Hilbert's lecture assistant, and I suppose I had some business with von Laue in this capacity. He took to me and involved me in fascinating discussions on the problems raised in our lectures. He had already graduated Dr Phil. in Berlin under Drude and had come to Göttingen to attend some advanced courses. When he noticed my inclination to give up Voigt's lectures, he strongly opposed this and went so far as to fetch me from my digs and to convoy me safely to the lecture-room. I am still grateful to him for this service, for I learned a great deal in these lectures and still more in the practical course on optics connected with it. It was an excellent course, attended by a small number of advanced students who were introduced not only to the elementary facts of optics, but also to rather complex phenomena of crystal optics, electromagnetic optics and spectroscopy. What I learned there formed the root out of which grew, twenty-five years later, my own text-book on optics.

Felix Klein, though himself as pure a mathematician as could be, was deeply interested in all forms of applied mathematics and worked unceasingly in founding new departments representing special applications. Through his influence there came into being a Seismological Institute on the Hainberg, and a chair first occupied by Wiechert; an Electro-technical Institute directed by Simon; an Institute for Applied Mechanics and a chair still held by Prandtl, and others. I occasionally attended lectures by all of these and I came into closer contact with Professor Simon by taking part in several technological excursions to centres of industry which he organized for students. The first of these was at the end of my first term in Göttingen (August 1904) and took us to the Rhenish-Westphalian Industry. We visited all the great factories and saw many important branches of production. The journey culminated in Essen where we visited the Krupp works and were treated as the guests of the owner, staying in the firm's splendid hotel, the 'Essener Hof' and enjoying lavish banquets after the day's strenuous work of inspecting. There I had the first glimpse of modern German industrialism. How insignificant compared with these gigantic enterprises were our Silesian textile works, and how different the men directing them! But at that time I had no sinister foreboding of what this might lead to; I was just deeply impressed and proud as any ordinary German student was bound to be. Most of the participants were Simon's pupils, students of some branch of technology, and I was a kind of outsider who asked ridiculous questions which amused Professor Simon. He liked to explain things to me and we became quite good friends. He was a youngish man, with fine features and a friendly expression, in whose hospitable house I later spent many a lively evening. He easily persuaded me to take part in other excursions; in this way I saw a good deal of German industry, for instance the Zeiss works in Jena, the shipyards and the harbour in Bremen.

At the beginning of my second year, Toeplitz appeared in Göttingen, and I took this opportunity of disengaging myself from Alexander by joining Toeplitz. We took rooms in the same house, Kirchweg 4, belonging to a gardener called Rieke. This street, at the other end of the town from my former abode, was much nearer to the auditorium building containing the mathematical lecture-rooms, together with those of law, philosophy, history and other arts subjects. Toeplitz had a front living room and a back bedroom downstairs, and I had the corresponding rooms upstairs. We got on quite well for a time; but as time went on a certain tension developed which worried me a lot.

Toeplitz had graduated Dr Phil. in Breslau, with a learned dissertation on some algebraic problems, and I had attended the graduation ceremony at the end of the previous term. This was a strange mediaeval function, performed in the beautiful baroque *Aula* of Breslau University, partly in Latin, before an audience not only of students, but of anybody from the street who dropped in to enjoy the picturesque proceedings. The Rector and the Senate (a small number of selected professors) were present in magnificent robes. Half a dozen candidates for the doctor's degree—the only one given by the University—had to defend their theses against opponents. These opponents were supposed to be members of the audience, but were in fact the candidate's friends who had been well instructed about their questions. I acted in this capacity for Toeplitz, and we had several rehearsals of our learned dispute in which I, the opponent, was duly defeated by the doctorandus. All went well in the *Aula*, and when the candidates had victoriously emerged from the attacks of their adversaries, the Dean of the Faculty of each man praised his merits in Latin and the Rector decorated him with the doctor's hat. It was a charming remnant of olden times, yet just a little ridiculous, and it was given up a year or two later.

Now Toeplitz came to Göttingen as a proud doctor of philosophy and at once tried to reassume the patronizing attitude to me which he had used in Breslau. But alas, things had changed. I was by now Hilbert's 'private assistant'—he had offered me this unpaid position at the end of my first year—I was on easy terms with Minkowski, acquainted with all the lecturers and more distinguished students in mathematics and physics. Toeplitz, on the other hand, was the newcomer and depended on me to introduce him into these magical circles. This was a minor source of friction, though I do not think that I took undue advantage of this situation. More serious was the fact that in spite of my attachment to Hilbert, and even with his approval, I began to move away from pure mathematics and began to be absorbed into natural science. Toeplitz, strangely enough, took this as a rebuke and a criticism of the education which he had volunteered to give me in Breslau. I think he really believed that I had a specific gift for mathematics and ought to follow my vocation. But there was also a certain narrow-mindedness; he had learned from his father to consider mathematics as the summit of human attainment and he simply could not understand how anybody initiated in its mysteries could give it up lightly for other pursuits.

Now these pursuits brought me into contact with many other people and affairs which were not accessible to Toeplitz and that made him jealous, for he really loved me. I was loath to hurt him, but I had to free myself from this strange bondage. This situation was still more complicated by the arrival of other mathematicians from Breslau. First came Hellinger, a year after Toeplitz; and a year later Courant, whom I have not mentioned so far—for I hardly knew him in Breslau. He was several years my junior and had attended the same school but left it because his parents were not able to afford such an expensive education. He then showed amazing perseverance and industry, in studying privately without much help, for lessons were expensive. He earned his livelihood by giving lessons, and he even assisted his relatives. The only help he obtained came from my Aunt Käthe Mugdan, Franz's mother, who took him into her house for a considerable time. He succeeded in passing the school-leaving examination as an 'external', a year ahead (if I remember rightly) of his previous classmates. Then he studied mathematics, always relying on his own resources, first in Breslau, then in Göttingen. There he joined our group—four men from Breslau; it was only natural that we formed a group, and that we were acknowledged by others as a group. Hilbert in particular did this quite deliberately. When I decided to ask him to release me from my assistantship because I had to concentrate on my doctor's thesis, I recommended Hellinger as my successor; and when Hellinger too had to give up to prepare for an examination, Hilbert insisted on having Courant, 'since those fellows from Breslau were so useful'. Courant was useful indeed, and fifteen years later he became Felix Klein's successor.

There were other groups among the young scientists in Göttingen. The most important of these circles was '*der mathematische Stammtisch im Schwarzen Bären*'. '*Stammtisch*' means a table in a beer or wine restaurant reserved for a group of people who would meet at certain hours for a drink—or drinks—and a chat. The 'Schwarzer Bär' was an ancient inn with a small beer-cellar in a dark side street of the old town, where an excellent Bavarian beer was served. There were several '*Stammtische*', some of students' corporations, others of burgers, or 'philistines', as the students called them, and lastly one of young learned men, of all faculties, but mostly mathematicians and scientists. They were by no means a club or society; but they had the knack of treating an unwelcome intruder in such a polite but icy way that he would never come again. Therefore it was considered an honour to be an acknowledged member of this table. It was not sufficient to be a good scientist or teacher; you had to produce a particular kind of social gift, wit or originality, to be admitted. Not that the talk was permanently on a high level of *esprit* and entertainment, but almost every night there was at least one burst of amusing dispute or fascinating soliloquy which made it worth while to attend, quite apart from the good beer and cosy atmosphere of the place. Sometimes silence reigned, and everybody would listen to the talk of the neighbouring table occupied by a group of elderly citizens, mostly well-known characters. For example, one night two old fellows were sitting there sipping their beer, only muttering a word now and again. One says: 'Tscha, neighbour Maier's son has now set off for Africa too.' Long silence; then the other: 'Oh, then he'll have to change trains at Eichberg' (which is a junction just twenty minutes outside Göttingen).

As long as I was a student I could not, according to unwritten laws, become a regular attender at this *Stammtisch* but only an occasional guest of one of the members. Carathéodory introduced me, and as I was soon on friendly terms with others I frequently joined the *Stammtisch*. They were a strange lot indeed. The strangest was perhaps Zermelo, whom I have already mentioned with regard to his lectures on Cantor's theory and his attacks on statistical mechanics. He was a logician of the purest breed, always sensing offences against the laws of reason in the most harmless remarks which he then analysed with absurd seriousness. By this hunting for flaws in his neighbours' logic, he had acquired a general suspiciousness. When we had ordered our evening meal and the waiter brought our dishes, Zermelo looked from one to the other, and pointing with his finger at one of them he exclaimed with a sad and wailing voice: 'There, of course, you have a much bigger portion and a much better slice than I have—they always treat me badly'; whereupon the other had to offer an exchange which was graciously accepted. But though he was grotesque, he was respected not only for his mathematical talent but also for the integrity of his character. Then there were the two Müllers, Hans and Conrad. These two were not brothers and were opposite in every respect. Hans Müller, a man of the world, elegant not only in his dress, but also in his mathematics, was always involved in love affairs and would be seen on Sundays in the society of pretty females in the neighbouring cities of Hanover and Kassel. Conrad Müller was a sturdy fellow with a rather boorish face under a colossal skull, which enclosed one of the most prodigious memories I have ever come across. He was a living encyclopaedia and remembered everything he had ever read, mathematics as well as anything else. He also had a sound mathematical judgement but no productivity. Felix Klein made use of Müller's specific gifts for his great enterprise, the *Mathematical Encyclopaedia*, and for the administration of the library of the Mathematical Reading Room. Later on Müller became a librarian in Hanover, a position which suited him.

A very conspicuous member of this group was Max Abraham, a theoretical physicist, also well known outside Germany from his text-book on electricity and magnetism which was one of the first systematic presentations of Maxwell's theory in German and is still, in its modernized edition by E. Becker, one of the best books on the subject. He was a tall, slim, rather Jewish-looking man with a caustic humour, unable to suppress a sarcastic remark even if its victim was present. In this way he acquired many enemies, amongst them influential scientists, and he was never promoted. Once the Göttingen Academy of Science accepted a paper which Abraham considered wrong; when he met the young author in our circle he greeted him with the words: 'Hallo, you have cheated our learned Academy nicely by persuading them to publish your paper—I congratulate you', thus compressing into one sentence his contempt for the academicians and for the paper in question. Abraham was wrong in this case, as in several others. Later he opposed Einstein's theory of relativity and got himself into an awkward position as he, the champion of progress, now appeared to us youngsters as the defender of antiquated tradition. In the end he left Germany and got a chair of mathematical physics in Milan, Italy, which he lost in the first world war. His end was tragic. Just when for the first time in his life he had been offered a professorship in Germany, he fell ill and died on the way to his new post.

The man who impressed me most in this group was Erhard Schmidt. He came from the Baltic provinces, which at that time were still a part of Russia, yet dominated by a German upper class, the well-known Baltic Barons. Erhard Schmidt was no baron, his name being as common as Smith in England, but he was an aristocrat by conviction and by looks; his features were noble and expressive, somewhat like those of a young Roman emperor. If you had to introduce him to somebody, you had to say 'Herr Erhardschmidt', in one word; he explained to me that this Christian name, hereditary in his family, made all the difference as it distinguished them from the ordinary Schmidts. He had studied mathematics in Berlin and came to Göttingen with the reputation of being one of the most brilliant students; he had already published some papers, but had not yet acquired his doctorate, nor did he work systematically for it at Göttingen. His enemies said: 'Schmidt is waiting until they have to give him an honorary degree.'

Schmidt had great charm and sparkling wit, and a lively intellectual curiosity. He was soon aware of the fact that I admired him, and he accepted my homage with friendly grace. He liked an audience for the display of his wit, but he was also quite content with one attentive disciple like myself. We used to go for long walks on the Hainberg, when he began explaining the progress of his own work, at that time concerned with integral equations. His mind was as clear as crystal and his attitude to mathematics that of an artist. It was a great pleasure to listen to him as he was often carried away by the beauty of a train of thought to a rapture, from which he returned to the ordinary world with a sudden laugh. Then he began to ask questions about my own work, or about my life, my parents and my relatives. I think he considered my background just as strange as his own appeared to me. He was obviously impressed by the fact that I came from an industrial family of some standing; this and the fact that I was keen on riding and knew something about horses, appealed to his aristocratic taste.

He often told me anecdotes about the German or the Russian court; what His Majesty the Emperor Nicholas had said on some occasion or how His Majesty the Emperor Wilhelm behaved on another occasion. He took all this very seriously, having a genuine veneration for high-born people, and assumed naïvely that I shared this feeling. In fact, it puzzled me. In my parental home there had been no monarchical inclinations. In particular, the Russian tsars were considered evil tyrants. I found it very difficult to reconcile Erhard Schmidt's exceptional clarity in mathematics and many other things of science and life with his blind acceptance of traditional values in politics and society, which I considered rotten. It was hopeless to discuss such things with him. Once when the newspapers reported the assassination of some Russian duke or governor I ventured the remark that there might have been good reason for getting rid of him. This was to Schmidt an outrage, and I had great trouble in appeasing him. After that I avoided political discussions and we got on very well. I owe him a great deal, not so much as regards mathematical knowledge but as regards the way of tackling a problem by assuming the proper standpoint.

Schmidt eventually finished his paper on integral equations, which is now considered a classic, and graduated Dr Phil. Soon afterwards he was appointed to a chair in Berlin. He did not quite fulfil our expectations with regard to mathematical productivity, but I have been told by many people that as a teacher he was unrivalled. At the beginning of the first world war I became professor at Berlin and Schmidt's colleague. I did not see him very often; we disagreed of course intensely over the political situation and the war. He was my first example of the type of mind which is divided into watertight compartments: one for the free play of reason in a restricted field, the other for general use subject to strict rules of tradition. Later I came across many people with this cleavage in the brain. At that time it worried me a lot, and it made me look about for other guidance.

This I found for a short period in another man and another group of people: Leonard Nelson and the Friesian school of philosophy. Fries was a philosopher of the post-Kantian period, and Nelson a young lecturer at our Alma Mater. With regard to their teaching, Fries was Allah, and Nelson his prophet. I attended one of Nelson's lectures and later his colloquia, and I was deeply impressed by his sincerity and enthusiasm, also to some degree by his philosophy. The latter was a modified 'Critique' in Kant's sense, i.e. a rationalism which rejects the dogmas of the older philosophical systems, including those of the Church, but accepts principles *a priori* as introduced by Kant, namely the 'conditions for the possibility of experience'. Now at that time I was still under the spell of Kant, and my mathematical studies had so far strengthened the conviction that there are fundamental facts (like the axioms of Euclidean geometry) which are beyond empirical proof but nevertheless evident, by a kind of introspection. Nelson used all these arguments in a clever way, but his main interest was—like that of Kant in his later years—the application of the same ideas to ethics and politics. His political views were those of liberalism, and he maintained with all sincerity that a rational being could not have any other views since they could be derived from *a priori* principles. This again was a shock to me. For I knew people whose sincerity was as genuine as Nelson's, yet who had arrived at quite different conclusions. There was Dr Lachmann, the socialist, and Erhard Schmidt, the conservative. It was all rather bewildering, but it made me think. Kant's teaching has great attraction for one who is inclined towards rationalism. It provides a strong basis for belief and a well prepared arsenal of arguments. But it did not work with me. However, it was a slow process which made me acknowledge this failure and led me to a standpoint which might be classified as a kind of empiricism.

The people of the Friesian school assembled in Nelson's room for discussions. Some of them were sincere, clever and straightforward like himself; for instance the mathematicians Hessenberg (professor in Bonn) and Grelling (from Berlin). But there were a number of cranks and strange creatures who did not do much thinking for themselves but took Nelson's wisdom as dogma. Some of them were interesting and attractive, such as Hans Mühlestein, a former teacher from Switzerland, who was also a poet of some standing, and an enthusiast of typical Allemannian features, gestures and pronunciation. There was a majority of females, who gazed at Nelson with admiring eyes as the High Priest of their sect. Some years later, when I was already a lecturer and my views had been consolidated, I ventured to contradict Nelson or his disciples in these discussions, and one day he asked me politely not to continue my attendance since my arguments against his philosophy were bewildering for his followers. This angered me and still more my friend von Kármán (about whom I shall have more to tell you), and we did a foolish thing: we challenged the whole Friesian school to a public discussion on the principles of their philosophy.

This dispute took place in an overcrowded lecture-room of the University, and it ended with a complete defeat for us. We, that is Kármán and I, and a few friends of ours, had prepared some attacks on what appeared to us to be weak points of Nelson's system; but we were not prepared to propose a system of our own, nor even to defend systematically our empiricist standpoint. In fact we had no very definite and rigorous standpoint; we were just sceptical of all rigid formulations. Now Nelson spotted our weakness at once and applied all his wit and dialectics against us. We had, by courtesy, agreed that he himself should be in the chair, and he used this advantage without hesitation. He made fun of us 'harmless and naïve' scientists and had the laughers on his side. I think he was quite right to fight for his idealistic views with all possible means, for we had nothing better to offer. I felt this in my innermost heart, and overcame my initial resentment, remaining on friendly terms with Nelson and his followers. But from that time on, our relations were of a more private character.

He had meanwhile married a woman of great charm and wide interests who gave evening parties, where good music and entertaining discussions could be enjoyed. Nelson himself hardly ever appeared at these gatherings. He became more and more a moralist who considered even these harmless parties as frivolous and time-wasting. Many people began to regard him as a crank. Yet I had some experiences which increased my respect for his character. It was some years later when I had returned to Göttingen in order to work with Minkowski. I met Nelson on the ice-rink—skating was one of the few worldly pleasures in which he indulged; and he was very efficient, at least in comparison with the average inhabitant of Göttingen, where the climate is too mild to count on ice every year. As he found in me one who could appreciate and even execute some elegant figures on the ice we renewed our old acquaintance.

On one such occasion he told me the story of Walter Ritz. He was a young Swiss mathematical physicist (four years my senior) who was considered a rising star and just admitted to 'Habilitation' (i.e. to become a lecturer), but who was terribly ill, suffering from tuberculosis of the lungs, which had been neglected as he had not had the means to go to a sanatorium in the Alps. Nelson was deeply impressed by the man's genius and worried about his fate; he urged me to see him and I did so. I found Ritz in a small, simple room in an old house; his face was that of a martyr, thin and pale, the skin sharply drawn over the bones; beautiful, kindly eyes. He was sitting at his desk, coughing and working restlessly at his great paper on the vibrations of a rectangular elastic plate, a paper which contains the method of approximation known today on the Continent as Ritz's method. (In Britain it is usually called the Rayleigh–Ritz method, and in fact it is already contained in Lord Rayleigh's theory of sound, but without the rigorous proof of convergence given by Ritz.) He received me very kindly and sacrificed his precious time to discuss problems of physics with me. He was one of the first to attempt a theoretical derivation of the laws of spectral series and discovered the combination principle for spectral lines which became one of the fundaments of quantum theory. He was also deeply interested in electrodynamics of moving bodies and had worked out a comprehensive theory based on the hypothesis that the velocity of light, in spite of the wave character of the propagation, depends on the velocity of the source. I was at that time fascinated by Einstein's first papers on relativity which treated the same problems from an entirely different standpoint, and so we had some interesting discussions.

Yet that is not the story I intended to report; it concerns the human side of this encounter. I shared, of course, Nelson's feelings about Ritz's situation and we decided that something ought to be done. Nelson suggested we should collect a sum of money which would allow Ritz to go to Arosa for a cure. We estimated the costs to amount to some thousand marks. I wrote letters to some of my well-to-do relatives, of the Kauffmann and Lipstein tribes, and the response was quite satisfactory in view of the fact that none of these people knew anything about Ritz nor of mathematical physics in general. But when Nelson and I met again it turned out that he had already got together his whole share (he was related to one of the big banking families in Berlin) while I had hardly half of mine. So I simply added the rest from my own pocket, quite a considerable sum, almost my yearly allowance. Then we approached Professor Voigt and asked him to hand the money over to Ritz as a prize given by an anonymous donor for his scientific achievements, with proper instructions for its use. But this help came too late. Ritz died in 1909, thirty-one years of age. Nelson was despondent and relieved his feelings by violently abusing our society in general and the Göttingen professors in particular. I am sure that this accident was only one in a long chain which drove him into opposition to everything and everybody and made him an outsider and crank. He owed his career as a lecturer mainly to the mathematicians, who appreciated his reasonable attitude to mathematics in his philosophy; but he had a row with Hilbert. There was a permanent feud with the other philosophers, in particular with the head of the department, Professor Husserl, whose 'phenomenology' Nelson despised and publicly derided. Even my father-in-law, Viktor Ehrenberg, a man of mild and kind character, told me many years later that during the time he was Rector of the University he had a clash with Nelson.

Nelson's end was tragic. First his marriage went to pieces. He became lonely, badly looked after and neglected. He worked at night and began to suffer from permanent insomnia. This undermined his strength, so that when he fell ill, his body gave way and he died. I have told you his story as I think that he was one of the few who had a vision of the coming catastrophe; he suffered because his uncompromising liberalism and rationalism was offended by the reactionary tendency of the time and his noble heart could not bear injustice and selfishness.

I have just mentioned Husserl and his 'phenomenology'. He was the centre of another rather exclusive circle to which I had no real access. Near the end of my study I attended a course of Husserl's lectures, together with Hellinger, but found them so dull that I gave them up. Then we tried Husserl's seminar, which was dealing at that time with philosophical problems of mathematics, and this we found rather interesting. Husserl had come from mathematics and was well acquainted with the fundamental problems. The subject of our discussions was the epistemological validity of the mathematical axioms, and

Husserl tried to lead us to his phenomenological solution. This consisted of the conviction that by a proper kind of contemplation and reflection on the meaning of a notion you can approach the 'phenomenon' itself, obtaining in this way perfect evidence. This sounds rather obscure, but I have never succeeded in finding a clearer explanation of phenomenological reasoning. Under the spell of Husserl's dialectics we were for some time impressed by this idea. But the hollowness of it was soon obvious. It is a kind of 'a-priorism', not a rational one, like that of Kant, but a mystical one. I can quite well grasp Kant's idea that there are principles or categories of thinking which are the conditions of actual knowledge and which you can discover by investigating the structure of this knowledge. That is what we theoretical physicists are really doing—with the sole difference that we do not claim that our latest analysis is a final and irremovable law, but just another step to a remote truth. To look for ultimate evidence by introspection, contemplation and verbal analysis, is irreconcilable with science, and seems to me in practical life the source of all ideological struggle. For anyone who believes he has obtained such an internal evidence easily becomes a fanatic, a mystical believer unapproachable by reason and argument. If science stands for anything it has certainly no use for Husserl's philosophy.

As a matter of fact, the phenomenologists in Göttingen were a little group of conceited fellows who did not appeal to me. One of them, Heidecker, tried for some time to convert me to their creed, but in vain. I disliked not only his philosophy but also his personality. He became later Husserl's successor in Freiburg and wrote a book which appears to me the summit of senseless accumulation of words.

After all these accounts of scientific and philosophical circles with whom I was in contact you might get the impression that my life at that time was keyed to a high pitch of intellectualism and devoid of all simple human relations and amusement. But that is far from true. I have already mentioned that I had taken up riding in Göttingen.

There was an old Academic Riding School at Göttingen, founded simultaneously (1737) with the University by the Hanoverian Chancellor and Curator Baron von Münchhausen; a very necessary institution at that time when the Alma Mater was destined to attract young noblemen in whose education riding was certainly of higher importance than lectures on learned subjects. This school was housed in a spacious old building opposite the *Auditorien-Haus*, and the University riding master in my time was again a Baron von Münchhausen.

This name is well known in Germany (and I presume also outside) from a novel by Immermann in which a gentleman of this family is depicted as arch-liar and boaster. These Münchhausen anecdotes have become very popular through numerous children's books. Well, our Münchhausen bore no likeness of any kind to his noble relative, the founder and first Curator of the University, but a striking resemblance to that celebrated fictional figure, the braggart. He was a boorish fellow with a long grey beard, a good horseman and patient teacher, in particular if the pupil was a pretty young lady. He was extremely proud of his title and noble family, but devoid of any education, rude in his manners and given to boasting and lying as if he had inherited it from his namesake in fiction. We used to ride out in large parties, early in the morning, over the fields across the hills and valleys. In this way I learned to love the country around Göttingen, the lovely beech woods, the wide views over fertile plains, the snug villages and the inviting inns, the ruins of the proud old castles, Plesse, Hanstein, Gleichen, Hardenberg. As I was soon found out to be a fairly good horseman I was trusted with the better type of mares and sometimes honoured by being invited to ride beside the Master. Then he revealed his talent for lying. His stories were mostly about the families of the gentry whose manor houses we passed, and were often extemely indecent. I cannot remember a single decent one to write down. But he at once stopped short when a lady was within earshot. Later, when my horsemanship was well established I was sometimes allowed to ride alone, and this I enjoyed even more.

In winter we had riding lessons in the long hall. There I came into contact with all kinds of people I had never met before, students of other faculties, officers of the army, daughters of landowners and of professors. I discovered that most of them belonged to the Welfish party, which resented the annexation of Hanover by Prussia and professed allegiance to the old Welfish House of Hanover-Brunswick (from which the British Hanoverian kings had sprung). This was an attitude quite new to me. In Breslau there were also anti-Prussians, namely liberals who disliked reaction and militarism. Here I found people who hated the Prussians not because of their character, but as 'foreigners'. I could not see much to be preferred in the old Hanoverian government which was reactionary and backward. But I soon learned to hold my tongue in political disputes. The 'prussianized' university had not a good name amongst these Hanoverians. Therefore the university people kept together. I came on friendly terms with some young ladies, daughters of a professor of medicine, and was introduced into the Göttingen 'society' proper. But I did not enjoy it. The parties and dances were rather insipid and dull, compared with those at the Neisser house in Breslau, and I soon gave up frequenting these circles.

VII. Doctoral Thesis and Graduation

It was in my second year in Göttingen (1905) that my scientific future was determined, not by my own deliberate choice, but by a series of events in which I was involved. At that time Klein's interest in the physical and technical application of mathematics was at its height and infected the other professors, even mathematicians as 'pure' as Hilbert and Minkowski. One afternoon a week was fixed for what we called the '*Bonzen-Spaziergang*' ('walk of the mandarins'—if this word can be taken as a translation of our slang expression '*Bonze*', which in German means a Chinese priest or official of unapproachable dignity). Klein, Hilbert, Minkowski and Runge were the '*Bonzen*' who used to go for a walk on the Hainberg, discussing current events in science and in the life of the University. Once Hilbert told me that on such an occasion Klein had been very enthusiastic about modern progress in mechanics and physics which he discussed with Runge (professor of applied mathematics), rather over the heads of the other two. But after a while they also became interested, in particular in the work on electrodynamics of H. A. Lorentz in Holland and Henri Poincaré in France.

The result was that in the same year (I have forgotten whether simultaneously or in consecutive terms) two advanced seminars were held on mathematical physics: one by Klein and Runge on elasticity, the other by Hilbert and Minkowski on electromagnetic theory. It was the latter which fascinated me. We studied the papers of Lorentz, Poincaré and others on the difficulties which the theories of the electromagnetic ether had run into as a result of Michelson's celebrated experiment. This experiment showed that the motion of the Earth through the ether did not produce an ether-wind as common sense and the current theory predicted. At that time the ether was considered well established, and some people said that its properties were better known than those of matter. Ether was the word most used by Voigt in his lectures on optics. And now all crucial experiments to detect the ether were giving negative results. That was indeed exciting, and I felt the desire to concentrate on this line of research. Minkowski must have felt the same; for this seminar was the starting-point for his celebrated work on electrodynamics of moving bodies. Einstein preceded him in the formulation of the principle of relativity, but to Minkowski is due the discovery of the mathematical structure behind the physical phenomena, which is the backbone of modern electromagnetic theory.

However, I had to wait several years until I was in a position to do fundamental research on that subject, and that was the consequence of the other seminar, on elasticity.

I was not very deeply interested in elastic problems and intended to listen only to the lectures and discussions without taking any active part. But I could not avoid participation. For each problem a speaker and a deputy speaker were appointed, the latter mainly in case the former was prevented by illness from delivering his lecture. Now I became deputy speaker to a student named Arthur Erich Haas, and our problem was the elastic line, or elastica, and its stability. I read some sections in the current text-books on our subject, but left the study of the special literature to Haas. However, about a week before our report was due, Haas appeared in my room and declared he could not address the seminar as he was overworked and ill. In fact he did not look well, but I had my doubts about his illness and wondered whether it was not a kind of stage fright—quite a common phenomenon in performances under the direction of the awe-inspiring Felix Klein. I tried to get some information out of him about the literature but found his knowledge just as scanty as my own.

There followed a hectic week. I tried to read several papers, some in English (at that time a handicap for me), but I could not master the details. However, I understood that the problems of the elastica, i.e. the shape of a bent wire which was originally straight, was an example of a minimum problem of the kind treated in the so-called 'calculus of variations'. I had just attended a brilliant course by Hilbert on this subject which I had well digested. So I set out to apply the theorems I had learned there to the elastica problem and succeeded in formulating the stability conditions for a very general case in a mathematically rigorous way and in showing that some of the known examples (Euler's stability formulae for straight wires) were special cases of my general conditions.

I presented my results to the seminar, where they were received with some interest; yet Runge regretted that, owing to Haas's breakdown, the seminar was deprived of learning the many interesting cases to be found in literature. (As a matter of fact there were not very many.)

After having fulfilled this task I thought it was the last I was ever to hear of bending of wires. I was wrong.

When I was at home during the Easter holiday (1905), I received a letter from Felix Klein, written in his beautiful clear hand, which gave me a surprise. He said that my impromptu lecture had pleased him and that he liked the way I tackled a problem, not with the help of special cases such as those found in literature, but from a general mathematical standpoint. Therefore he had suggested to the Philosophical Faculty that they choose as a subject for that year's academic prize 'The Stability of the Elastic Line' and he expected me to compete for this prize. Now this prize was considered rather a high distinction; for there was only one prize given each year which was allotted to the different departments in rotation, and as the number of these was considerable, the turn of mathematics came very seldom. The problems were always chosen in such a way that one could expect some students to enter the competition. This time Klein had selected me particularly, and I ought to have felt honoured and pleased. But I did not. I was not much interested in elastic problems and did not wish to specialize in this line. There was also in our circles—my friends from Breslau as well as the *Schwarzer Bär*—a strong dislike of Klein's predilection for applied mathematics. We believed that this hobby of his was not in conformity with his real gifts and was only produced by his ambition to play a role in industrial and political circles—he was at that time a member of the Prussian *Herrenhaus* (House of Lords) for the University of Göttingen, a position held some years later by my father-in-law. We had also observed that students of minor mathematical gifts but greater adaptability were preferred and promoted by Klein if they flattered his partiality for applications.

All this went through my mind, and without asking anybody's advice I replied to Klein that I was very grateful for his kind suggestion, but that I had no deep interest in elastic problems, that I wanted to work on electrodynamics or optics, and that I preferred not to compete for the academic prize.

I received no reply. A few weeks later, when I arrived in Göttingen for the summer term, I found my friends, Toeplitz, Hellinger, Carathéodory, Erhard Schmidt, all in a state of considerable excitement, wanting to know what I had done to Klein. He had thundered crushing verdicts on my character and my stupidity, and even Hilbert had shrugged his shoulders when he heard of my offence. I told them what had happened and after a burst of laughter at my foolishness they were really worried. For having Klein as an enemy meant the end of a mathematician's career in Germany, and not even Hilbert could save me. So they persuaded me to approach Klein humbly in order to surrender.

My interview with the great Felix was an ordeal, but totally unsuccessful. He treated me like a naughty schoolboy. When I said I regretted my letter and my decision, and was prepared to do the work proposed, he said in an icy voice that he was not interested in what I was going to do. Then I braced myself and asked him whether, should I succeed, he would accept my work as a doctor's thesis. He replied: 'Why not? But I doubt if you know enough geometry to pass the oral examination. You have never attended my lectures.' I was stunned, for that was true, but how could he know this, as his classes always numbered about a hundred students? I said I had read books, whereupon he shot a question at me: 'How many inflexion points has the general curve of the third order?' Even if I had known I would hardly have been able to answer, shattered as I was by his dictatorial manner. He nodded and I left the room in deep despondency.

Then I held a council of war with my friends, and at last a way out was discovered. The difficulty was this: had Klein agreed to be my supervisor I would have been considered a pure mathematician, and consequently would have been examined, apart from the supervisor, by one other mathematician and one scientist; that would have been, in my case, Hilbert and Voigt. But now that Klein had refused me I had to ask Runge to be my supervisor and was accordingly classified as an applied mathematician; these however had to be examined by another mathematician and by two scientists. Therefore two questions arose: Would Runge accept me? And what other science subject should I take, apart from physics? The first question was soon satisfactorily answered by Runge's kindness, for which I have been grateful all my life. Concerning the second, the only science subject I had ever studied with much interest was astronomy. So I approached the newly appointed young astronomer Schwarzschild and frankly explained my situation to him. He said there would be no difficulty; I should attend his seminar, read Moulton's *Celestial Mechanics* and brush up my former courses of old Franz's—that would do. This seminar was most fascinating; it was the first time I came into contact with modern astrophysics and I half regretted not having chosen this as my main subject. I had to give a talk on the atmospheres of the planets and learned on this occasion the elements of the kinetic theory of gases and statistical mechanics. My lecture seemed to please Schwarzschild, for afterwards he always took an interest in my progress and my personal affairs. When he heard that I played tennis tolerably well, he invited me to become a member of a tennis party consisting of lecturers and young professors, among them himself and Prandtl, the professor of mechanics and hydrodynamics. They were poor players, but full of enthusiasm, and when I trained them in the sport they gave me in exchange valuable scientific information. Thanks to this personal contact, all went well with my astronomy, in spite of the shortness of time.

But now I had to sit down and develop the theory of the elastica, and that did not go at all well. The inspiration which had led to my first results, presented in Klein's and Runge's seminar, failed to reappear. I brooded over my papers but nothing came of it, because I was disgusted with the whole affair and could not concentrate on it. There were so many distractions. One could attend innumerable interesting lectures in other subjects. I took, for instance, a short course in anatomy, given by Dr Stolper, whose wife I have mentioned as the sister of my schoolmate, P. the budding Nazi; there I, the son of an anatomist, saw my first dissections and learned something about my own body. Then there was the surrounding country which I roamed on foot and on horseback, with friends or alone, and which I learned to love more and more.

I had a piano in my room and played not only sonatas and other piano music, but also transcriptions from operas, Bizet, Wagner and others, much to the annoyance of Toeplitz, who lived in the room below mine and disliked Wagner intensely. For a short time I also had a trio with two other students whose names I have forgotten. There were many concerts, mostly solo recitals, but also some orchestras, the famous 'Meininger Hof-Kapelle' or the Berlin Philharmonic. The municipal theatre was not bad; they played drama and comedy on week-days and operettas on Sundays—a frivolous thing to the British—and in summer there was a real opera season, with guest singers from Kassel and Hanover.

Often we spent the evening in the municipal park, where a small orchestra played light music for people sitting at tables under old trees drinking beer. I frequented also the *Kneipe* of the Mathematical Students' Society, which was

101

of the same kind as that to which I had belonged in Breslau. They arranged periodic excursions to some beer-garden in the country, for instance to the famous Maria Spring, where some of the old romantic students' life still survived. In a small valley at the foot of the hill which was crowned by the ruins of the Plesse castle, was a wooden dancing platform and music stand, surrounded by innumerable tables and benches which rose like an amphitheatre on the slopes, shaded by gigantic beech trees. Every Wednesday afternoon trains to the nearby station of Bovenden and innumerable buses and vehicles of all kinds took the families of the Göttingen citizens there in their Sunday best—the mothers going as advance parties long before the rest of the family, to procure good seats where their daughters would be seen to advantage. Students in their coloured caps from Göttingen gathered, and many people from the surrounding country, farmers' families and landed gentry as well, and a great deal of dancing, drinking and flirting began. It was a lovely picture, that gay, colourful crowd in the bright green light under the natural dome of magnificent trees. On Sundays there was a similar dance which however was considered of a lower class; the snobbish students' societies, the 'Corps' and the *Burschenschaften*, did not make an appearance *in corpore* but there were pretty peasant girls and maids from the town who attracted many students in mufti—that is, without their coloured caps.

Once every term the members of the Mathematical Reading Room gathered for an excursion under Felix Klein's leadership. I remember some of these as very gay and entertaining.

In this way the summer semester of 1905 passed pleasantly without much work. It was interrupted by the Whitsun holidays which I spent in Eisenach attending a Bach festival. Apart from several secular concerts, I heard the two big Passions of St Matthew and St John in the lovely parish church. My piano teacher Auerbach had introduced me to the 'Well-tempered Clavichord' and I had previously heard the St Matthew Passion in a concert hall. But it is a very different thing to listen to this music in a church, and it was there that I learned to love Bach. From Eisenach I set out for a lovely tramp through the Thüringer Wald, following mostly the 'Rennsteig', the remnants of the old Roman road which runs along the crest of the whole range. I came to Weimar and Jena, and visited all the places connected with Goethe and Schiller. Then instead of returning to Göttingen, I went to the Harz mountains and crossed them, starting from the Bodetal, then up the Brocken and down the other side to Goslar.

When at last I returned I still did not concentrate on my work. A new interesting figure had appeared in our circle, a Mohammedan Indian, named Zia Uddin, a slim dark fellow with sparkling eyes and ready laughter. He worked under Schwarzschild on a doctor's thesis, something about the history of Indian astronomy, I believe. But he also had to read modern European papers on mathematics for his orals, and this he found very difficult. So he came to Toeplitz and me and asked for our help which we gave him, since he was a new and interesting human phenomenon. As people could not remember his real name, he called himself Ahmad for simplicity.

I spent the summer of that year with my sister Käthe in Sills Maria, Upper Engadine. My Aunt Luise Kauffmann and her two daughters Alice and Claire were also there, and we visited the Bernhardt's in St Moritz. It was the first and last journey I made with my sister; she married the following year.

My thesis made hardly any progress during this summer and autumn, and after my arrival in Göttingen for the winter semester my friends were seriously worried when I continued my lazy life. Sometimes I sat down to work, but I could not become absorbed in my problem and therefore made no progress.

But a change came all of a sudden. All my friends, even young Hellinger, had become members of the Mathematical Society. For this one had to apply to Klein's assistant. When I spoke to Toeplitz of my intention to do this, he warned me not to risk it; he said that Klein would certainly be informed of my application and very likely veto it. If I consider this now in the light of my later knowledge of Felix Klein's character, I am inclined to think that this warning was a trick by the astute Toeplitz to rouse me from my state of torpor and indifference. It worked all right. I was too scared to try to become a member of the famous Society. Every Tuesday afternoon all my friends went to the meeting without me, and after their return I had to listen to their accounts of the proceedings. It was quite unbearable. Toeplitz said: 'If you got the prize and your degree, I don't think Klein could object to your admission.' So that was the way out of this humiliating situation. I sat down to work with a vengeance.

And as soon as I had really grasped the problem I came under the spell exerted by any productive work. You may ask 'How on earth can the way in which a wire bends be of any interest?' Well, I would never pretend that the bending of wires as such had any fascination for me, but as a stability problem it is attractive indeed, as it is by no means trivial or simple and yet it is still within reach of exact methods. Stability problems, however, are the clue to understanding not only natural structures such as atoms, molecules and crystals, man-made structures such as bridges, buildings, machinery and even cosmic structures such as stars and galaxies but also the catastrophes which may occur in such systems. If for instance you take a piano wire and slowly set it oscillating, what will happen? First it will respond steadily to your effort and pass through a continuous succession of graceful curves; but suddenly it will break and change over to quite a different state—this is one example of such a 'catastrophe'. If you play a while with the wire in order to find out on what conditions this catastrophic instability depends, you will be puzzled by the variety of possibilities which defeat any attempt to solve the problem by trial and error. I could not solve it even with all the apparatus of higher mathematics, but had to simplify it. I did not use round wire, but steel tape like those used for making tape measures; they can be bent easily in a plane, and I studied their stability in this plane only. The necessary formulae are then easily written down in terms of so-called elliptic functions, but discussion of these seemed to be hopelessly complicated. After brooding for a while, I hit on the right idea to obtain practical results. It consisted in transforming a method used in pure mathematics (calculus of variations) for proving abstract theorems, into a practical device which, with the help of graphs, allowed me to determine the solutions of my complicated equations, not only in special cases but for all possible configurations over a wide range. But now these graphs actually had to be made, which meant a great deal of work and many tables to be computed and curves drawn.

Meanwhile the New Year (1906) had arrived and there were only three months left, as the manuscripts had to be delivered before 1 April. Never in my life have I worked harder than in those three months. I was no good at calculations and had no practice in drawing; both had to be learned and applied to examples of an uncommon type. But by the end of February final results had emerged, which predicted the conditions under which those instability catastrophes would occur. This, of course, aroused my curiosity, and I began to play with tapes of stiff paper or cardboard; but they were hardly any use since they had a regrettable tendency to break. Suddenly I decided to try real experiments. During a sleepless night I designed a simple apparatus; in the early morning I made some drawings of its parts, and when the shops opened I was entering the workshop of one of the mechanical firms (Spindler and Hoyer) which had a high reputation in the scientific world for skill and accuracy. I explained my case, the problem and the urgency of a quick solution, and I found interest and sympathy. They made my little contraption in a few days, and while I was still working at the theory in the evenings, I was busy the whole day experimenting and measuring. And lo, my theoretical predictions of stability limits turned out to be correct.

It is always a moment of deepest satisfaction for a scientist when he is able to check a theory by experiment (even when done by somebody else). This first time I felt that it made me a scientist; it was the crucial test of my ability, and I was happy. But I had only a few days left to write it all down, to have it cleanly copied and bound. So I had no time for much happiness, until the manuscript was duly delivered to the Secretary of the University; and after that I was too tired.

I do not remember what I did in the Easter vacation. Presumably I went home and worked all the time for the orals. The last term in Göttingen (summer 1906) I remember as rather unpleasant. I had to work hard, in particular at modern astronomy. Toeplitz coached me in mathematics and made me feel that I knew nothing. I did not dare to apply for admission to the Mathematical Society for fear of Klein's wrath and my friends did not refrain from teasing me about this inferiority complex. I attended several meetings as a guest of Toeplitz or Schmidt, sitting modestly on a back bench. But I felt excluded from the inner circle to which I had belonged as Hilbert's assistant. I had the same feeling of being an outsider with respect to the social life of the town. I had many friends, but no friend—no one in whom I could confide. On 13 June 1906 there was the academic function where the prize-winners were announced. After the usual ceremony and a long speech by the Rector the moment came when he opened the sealed envelopes, marked by the motto of the thesis, and read the name of the author. I was standing in the doorway to the anteroom and heard my name pronounced as the winner of the prize of the

Philosophical Faculty. Though I had been pretty sure of the soundness of my work, it was a great relief. But there was not much time for celebration, as my reading for the examination had to go on. This took place a few weeks later. As I have said before, it was the only examination in all my university career, and it was not formidable at all. With Hilbert I just had a nice mathematical chat. Runge enquired about my graphical method used in the thesis and put some harmless questions. Voigt examined me on Maxwell's theory and optics, and I do not think I missed a single question. When it came to astronomy I had become so confident to such a degree that I committed a considerable impudence. Schwarzschild began by asking: 'What would you do if you saw a falling star?' Now there is a superstition in Germany that if you think of a wish at that moment it will be fulfilled. So I answered rashly: 'I'd make a wish.' Hilbert, who was listening, laughed heartily, but Schwarzschild remained earnest and said: 'So you would, would you? But what else?' Meanwhile I had had time to collect my wits and said, as required: 'I'd look at my watch, note the time and the place of appearance and disappearance of the meteor in relation to the stars'—and so on; then he made me develop a method for determining the approximate orbit from these data, which I did with some assistance from the professor.

I did not get the highest mark, *summa cum laude*, but the second one, *magna cum laude*; and I thought it was a fair decision since I never claimed to be a real mathematician, like Schmidt or Carathéodory or Toeplitz, but just a fellow with common sense who could learn obvious things like mathematical theorems from lectures or books, without claiming any originality or productivity. If later something like that appeared in my work, it was in a rather different field, in the quasi-magical process of theoretical physics, of abstracting natural laws from experimental evidence.

The following day we celebrated my success in Münden. And then I left Göttingen, rather fed up with the place, and decided never to return. Toeplitz, however, said: 'You will come back, not only once, but twice: first as a lecturer, later as a professor.' I was not pleased, but his prophecy turned out to be correct.

The formal graduation was six months later (14 January 1907), *in absentia*, as I was then serving in the army and could not attend.

There I was, a young doctor of philosophy, a young man of means and prospects—the world lay before me to be conquered. But I had hardly any plans. I wanted to learn more and to experience the thrill of discovery, of which I had a foretaste by my rather trivial elastic experiments.

VIII. Military Service

Since I finished the last chapter a few months ago (Christmas 1944) terrible things have happened. The places of my childhood which I have described in these recollections are in the war news. Breslau is being besieged by the Russians, who are blasting their way through the southern suburbs where the Kleinburg garden was situated. It may well be that the guildhall and university, cathedral and *Kreuzkirche* are in ruins. How often I have dreamed of seeing the city of my youth again, but now there may be nothing left to see. It is a strange and sad feeling, and I sometimes wonder how the soul can bear it, after all it has gone through. But then I ponder: what does Breslau matter when the whole of Europe is being destroyed? The old world is gone, and we have to go on living.

That same summer (1906) when, with my graduation, my student life came to a close, my sister's girlhood ended with her engagement and marriage. My new brother-in-law, Georg Königsberger, was a municipal architect. He had just accepted a new job in a suburban district of Berlin. I met him and his parents and brothers at Schandau, on the Elbe, near Dresden, where the engagement was celebrated. My sister was obviously happy, and I liked her fiancé. They were married in October and settled in a nice old-fashioned house in Grünau on the Dahme, one of the many tributaries of the river Spree, near Berlin.

I spent the summer vacations with my family, Mama, Käthe and Wolfgang in Swinemünde on the Baltic Sea; it was the last time we were all together. I found Mama rather depressed by the idea that she had forthwith to live alone. So I suggested to her an excursion to Sweden to distract her mind. We went to Göteborg and through the whole Göta Canal to Stockholm. This journey on a tiny steamer through rivers, canals, and lakes was enchanting; so much so that many years later I suggested the same trip to my bride for our honeymoon, with rather sad results. For the big Swedish lakes, which had been as calm as ponds on my first journey, turned out to be rough and nasty, with dire consequences. I was terribly seasick and Hedi—let us say—was not very well. But that belongs to a later chapter of my life.

I had now before me an unpleasant prospect: military service. Most of my contemporaries had entered the army immediately after leaving school and were by now NCOs of the 'reserve'. I had also undergone a medical examination, but had been rejected because of my asthma. After I had begun my studies I applied every year for postponement of the call-up because I did not want to interrupt my work. But now, after my graduation, further postponement was impossible; I had to report at a recruiting office, and I was accepted, to my great annoyance, for I had rather expected to be rejected again. I was twenty-four years old and accustomed to the free life of a scholar. So the prospect of a year's sharp discipline was not attractive. You will have gathered from my description of our parental home that we were not imbued with military spirit and admiration for the Prussian army. But on the other hand the duty to serve one's country was never questioned. Everybody did it; many enjoyed it, others were indifferent, nobody objected.

I belonged decidedly to the indifferent ones. This was not so much the consequence of my father's tales from the wars—for at that period nobody believed in the possibility of another European war, and the armed forces were considered more an adornment of the monarch than his last resource in the case of danger—no, my attitude was a reaction to the special position of the Jews in the army. The Jews of my father's generation had enjoyed complete equality of rights in civil life and in the army as well. My uncles had all been officers in the reserve army. But in the last decades of the nineteenth century the first strong wave of anti-semitism had swept over the country, produced and directed by the 'Oberhofprediger' (court chaplain) Stöcker and by a demagogue, Ahlwardt, who became a member of the Reichstag. Slowly these ideas spread and infected the officers' corps of the army. It was against the law to exclude Jews from promotion in the service; but laws can be evaded. Jewish soldiers were promoted to the various ranks of NCOs and finally admitted to examination for a 'commission', as it would be called in Britain. If this was passed, they had to apply to the officers' staff of a regiment for admission, and the result depended on a ballot of the officers. Here not only the military merits of the candidate counted but also his character, family, fortune and so on, and there were no restrictions on selecting or rejecting. In this way some regiments, mainly those of the Royal Guard, were reserved for the aristocracy; others admitted members of the wealthy middle class, while the ordinary provincial regiments were less particular. But Jews in general were not admitted. However, this was not strictly a racial prejudice, for baptized Jews of some wealth and social standing had no difficulty in finding a regiment where they were admitted.

When this practice developed, many of the Jewish officers of the reserve resigned and returned their swords, including my uncle, Berthold Schäfer, cousin Hans's father, and others of my relatives. I strongly resented this discrimination against Jews, and my reaction was a decision to give no more of my time and energy to the army than was absolutely necessary. Therefore I looked out for a regiment where the service was easy and leave plentiful, and I found one with the help of Mama's younger brother, Ernst Lipstein, who had just finished his year's service in the 2nd Dragoon Guards Regiment, stationed in Berlin. He told me why it was that this noble regiment attracted many young men from well-to-do families who liked to spend their one year's service in an easy way. The officers were all high-ranking aristocrats who did not take their military duties very seriously. Therefore the NCOs and sergeant-majors had a free hand to handle things as they liked; and what they liked was to be on good terms with the one-year volunteers who had money to spend. It was a harmless kind of corruption. The main advantage was that one could go into town in mufti; it was not officially permitted but no officer or NCO ever 'knew' you if you had the misfortune to meet him in a tramcar, restaurant or theatre. Ernst Lipstein also recommended to me his digs, where he had been very comfortable under the care of a fat widow named Palm. All this was just what I wanted, and as I liked horses and riding I did not hesitate to become a dragoon of the Prussian Guard.

Berlin had many attractions for me. My sister and her husband had just settled in Grünau, a rather distant but lovely suburb; their house was soon a second home for me. I was a welcome guest in the house of all the numerous relatives, Kauffmanns and Lipsteins. After having spent several years in a small provincial town I looked forward to the pleasures of the capital; splendid theatres, operas, concerts, parties and dances. My friend James Franck was serving at the same time in an engineers regiment stationed in Berlin. And last but not least, there was a new affliction of my heart, the cause of which lived not far from the Brandenburger Tor.

So I moved into Frau Palm's little flat, Fürbringerstrasse 21, a bleak and gloomy suburban street leading directly to a back entrance in the wall surrounding the barracks. The first weeks of my military life were the usual nightmare of fatigue and exhaustion caused by the unaccustomed and ceaseless physical activity: out of bed at 5 a.m., after hasty dressing in a dirty uniform off to the stables, two hours of cleaning horses, dashing home, breakfast, re-dressing, back to the stables, riding lessons, marching, lance fighting, drill, horse cleaning again and so on until 6 p.m. when I was hardly able to eat any of the ample dinner provided by Frau Palm, and then to bed. The worst of all these exercises was lance fighting. The lance was a long and heavy iron tube with a sharp point, and we had to learn to handle it with such efficiency that we could later be trusted with it on horseback. The hardest exercise was to hold the spear with an outstretched arm as long as the sergeant thought possible, and woe betide the recruit who wavered or let it sink down; he was sure to get an extra hour of drill.

The riding lessons, on the other hand, were quite pleasant, at least after an initial incident. Our lieutenant, a young conceited nobleman, used to stroll in in the middle of a lesson and watch our performance with great amusement. One day we had for the first time to jump a 'hurdle' consisting of a pole fixed a few inches above the ground. We had neither saddles nor stirrups but were sitting on blankets. Many of the recruits fell off at once, and for the rest the height of the pole was increased until they also fell off. In the end I was the only one who refused to be thrown off. Now the young officer disliked the one-year volunteers particularly, as he had to treat them as gentlemen when off duty. So instead of the expected appreciation of my performance I found him getting angry: 'one inch higher' came his command, and again 'one inch higher'. In the end the horse refused the jump, and I remained on its back. Then the lieutenant got into a rage and was just reaching for a long whip to drive the poor creature over the pole when a sharp voice was heard: 'Lieutenant Count X, no whip please during riding lessons.' It was our *Rittmeister* (captain) Baron von Schönaich, a man whom I soon learned to respect. He was different from the average aristocratic officer, a good soldier, well educated, stern but human, conscientious in his duties and interested in the well-being of his subordinates. He was aware at once of what had happened and took over command. When he found out that I knew how to handle a horse he sometimes dropped in during a riding lesson and ordered me to come with him for a ride over the Tempelhofer Feld, the immense parade-ground adjacent to our barracks. He was very friendly on these rides, asked me about my family, profession and interests, and gave me hints about how to become a good soldier. He was immensely popular with the whole 2nd Escadron (squadron) to which I belonged, and the men of the other squadrons envied us for having him. Years later, after the first world war, I read in the papers about a General von Schönaich who had distinguished himself in battle but was so disgusted by the slaughter that he became an ardent pacifist. He fought with courage for reason and peace, and suffered, as far as I know, the usual penalty of being murdered by the Nazis. I wonder whether this general was identical with my captain of the 2nd Dragoons.

After a few weeks we were allowed to go out in uniform (if we liked; for as I said before, we went in mufti without getting into trouble). For those of the volunteers who were students this was important as they could now go to the university to matriculate; the military year was counted for examination purposes even if no lectures were attended. So one morning quite a number of the volunteers took a day's leave and drove in taxis to the university. As I was already a graduate, I was not among them, but did my stable duty and horse-cleaning as usual. When I returned for lunch I found Frau Palm highly flustered; something terrible must have happened. Her friend, the wife of the sergeant-major of our squadron, had come and told her there was great excitement in the officers' mess about some affair concerning uniforms. There would be an inspection by the colonel himself, and Frau Palm should see to it that the 'extra-uniforms' of the two volunteers lodging with her were exactly in line with the regulations.

We volunteers had to wear during service the old, dirty uniforms provided by the regiment, just like all the other recruits. But we had the privilege of buying 'extra-uniforms' of our own for going out. These were made by the same tailors who worked for the officers, and were as elegant as officers' uniforms, of good material, with silk badges and stripes, patent leather belts and boots, etc. I was rather proud of my new outfit, which I had never worn in public yet. When I came back from my drill at night, how awful this uniform looked, for all the silk and other embellishments had gone. But Frau Palm was quite adamant that this was the right thing to do, for by now she knew what had happened that morning, and told me.

When two of our student volunteers, dressed in elegant 'extra-uniforms', had stepped out of their taxi in front of the university building, Unter den Linden, there stood an officer who, after a searching glance, approached them and asked them to follow him. He led them to the garrison headquarters where they were meticulously inspected by some officers and then dismissed. Even before they had returned to our barracks, the Regiment was informed by telephone that two of its volunteers had been caught wearing uniforms which were contrary to regulations; it was to punish the culprits and ensure that its men were properly dressed. Great excitement; why the devil this sudden interest of the garrison command in details of dress? The colonel telephoned and soon learned of the strange events which had led to this ridiculous interference, and we read the story in next morning's newspapers. It was the famous affair of the captain from Köpenick.

There was a fellow named Vogt, a cobbler by profession, who had served in the army and was discharged in consequence of an accident. He had claims against the war office but could not make them effective. He bombarded the authorities with letters, besieged them and carried on law suits, all to no avail. Meanwhile he had become an elderly man, shrunken and a little stooped, with white hair and a long white moustache. This affair had become a mania with him, and he believed that everything would be all right if he could only get hold of certain papers which the authorities were withholding from him. One day he made up his mind to get these forms by fraud or violence. He bought a captain's uniform from a second-hand shop, took the suburban train to the little town of Köpenick, disappeared into the station lavatory and emerged after a while as an elderly captain, dressed not quite according to regulations, but sufficiently convincing for the lance-corporal of a platoon of grenadiers whom he met on the main street. When he saw that they gave him the military honours, he stopped them and ordered them to follow him. This they did with the Prussian soldier's traditional obedience to the officer's uniform, although they were on some definite commission. Old Vogt led them to the guildhall, ordered a picket to watch all entrances and marched with the rest of the men into the building, straight up to the room of the mayor—Langerhans was the unfortunate fellow's name. 'You are under arrest,' he thundered at the bewildered mayor who handed over on command the keys of the safe. Vogt opened it and searched for the papers he desired, in vain of course, for how could military papers be found in a municipal office? But he did find a considerable amount of money and pocketed as much of it as his claim amounted to. Then he told the corporal to watch the mayor while he telephoned the authorities; he left the room, went down the stairs, passed his sentinel on the main entrance and disappeared. An hour or two went by, until the flustered officials began to get suspicious and to use their telephones. Soon a police car arrived and it became clear that a practical joke of colossal dimensions had been carried out by one who understood the magical power of the officer's uniform over the Prussian mind. But that uniform, discovered in the lavatory of Köpenick station, was the only trace left by the perpetrator of this outrage.

When the papers published the story the whole country roared with laughter. For that magical spell inherent in the uniform was also coupled with a strong resentment against its bearer, the proud and arrogant officer. The strangest thing was that the Emperor, Wilhelm II, joined in the laughter, very much to the embarrassment of the generals who, though furious, were forced to smile. They could do nothing but issue an order for an immediate tightening of the regulations concerning uniforms, for it turned out that Vogt's uniform had been very faulty and that the soldiers from Köpenick ought to have recognized the swindle at once had they known the correct attire of a captain. So it came to pass that we private soldiers were the only ones who soon had nothing to laugh at.

In the early morning of the day when these events were published, we stood in the yard, dressed in our 'extra-uniforms', and waited for an hour or more. (Waiting is the chief occupation of the soldier.) Then the colonel appeared with his staff. The two culprits whose elegant uniforms had initiated the trouble were quickly sentenced to two weeks' hard detention. Then the officers walked along the line of volunteers and every man got three days' detention and in addition four weeks' stable duty (i.e. night work in the stables, horse-cleaning, etc.). Then they approached us three men from the 2nd squadron. Rittmeister von Schönaich saw us first and exploded with laughter—we looked funny enough indeed, in our adjusted uniforms. Even the severe colonel grinned. He could not punish us but gave us a week's extra stable-duty in order to dampen our hilarity. The saddest consequence of the affair was that the whole regiment was confined to barracks for many weeks: we volunteers had to stay in our digs and were checked to make sure we did not go out in mufti. There was the city with all its attractions, and we had to stay at home. But the controls soon began to slacken and we could at least visit one another. We had some hilarious nights with a lot of wine and beer, occasionally also with the girl-friends of some of the fellows.

As to the captain from Köpenick, the old cobbler Vogt, I can narrate a happy ending to the story. He was, of course, caught all right, prosecuted and sentenced to a year's imprisonment. In jail he had a good time since innumerable admirers sent him presents and the jailors respected him as a celebrity of world fame. A collection was made for him, and when he left the prison he found himself a man of means. Literature, stage and film have contributed to his glamour as the man who ridiculed Prussian militarism.

But mayor Langerhans of Köpenick had to quit his post and disappeared from public life, as far as I am aware.

The rest of my military career was highly inglorious and painful. My asthmatic disposition was noted in my military documents; the army doctor who examined the new recruits had seen it and decreed that I should report to him at regular intervals, every fortnight I think. During the first months everything went well. I had slight asthmatic trouble, particularly during a night's watch in the stables; the exhalations of some 120 horses, and the dust from straw and hay, are the worst atmosphere for an asthmatic bronchial system. Therefore I never attempted to sleep on the straw between the legs of my horse as other men did, but spent the long hours of the night reading the proofs of my doctor's thesis—much to the amusement of my comrades. The next morning I was never very fit, but by good luck the doctor never saw me in that state of breathlessness. However, after the adventure of the captain from Köpenick, we had got a full week's stable watch, and during this time I had to report to the doctor. He kept me for a week in the most unpleasant and uncomfortable sick-room of the barracks, and after that I had to report every week.

All went well until the Emperor's birthday, 29 January. The main celebration consisted of a grand parade of the whole garrison of Berlin, in front of the imperial palace. It was a clear, extremely cold winter's day when we marched in our splendid uniforms through the endless streets to the centre of the city. As cavalrymen we were not accustomed to marching, and we were pretty tired when we arrived and took our position on one of the terraces surrounding the palace. There we stood waiting a long time—one hour or more, I do not remember. It was terribly cold and we were not allowed to move. Then music, the usual wave of excitement, the commands 'Attention!'—'Present arms'; and along came Wilhelm II on horseback with his suite, slowly moving along the long front of his Guards regiments, perhaps 20 000 men altogether. It was what Mr Churchill would describe as 'a magnificent spectacle of military splendour and power' like those which he thoroughly enjoyed in India when serving in the 4th Hussars (see his *My Early Life*). However, I did not enjoy it. That our feet, hands and ears were frozen stiff did not matter so much; we had experienced that often enough on our barrack parade-ground. But to be a little cog in a little wheel in the big military machinery was abhorrent to my mind. I was not devoid of patriotic feeling and had I watched the same parade from outside I might have felt that well-known wave of patriotic elation which such military displays are intended to produce. But standing there motionless in rank and file, I remembered what my father had told me about his experiences during the Franco-Prussian war, and suddenly I felt the same degradation of human dignity which he had described to me as his reaction. I still remember this moment as vividly as if it had happened yesterday. It determined my attitude to militarism for the rest of my life.

This same parade also determined my personal fate in the army. After our march back, which took hours because of the crowds in the streets celebrating the monarch's birthday, I had a bad cold, and the next day I was seriously ill. I was sent to the garrison hospital in Tempelhof and spent some very uncomfortable weeks there with an asthmatic bronchitis and high temperature. After having recovered I returned to my regiment for another few weeks; then suddenly I was discharged as unfit for military service.

I felt no regret at leaving the army, and I was convinced that I would never again have anything to do with it—but that was a great mistake. It was my fate to serve again, and more than once, in different regiments: curassiers, air force, artillery; in peace and in war. The Prussian militarists have given me ample opportunity to see them at work, but no love has sprung from these experiences. You may ask: why did you not become a pacifist, a conscientious objector? Why did you obediently surrender to a power which you disliked and despised? The answer is that at that time the idea of refusing military service was considered absurd, impossible. If you did so you only had the choice between death by shooting or a lunatic asylum. I do not think the idea even entered my head. Some pacifist societies made propaganda for abolishing armies and war. But there was something cranky about them. Moreover, in this period between 1900 and 1910, there was a general conviction in the bourgeois world to which I belonged that no European war would ever happen again. We believed in perpetual progress. Had not the Tsar Nicholas of Russia convened a general peace conference at The Hague? If even the supreme commanders of the greatest armies, if emperors and kings began to repudiate the possibility of war, why worry? One had to sacrifice some time and to submit to some discomfort until the armies were reduced or abolished, so let us take it easy. My comrades in the Dragoons certainly took it easy and I tried to as well.

Yet I remember some moments when I was suddenly aware of the sinister background of all that merry exercise, of riding, shooting and fighting. One day we began to learn to use our lance on horseback; at a brisk canter we had to hit a bag filled with straw. It seemed to be an amusing sport, and I was not bad at it—until it suddenly dawned on me that this bag represented a human body. What would you do if you were in a real battle and had to charge with your lance against living men? The thought seemed to me so absurd that I simply brushed it aside. Such things have happened, they are described in books, but they will not happen again; anyhow not to me, Max Born, a peaceful scholar. When a few of us volunteers had a little supper together I directed the talk to this subject in order to find out their opinions. To my amazement, not one of them understood what I meant: of course they would pierce an enemy when ordered. A soldier has not to think but to obey. Even two Jewish boys accepted this as unquestionable. There I was, an outsider again, as so often before. Only half a dozen years later I watched the real thing, in the first world war. All the young men did what they had so eagerly professed. It must be deeply rooted in human nature. For today they are busy again with a murderous war. How I went through these two wars I shall tell you in later chapters. Here we are still in the happy period which we believed to be the beginning of eternal peace on earth.

Before I close this chapter I must mention that my friend Franck was discharged from his engineers regiment about the same time as I, also because of illness. But his case had been far more serious than mine. He had been hurt in the leg and was limping; but he continued to take part in the drill—whether from his strong sense of duty, or from military ambition, or compelled by a savage sergeant, I have forgotten. Anyhow, his leg swelled to an immense size and he was taken to hospital very ill. The military doctors gave him up, but his father succeeded in getting him home and into the hands of a first-class specialist. So he was saved. When I saw him some weeks later he still looked like a ghost and had to lie motionless in order to avoid a clot of blood entering the heart. After his recovery he became an assistant to Rubens, the professor of physics at Berlin University, and began his scientific career. Our paths seemed to diverge; he was from the beginning a modern experimental physicist while I grappled my way to my goal via a long detour through a mathematical jungle. Yet in the end we came together all right.

IX. Cambridge

When I returned to Breslau after my discharge, I found my cousin Hans Schäfer busy preparing for a journey to England. He had obtained his Ph.D. in chemistry, entered the Kauffmann business as an apprentice and was now to be sent to some textile-works in Bradford, Manchester and other places in the English industrial North in order to learn up-to-date methods of spinning, weaving and dyeing. He easily persuaded me to go with him to England for I had no other definite plans and wanted to see the world.

I had at that time little idea about the leading part which Britain has played in science, still less about what was going on there at that time. Toeplitz, who had formed my first notions of the mathematical world, did not think very highly of the contemporary British mathematicians. In Göttingen, English names had frequently appeared in my lectures; Sylvester, Cayley and Hamilton were often quoted by Klein, and Hamilton's dynamics also dominated several of Hilbert's courses. As I have mentioned before, Maxwell's electromagnetic theory was presented to me rather early by Clemens Schäfer, and I had studied it more thoroughly with the help of Abraham's book. My friend Franck was strongly influenced by the English school of experimental physics; he spoke of Townsend, J. J. Thomson, Sir Oliver Lodge, Rutherford and others with great admiration, and he made me curious to learn something about their work. I had also heard of Willard Gibbs from cousin Hans, who quoted terms like 'phase rule' and 'heterogeneous equilibria' with mystical reverence. I bought Gibb's great work and tried to study it, but found the language a great handicap. So I decided to see England and to learn English.

The latter was begun at once by taking lessons. My teacher was a pleasant young lady, very good to look at but not very efficient in her job. Anyhow, my linguistic talent is not considerable, and a fortnight not sufficient to master a language. Hence when Hans and I arrived in London (20 April 1907), I understood nobody and could not make myself understood. At first it did not matter much, as we stayed at a big international hotel, De Keyser's—it seems to have disappeared since. We went sight-seeing together, and Hans visited some businessmen to whom he had introductions. I do remember that we were both impressed in some vague way by the splendour of the city and of our hotel in particular. We liked to sit in the lounge with a glass of port and watch the elegant crowd, and we enjoyed in particular the little messenger boy in uniform passing along and shouting 'number 267, please' or 'Mr Brown wanted on the telephone, please', and we tried to imitate his cockney accent. We went to a theatre and saw Beerbohm Tree as F.. -ff in *The Merry Wives of Windsor*, and we were most astonished when, at the ... of the performance, he came to the footlights and addressed the audience—something unheard of on a German stage.

But De Keyser's Hotel was expensive, and so we moved into a boarding-house somewhere in Bloomsbury. There life was more difficult, for if you wanted anything you had to make people understand you, and if you wished to join in the conversation at dinner you had to grasp what they were talking about. I cannot remember any particular incident; but one thing has stuck in my memory, namely that, compared with our standards, the place was dirty and ugly. We learned a fact which has hardly improved in the thirty-eight years since: you can have comfort and cleanliness in Great Britain if you can afford a first-class hotel, but almost everything below this level, with rare exceptions, is uncomfortable and not very clean.

After a few days at this place Hans and I separated—he went to some textile-works in the Midlands and I proceeded to Cambridge. Following the advice of my Baedeker I stayed at the University Arms, which was almost as expensive as De Keyser's in London. The next few days I admired the sights of the old town and enjoyed the lovely college courts, King's Chapel and the Backs. I also tried to get into touch with the University, but in vain. Nobody seemed to understand what I wanted. They took me for a sightseeing foreigner, and when I asked for the University offices they showed me the lovely façade of the Senate House. It was exasperating and I became quite depressed. One day I was walking along Trinity Street behind a group of students in cap and gown, and to my amazement heard them speaking German. So I pulled myself together and addressed them in German. I told them who I was and what I wanted, and I asked them to help me. It turned out that all three of them were German students, attached to different colleges and all from distinguished families. One was a son of the director of the Deutsche Bank, von Gwinner; the second was a son of the celebrated chemist, Emil Fischer; and the third, Ernst Philip Goldschmidt, a member of a wealthy Viennese family, related to the Rothschilds.

These fellows laughed at my difficulties and promised to help me. But it was not easy, for term had just begun and all the colleges were full up. Goldschmidt was at Trinity, but informed me that no more students could be admitted there; similar reports came from the other men. I was despondent and was beginning to think of leaving Cambridge and returning home when Goldschmidt appeared and told me that he had met a man from Caius who said that they would probably take me. So I tried and was accepted as an 'advanced student' of Gonville and Caius College.

I did not get a room in the College—this was long before the new buildings were even planned, and many students had to live in digs. I found quite comfortable rooms in Bene't Street. The landlady was a nice woman who tried hard to understand me and to teach me English, or what went for English with her. I imitated her as well as I could and caused a great laugh one night at dinner in Hall when I pronounced 'sta-ition' and similar words in the local Cambridge accent. Then I decided to take proper lessons. My teacher was a man with a German aristocratic name, von Glehn, whose ancestors had immigrated from the Baltic countries, I believe. But he was one hundred per cent English, a schoolmaster at the Perse School and a very efficient teacher. Though he spoke German fairly well he never used a German word in the lessons but gave me the Oxford Dictionary to find the meaning of any expression. We read Kipling's *Stalky and Co*, which opened up to me the strange new world of an English boarding school, and he made me learn by heart long sections of *Hamlet* and other Shakespearean plays. Alas, only a few lines have stuck in my memory. When I returned to Cambridge in 1933 as a refugee I found my old teacher still alive and happily living in Granton; but he was now Mr de Glehn—a little change due to the first world war, when a German 'von' was an unpleasant attribute for an English gentleman and schoolmaster.

These lessons helped me a lot in my intercourse with the dons and undergraduates at the college. How strange everything appeared to me! After the free life of a German student, it seemed to me almost absurd that the English students were subject to strict discipline which, though not very oppressive, nevertheless restricted freedom in many ways, in particular with respect to being out at night. The institution of Proctors and Bulldogs seemed to me ridiculous, and the students' duty to attend chapel a hypocritical remnant of mediaeval times, for none of them seemed to take it seriously. But I soon began to like this regulated life. I began to understand a little of the talk at dinner in Hall and take part in it. There was one student who spoke German fluently and helped me in my difficulties; his name was Gilliat-Smith. He was a nice fellow with a remarkable gift for languages; I do not remember how many languages he spoke, certainly all the European ones, but his main interest was the gipsy language, which he had studied to such a degree that he could tell the district from which a gipsy came—so he said.

The Master, Mr Roberts, was very kind to me and invited me frequently to tea-parties in the lodge. My tutor was the pharmacologist Mr (later Sir William) Hardy; I remember very well his lovely house and garden, and his beautiful wife. Hardy died just as I returned to Cambridge in 1933. Another house which gave me pleasant hospitality was that of Professor Ruhemann, the chemist, of German origin.

All the same I often felt terribly lonely and homesick. Neither at College, nor at the Cavendish did I make a real friend, whether personal or scientific. I took two courses, one by Larmor on electricity and magnetism, and one by J. J. Thomson, on modern experimental physics. Larmor's lectures were almost incomprehensible to me, owing not so much to his abstract and abstruse way of presenting things, but rather to his strange pronunciation (I think it was an Irish accent). Thomson's experiments were fascinating, just as I had expected from Franck's tales of English physics; but the theory was often beyond my understanding. I knew at that time very little about ions, electrons and the conductivity of gases. I had intended to do some experimental work in the Cavendish; but after one or two of Thomson's lectures I decided that I could not possibly apply to him for a place in his laboratory. I lacked even the simplest knowledge of technique in this field. So I humbly took an elementary course in electricity by Searle. This however was also a mistake, for I had already tried all these simple experiments and measurements. Searle found this out and asked me whether I would prefer to do some more advanced work. I suggested to him that I should like to build up a system of absolute electric and magnetic units from scratch, that is with the help of simple coils, batteries and other things available. He agreed, and I learned a lot. Together with me in this work was another student, a pretty girl from Newnham, who was very shy and aloof—or seemed so to me; in fact I suppose I was the shy one. One day we had some difficulty with an instrument, and when Searle came to our table I asked him: 'Dr Searle, something is wrong here, what shall I do with this angel?' (meaning of course 'angle'), whereupon old Searle looked at both of us over his spectacles, shook his head and said: 'Kiss her.' After that the shyness between us increased still more and was never overcome—a great pity, for she was a lovely creature.

I was, however, not quite without female society. My old friend Lore was at that time governess in a family who spent the summer at Ventnor on the Isle of Wight. I went there for one weekend, and she returned the visit at Cambridge some weeks later, accompanied by a friend, a Miss Zeiss, grandchild of the founder of the celebrated optical firm. We had a good time; I showed them the sights and took them on the river in a punt, in perfect Cambridge style.

When May Week came I felt particularly lonely, for I had no partner. I saw the boat races with a party of German students, and I went to my college dance, but danced only a few times with Mrs Hardy, who discovered my loneliness and tried to cheer me up. In Germany at that time one did not go to a dance with a definite partner but danced with any girl present. The English custom seemed to me rather stiff and boring, as did the whole atmosphere at these festivities. Anyhow, I thought it not worth the several guineas I had paid for it, in spite of the splendid surroundings and the opulent dinner. In Göttingen I had enjoyed myself much more at students' dances, like those in Maria Spring, which did not cost more than a sixpence.

After the end of the summer term I remained in Cambridge for the 'long vacation term' which lasted about four weeks in July. I got a room in College and came into closer contact with some undergraduates. In Hall I had an embarrassing experience; it was found out that I was the oldest 'advanced' student and as such I had to preside at the table of these scholars, supervising their behaviour—not an easy task for a foreigner with a restricted knowledge of language and slang. But though I was often the centre of hilarious outbursts, all went quite well.

Apart from my experimental course with Searle, I had nothing to do, and I spent most of my time lying in a punt or canoe on the Cam and reading Gibbs, whom I now began to understand. From this sprang an essential piece of progress in thermodynamics—not by myself, but by my friend Carathéodory. I tried hard to understand the classical foundations of the two theorems, as given by Clausius and Kelvin; they seemed to me wonderful, like a miracle produced by a magician's wand, but I could not find the logical and mathematical root of these marvellous results. A month later I visited Carathéodory in Brussels where he was staying with his father, the Turkish ambassador, and told him about my worries. I expressed the conviction that a theorem expressible in mathematical terms, namely the existence of a function of state like entropy, with definite properties, must have a proof using mathematical arguments which for their part are based on physical assumptions or experiences but clearly distinguished from these. Carathéodory saw my point at once and began to study the question. The result was his brilliant paper, published in *Mathematische Annalen*, which I consider the best and clearest presentation of thermodynamics. I tried to popularize it in a series of articles which appeared in *Physikalische Zeitschrift*. But only a few of my colleagues accepted this method, amongst them R. H. Fowler, one of the foremost experts in this field. Fowler and I intended, a few years ago, to write a little book on this subject in order to make it better known in the English-speaking world, when he suddenly died. That will, I suppose, be the end of it, until somebody re-discovers and improves the method.

One day, during this vacation term, I had to call at the German Consulate in London; the reason was a communication from the German military authorities ordering me to report to the nearest recruiting office for a new medical examination, and followed by the usual threats of punishment in case of disobedience or delay. After some waiting I was shown into the room of the physician; I entered with a friendly 'Guten Tag', but was received by a thundering 'Sir, how dare you? Here you are a soldier in front of his superior!' I suppose that my dress and poise, a fair imitation of a Cambridge under-graduate, offended this Prussian officer as much as my informal greeting. Anyhow, he was full of rage from the beginning and after a very superficial examination and a few questions about my military career I was dismissed with the words: 'You are a shirker; you will hear from me.'

This worried me a little, but I soon forgot it. Alas, I was indeed to hear from him.

I planned to attend the British Association meeting—I think it was held at Birmingham—where Lord Kelvin and J. J. Thomson were to discuss the principles of electronic theory. But when the end of the vacation term came I felt tired and went on a sightseeing trip, mainly to the cathedral cities of England. I do not remember where this sudden interest in mediaeval architecture came from; I suppose Philip Goldschmidt had told me of the beauty and grandeur of these churches and had roused my curiosity. I enjoyed this journey so much that I missed the British Association and the discussion between the old and the young generation in physics. I never saw Lord Kelvin, for he died a short time later.

The only one of the big cathedrals I knew was Ely, which is close to Cambridge. Now I started with Peterborough and went up the east coast to Lincoln, Durham and York. There I met cousin Hans, and we spent some days at Scarborough. From there I proceeded to Newcastle, not quite clear in my mind where to go next. This was decided for me by the fact that people in these parts did not like to take my cheques on a Cambridge bank; but I had a letter of credit to the Bank of Scotland so I took the next train to Edinburgh in order to get my purse filled. This was the accident which brought me for the first time to the city which is my present home. I knew nothing of it, apart from the fact that it was the capital of Scotland.

I arrived there one lovely morning with a few shillings in my pocket and went to a good hotel in Princes Street. Then I asked the porter to show me the way to the Bank of Scotland where I wanted to draw my money. He said: 'There it is, on the hill; but you cannot draw money today, sir.'—'Why?'—'Sunday, sir.' I had lived in a holiday spirit, not realizing it was the Sabbath. So I had to spend a day without money and discovered it to be quite easy if you are well dressed and in a good hotel. The days in Edinburgh have left a vague recollection of a lovely city with a picturesque castle just opposite the main business street, and of a magnificent bridge over the blue waters of the Firth of Forth. Little did I dream then that I should return to stay.

I made my way back via Carlisle to the Lake District where I met cousin Hans. We stayed in Ambleside and intended to climb the nearest mountain range and to descend into the Langdale valley, a tour quite small on the map with respect to distance and altitude compared with what we were used to in the Alps. So we strolled off in our city suits and boots, but were thoroughly beaten. After having reached Easedale Tarn, we soon lost our way in bog and mist, and had no choice but to return, sliding dangerously on the slippery rocks.

I think I visited Hans in Bradford, had a glimpse of Manchester, saw lovely Chester and spent a few days with him in Llandudno, in North Wales. In this way I took home with me an impression of a considerable section of Great Britain when I left for the Continent at the end of August 1907. After meeting Carathéodory in Brussels, as already mentioned, I went home to Breslau.

X. Breslau. Second Military Service. Experimental Physics

My mind was made up to give up mathematics, except in so far as it was needed for physical theory, and to become a real physicist. I planned to apply to the physics department of Breslau University for admission as a research worker and I tried to make contact with the physicists whom I knew from my student days. I made these preparations for my future professional work in a leisurely way and spent a great deal of time playing tennis with my cousins and friends in the Kleinburg garden or horseback riding on the outskirts of the city. Suddenly this idyllic life was shattered by a call-up order from the military authorities. It stated that the last medical examination—the one in London I described above—had proved me fit for service and it ordered me to report to a cavalry regiment on 1 October (1907), two weeks away. I had sold both my uniform and equipment of the Dragoons, and I hated the idea of returning to this regiment with the prospect of going through all the drudgery of a recruit yet again, constantly threatened with asthmatic attacks and confinement to military hospitals. I racked my brain to find a way out of this predicament, but in vain—until one day a friend and pupil of my father, Dr Rosenstein, an excellent physician, visited us and heard of my difficulty. He pondered for some moments and then said: 'I have the solution to your problem. Go tomorrow to the barracks of the Cuirassier regiment and report for service, and everything will be all right.' I thought he was crazy, for the 'Leib-Kürassier Regiment No. 1 Grosser Kurfürst' was, together with its sister regiment, the 'Garde du Corps' in Berlin, the noblest and most exclusive regiment in the whole Prussian army. The officers were all aristocrats of even higher rank than those in the Dragoon Guards, they had very few one-year volunteers and these were all young noblemen of the Silesian landed aristocracy. I had never heard of a simple citizen without a 'von' before his name serving as a volunteer in this regiment, not to mention a Jew. Yet Rosenstein insisted that I should try. He said the chief medical officer of the Cuirassiers was a Dr Scholz, who had been one of my father's pupils. Rosenstein knew him well and was convinced that he had no prejudices and would most likely accept me, just to play a trick on the snobbish regiment. I was to tell him that he, Rosenstein, would give him some hints and see to it that my military career would not last for too long.

Well, I did as Rosenstein advised. I went to the red brick Cuirassier barracks, which were quite near to Kleinburg, and entered my name in the list of prospective volunteers. On 1 October I appeared for medical examination and found two other men there for the same purpose. Dr Scholz himself examined us. Everything went as predicted by Rosenstein. When the doctor looked through my previous military record and heard my story, he shook his head angrily, and I heard him murmur distinctly: 'Idiot, that fellow in London.' We were all three accepted. Before I left the room, Dr Scholz took me aside and said: 'Don't buy an extra uniform, my boy, until I tell you; you may not need it.' I took a room in an ugly and filthy house, the only one near the barracks, opposite the main gate; a wooden box, apart from a bed and a chair, was all the furniture I had. I lived a most simple and provisional life, for I was convinced it would not last long. The officers and NCOs were not accustomed to having one-year volunteers at all; if they had one, it was a count or a prince, who had to be treated with the utmost consideration. So I was treated with the utmost consideration. When it turned out that I had served almost half a year with the Dragoons and knew how to handle a horse I was exempted from horse-cleaning, night duty in the stables and from part of the drill; instead, I got a good officer's horse to ride in the pleasant country around Kleinburg. We three volunteers had our meals in a small room attached to the officers' mess and supplied by the officers' kitchen. The other two fellows were members of the aristocracy, and the orderly who served us assumed as a matter of course that I belonged to the same class. We had to write our names on the serviette rings, and as soon as he had seen these he addressed the other two men as 'Herr Graf' and me as 'Herr Baron'—apparently misreading the name Born. I did not correct him, nor did the other two—poor lads, they were rather apathetic after the usual fatigue caused by the first weeks of military training. They considered me a kind of expert soldier whose advice on horses, recruits and sergeants was of some value. My duties ended at about 4 p.m., then I changed into tennis dress and cycled to the Kleinburg house for play, music and flirting. In this way I had spent a month in quite a pleasant though futile way, when one day, on entering the barracks in the early morning, I was told that I was to be released at once for health reasons. There was no further medical examination, but I was exempt from duty and discharged after a few days. Years later I met Dr Scholz at a party and thanked him for the fair treatment of my case. He laughed and said something not very flattering: that the Prussian army could do very well without such brainy weaklings as myself. I wish it had, but again I was to be disappointed; they could not do without me. I shall tell you later of my third military episode.

Now I was free to start on my scientific life. At that time physics at Breslau was represented by two men who were friends and just as inseparable as Hilbert and Minkowski in mathematics. They were Lummer and Pringsheim. They had been together at the National Physico-Technical Laboratory in Berlin-Charlottenburg, where they had worked together at the establishment of the laws of radiation. It was the period following Max Planck's announcement of his radiation law from which he derived the sensational result that radiation consists of finite quanta of energy $h\nu$, where ν is the frequency of the oscillation and h a constant (Planck's constant). Everybody felt that if this were true it meant the greatest revolution in physics—as in fact it turned out to be, not only in physics, but in the whole domain of science and even beyond it, in all applications of human reason; that will be the background of the story of my scientific life in the rest of these recollections.

The first step in this fundamental development was the task of confirming or rejecting Planck's law by experiment. He had found it by an amazingly clever interpolation between two well studied extreme cases: Wien's law for short waves and the Rayleigh–Jeans law for long waves; the question was whether it was correct over the whole range of wavelengths. Lummer's and Pringsheim's careful measurements were essential contributions to the final establishment of Planck's formula. When Lummer was called to the chair of physics in Breslau he accepted only on the condition that a new chair for theoretical physics be created and Pringsheim appointed to it—a case very similar to that of the mathematical Castor and Pollux of Göttingen; for Lummer was a blond Aryan, and Pringsheim a Jew, or at least of Jewish origin. A more dissimilar pair can hardly be imagined. Lummer full of life, highly temperamental, inclined to boasting and subject to fits of temper, a magnificent experimenter, whose big lecture was a sensational performance, attended not only by physics students, but by all kinds of people who wanted to enjoy an hour of intellectual entertainment. Pringsheim was a quiet thinker, elegant in manners and attire, cautious and reserved in his statements, modest and unobtrusive. But they got on very well together.

To these men I applied for admission as a research worker in the physics laboratory, and I was accepted. Being a trained mathematician, I was attached to Pringsheim, and he suggested to me that I should investigate some problem in his own field of research, black body radiation. The exact nature of this problem has disappeared from my mind if it was ever there, for the whole complex of problems concerning radiation was strange to me; I had to read numerous papers and books before I could even understand what I was supposed to do. Practically the first thing I had to learn was the handling of a so–called 'black body', that is, an electrically heated tube of porcelain or some such material with a small hole through which the radiation contained in the interior could be observed, and fitted out with instruments for measuring the temperature up to very high degrees. The whole thing was surrounded by a big double-walled metal box through which water ran, in order to shield the surroundings from the heat. It was this unfortunate cooling arrangement which was to play an essential part in deflecting me from my experimental efforts back to pure theory. But before I record these tragi-comical events, I have to tell you about other people and things surrounding me at the place of my activities. For some of these have had a lasting influence on my life.

Lummer's assistant, his right-hand man and friend, was Wätzmann, whom I knew from my student days. He had specialized in acoustics and was considered a great expert in this field. A nice friendly mild chap, he would do anything to help anybody. He was on bad terms with one man only, his colleague Clemens Schäfer, a lecturer and also assistant; between these two there was a permanent feud of little jealousies and intrigues. We others, mainly Ladenburg, Reiche, Loria and myself, tried to keep out of it as much as we could. I knew Rudolf Ladenburg from my childhood, as his father, a well-known chemist, was a professor at Breslau University, like my own. We had met at children's parties, and attended the same school where he was one year my senior, but no friendship had developed. Ladenburg had quite a different set of friends, mostly tall, elegant, sporting fellows like himself, who appeared to me a little snobbish (as I did to them, very likely). When we now met again it was quite different. Ladenburg had gained his Ph.D. in Munich under Röntgen and was not only an accomplished experimenter but a very learned and enthusiastic scientist who would forget everything else over an unsolved problem, just as I used to do. We very quickly discovered a general identity of interests and tastes, in science, in music, in art, in our love of nature. So we quickly became friends and have remained friends to this day. I admired Ladenburg's manly features, his well-trained body, elegant manners and dress, his ability for quick decisions and determined actions, qualities which I sorely lacked. That our friendship developed rather quickly is partly due to some tragical events in his life. He lost in quick succession his mother and his two brothers, all of whom he loved very dearly. It was a terrible blow to Rudolf who was left alone in a big and elegant house with an ailing and often bad-tempered father. Under these circumstances he found the sympathy and friendship which I had for him particularly welcome.

Fritz Reiche was one of the very few research pupils of Max Planck and had just graduated in Berlin with a thesis suggested by his master. He was a tiny delicate Jew who combined the typical humour of the Berliner with a deep melancholy and pessimism. At times he was most entertaining. I learned from him a great deal about radiation and quantum theory which he had studied at the source, in personal contact with Planck. Reiche later wrote one of the first books on quantum theory and became professor in Breslau, occupying the chair which during my time was held by Pringsheim, and after Pringsheim's death by Clemens Schäfer.

The last in our group, Stanislaus Loria, was a Pole from Cracow, hence at that time a subject of the Austro-Hungarian monarchy. But he was a great Polish patriot and hated the Austrians. To me he was a new type of human being. Poles, as I said in the description of my childhood, were in Breslau mostly domestic servants, like our nurse Valeska, who taught us the Lord's Prayer. Now I met a highly educated and elegant Pole who was proud of his nationality. He was an enthusiastic physicist and a most charming young man, with fine features and perfect manners. We got on very well, and the only point of friction between us was his exaggerated devotion to his professor, Nathanson, in Cracow, whom he declared to be one of the greatest physicists. We considered this to be a nationalistic overstatement, and contradicted him; but on the other hand this admiration for his master was rather touching, and we soon ceased our objections. Much later I learned that though he looked the prototype of a Polish aristocrat, he was not a thoroughbred but had a Jewish mother. This was, of course, his doom, when in 1939 the Nazis overran Poland; I heard that he was in a concentration camp, and I had no hope of seeing him again. But the latest information is that he not only survived, but is professor of physics in Breslau, now the Polish city of Wroclaw. What an amazing circle!

We four, and a few others who have vanished from my memory, met every weekday morning in the physics laboratory on the Dominsel and exchanged ideas about our work. Ladenburg was working partly in collaboration with Loria, on the experimental demonstration of anomalous dispersion in the first Balmer line of (excited) hydrogen. His darkened room with some capillaries glowing in a reddish light fascinated me like a sorcerer's den. He had an uncanny skill in glass-blowing and other techniques which I tried to learn from him, with little success. While he worked with his tubes and wires or adjusted his spectroscope, there came a steady flow of comments from Loria on the theory of the phenomenon, sometimes interrupted by Reiche with a sarcastic 'Oh, that's all rubbish'. Then a wild discussion would flare up. I shared their interest in optics; I knew the theory well enough from Voigt's lectures, but had to learn all the tricks of the experimentalist necessary to produce the phenomena represented by the formulae. But in this I was not very successful. My own experiments with the black body proceeded very slowly and were interrupted by a catastrophe already alluded to. One morning when I entered the laboratory some junior assistants and the mechanic received me with loud lamentations and reproaches; they took me to a cloakroom and lavatory in the basement where a stream of water was splashing down from the ceilings, which were already in a state of disintegration. Meanwhile, Ladenburg, Reiche and others appeared and led me in procession to my own room, which was just above those watery places in the basement. There I found the floor had turned into a pond, the water standing some inches deep, although charwomen were hard at work with buckets.

What had happened was this. As I said, my 'black body' was cooled in a metal box through which water ran. This water had to make its way through a long pipe-line of glass and rubber. When I stopped working in the afternoon, I had to wait about an hour until the temperature had dropped sufficiently for me to turn off the water. Now the previous night I had worked longer than usual, and when I stopped it was rather late. Perhaps I had some invitation or theatre ticket; anyhow I decided to let the water run throughout the night. But apparently one of the rubber tubes had found these long hours of work too tedious and had slipped off the water tap, with the result just described. Professor Pringsheim was mildly annoyed, Lummer was furious. I had to pay the costs, quite a considerable sum, and Reiche concocted a beautiful poem on the inundation which made the whole institute laugh at my expense, but also curbed the wrath of the terrible Lummer.

There were, however, other occasions for angering Lummer. He was a great expert in optics from the practical as well as from the theoretical side. Once or twice he visited me in my room and inspected in particular the optical part of my apparatus, playing with all the screws and handles, and spoiling my adjustments. I was never fast enough for him in getting them re-adjusted, nor did he relish my comments; I suppose quite rightly because at that time my theoretical knowledge was not well coordinated with my practical experience. We never got on friendly and confidential terms, and this fact became decisive for my future.

Outside the laboratory, our little group kept good company, and we had a merry time. Ladenburg had just bought a motor car, a great adventure, for in 1907 automobiles were rare, and not as reliable as today. Very often he called on me in the morning to take me to the laboratory, or he gave me a lift home. At weekends we went for long drives in the country, sometimes with young girls. My old friend Lore Jänicke and her sister Käthe often joined our party. In the winter Ladenburg and I went skiing together in the

In the Kleinburg garden, Breslau 1907.
Myself in the uniform of the 'Leibkurassiere', with some of the
Kauffman relatives (centre, Aunt Luise)

126

Riesengebirge. the highest part of the Sudeten mountains, easily reached from Breslau in two hours. I had been on skis before and had acquired some basic technique. Now we went for long tours over the endless expanses of the 'Kamm' (comb; the long flat ridge of the mountains), sometimes in blazing sunshine, other times in thick fog, spending the night in one of the numerous mountain inns called 'Baude'. We learned to speed down the slopes and to stop with a 'Christiana' or 'Telemark', or, more frequently, with a somersault in the snow. Though Ladenburg was the much stronger and more enduring, I became quite good at the tricks of this lovely sport, and it was he, not I, who had a bad accident. We were whizzing down into a valley over a steep slope covered by a slight mist when the first trees appeared. I slowed down and avoided one or two trees by inches. I stopped where the thick forest began and waited for my friend. But he did not come. Greatly worried I began to climb again, but it took me a considerable time, perhaps half an hour, to get to the point where I had lost him. There he was lying. He had seen the trees too late and run into one. His knee was badly hurt. It was too cold to leave him there in the snow and to get help from the next 'Baude'. He insisted on continuing the descent, though he could only use one leg. This descent into the Weisswasser-tal was a nightmare, which I cannot describe. In the end we reached houses and help, he in great pain and both of us exhausted. His leg was not broken, but badly sprained and he had some trouble with it for the rest of his life. Nevertheless we went skiing again and had no further accident.

At home Mama tried to make me as comfortable as possible and kept an open house for my friends. I took piano lessons with Max Auerbach and we played trios with Ilse Späth. Then there were the Neisser house and the Kauffmann villa in Kleinburg, the theatre, opera and concerts, dinners and dances. I remember with particular pleasure the parties arranged by the Neissers, after the concerts of the Breslau Philharmonic Orchestra; they took place in 'Hansen's', an old wine restaurant with an excellent cuisine and that comfortable atmosphere which you may know from 'Luther's wine-cellar' in Offenbach's opera *The Tales of Hoffman*. The guests of honour were the soloists of the evening, the conductor of the orchestra, Dr Dohrn, and its leader Herr Behr, brother of Theresa Behr, a celebrated singer who later became the wife of Artur Schnabel. There I met many musical celebrities and other famous men and women.

In this pleasant way the winter of 1907/8 went by. In spring 1908 I undertook a journey to Greece which began with the romantic episode in Venice described in Chapter III of these recollections. The first plan for this journey had sprung from an accidental meeting with Carathéodory in Berlin who told me that he intended to visit his relatives in Athens and Constantinople and asked me to join him. I took a small Austrian steamer from Trieste. After a pleasant journey (I was almost the only passenger on the boat) along the Dalmatian coast and a short stop at Corfu we steamed through the gulf of Corinth and entered the canal, which is a narrow and deep cut through the rock. And when we reached the other end which opens to the blue Aegean Sea there standing at the quayside was my friend Carathéodory in his usual black attire, a cigar in his mouth. But he told me to my great disappointment that he could not come with me (I think his uncle, who was director of the canal, was ill) but would join me later in Athens. So I had to spend most of the time without my friend on whom I had counted as an expert guide. But I hardly needed a guide; thanks to my classical education the whole journey was almost like a home-coming; all was well-known names and views familiar from pictures and descriptions, and everything beautiful beyond description, for Greek air and light can only be experienced.

I wish to recall a few incidents and impressions. I have already mentioned my visit to Cape Sounion. To reach it one had to take a train to a little town called Lavrion, on the top of Attica. In my compartment was a nice and loquacious man who on learning the purpose of my journey, declared that he could not understand why foreigners were so keen on visiting the old Poseidon temple instead of the interesting modern industry of Lavrion. He, Mr Levidis, mayor of the town, would show it to me. It was difficult to persuade him to let me proceed to the temple, but I succeeded by promising him that I would see the works in the afternoon. I was lost in beauty and sunshine when a carriage appeared and a footman asked me politely to follow him. What I was shown was not uninteresting. Two American engineers, whose English I understood better than Mr Levidis's French, explained to me that they were exploiting with modern methods the slag heaps of the antique and now exhausted silver mines which had played a decisive part in the classical history of Athens during the Persian wars and after. The Americans now extracted lead which had been thrown away by the ancient Athenians.

I went for a long trip to Delphi and the Peloponnese. At that time travelling in Greece was not as comfortable as in Central Europe. A steamer went once a week from Piraeus to the little port of Itea from which the travellers were taken by horse-drawn carriages up a steep, winding mountain road to a little inn near the ruins of the temple of Delphi. Here my classical education failed me. Whatever I had expected, it was not the wild, terrifying mountain landscape where the work of man is dwarfed by the colossal rocks and ravines over which eagles were circling. I can still see before my eyes the tragic loneliness and grandeur of the valley where the remnants of Apollo's temple are standing. For two or three days I rambled between rocks and broken columns in a state of enchantment. Then I had to leave in order to catch the steamer in Itea on its way back from Patras to Athens.

This return journey was not without its little adventure. I shared the carriage with two German ladies, mother and daughter. It was the last vehicle in a convoy of four or five, and soon we were left far behind. Our driver was not in a hurry. Several times he stopped at cross-roads to take parcels or letters from peasant women for delivery in Itea. When the little port came in sight deep down at the foot of the hill, we saw steam rising from the ship's funnel and heard its siren. The other carriages had already reached the quay and we were still high up. When we at last arrived the steamer was far out at sea. There we were stranded in a desolate dirty place, knowing that the next boat to Athens would arrive only in a week's time. My two ladies were exceedingly depressed and near crying, but help soon arrived. There was a motor launch from which a filthy looking sailor emerged; he approached me and offered to take us to the opposite side of the gulf, faintly visible in the distance, where we could catch a train from Patras to Corinth. He demanded an excessive price which I flatly refused. Then began a bargaining which lasted for hours, until the fare was reduced to about half. We then boarded the motor boat and while gliding out of the harbour, we could clearly see our coachman being taken in triumph to the local inn to celebrate his successful trick and it became obvious that it was an established performance, the spoils of which were shared between the two men. After my return to Athens, I went to Cook's office to complain about how we had been cheated; they were not in the least astonished or indignant but refunded my extra expenses at once.

The launch delivered us safely on the shore of the Peloponnese, at a small station, the station master kindly stopped the express train for us in the afternoon. I arrived in Corinth in the evening, instead of the morning as I had planned, for I wanted to climb the rock on which Acro-Corinth, the ancient fortress, is built. It was a lovely night; I decided to make the ascent in moonlight, and I found a boy with a mule willing to guide me. It was indeed worth the effort. The whole mountain is covered with the most fantastic ruins of buildings and fortifications. The ancient Greeks, Macedonians and Romans have left their marks everywhere, and after them the Crusaders, the Turks, the Venetians covered the rocks with their temples and chapels, mosques and bastions. Everything is overgrown—a green wilderness with vines and half-tropical plants of all kinds. We arrived at the summit just as the sun rose in incredible splendour out of the Aegean Sea. At that moment my guide shouted in order to direct my attention to something red jumping from a rock and disappearing into a hole—a fox.

We descended to a village at the foot of the ruins of a big temple on whose steps peasants were sitting, drinking their retsina. I was very tired and thirsty from the long climb and gratefully took the pumpkin gourd offered to me. I drank the bitter wine in one draught, said 'thank you', mounted my mule and proceeded towards the station. How I arrived I do not know exactly; I fell asleep on the back of the mule, and the boy must have had considerable trouble in keeping me in the saddle. In the station I continued to sleep on a bench in the restaurant. Now I had an appointment with a man whose acquaintance I had made in Athens, a German wine merchant named Hildebrand, from Rüdesheim on the Rhine, who was to meet me at the station of Corinth from which we intended to proceed to Nauplia. He found me in the restaurant lying snoring on a bench, and the waitress was very worried. 'He is ill,' she said. Hildebrand bent over me, sniffed and replied: 'No, drunk', and shook me until I was conscious again and quite well, to the great relief of the girl.

The onward journey brought me and my companion to Nauplia, Epidavros, Mycenae and Tiryns. But little has remained in my memory, and if I remember the Lion Gate of Mycenae, it may be just the impression of the innumerable pictures I have seen of it.

From Athens I went to Constantinople. On the boat I met a Greek family with a German name (since the time of the Bavarian kings quite a number of German families have settled in Athens and become Greek) with two lovely daughters who taught me Greek folk songs which I tried to accompany on the piano. Then we danced half the night, and I was sorry when the minarets of Istanbul appeared on the horizon and we had to part. But new impressions consoled me quickly. In the marvellous Serail Museum, I was most impressed by the Hellenistic sarcophagi with their somewhat sweetish and sugary reliefs—I think it would be considered bad taste today. I called on the German Embassy and got a ticket for the great Selamlik. It was still the time of old Sultan Abdul Hamid. The Selamlik was a parade of his crack regiments and well worth seeing. I remember a regiment of black men in red uniforms on white horses, a splendid spectacle. From Carathéodory I had introductions to some of his relatives and was invited to parties. In this way I saw something of the life of these Greco-Turkish families. They were rather grand people with lovely dark features, in particular the women, and they received me with great kindness.

I returned home by a slow train across the Balkans for my money had run short and I could not afford the Orient Express.

Back home I resumed my pleasant life and work, and you will be surprised to learn that I was not happy—had I not everything a young man could desire? Perhaps I had, but one essential thing was missing: decent prospects regarding my chosen profession. I had not succeeded in getting on good terms with Lummer. It was obvious that he did not consider me a physicist because of my deficiency in the art of experimentation. But how could I have learned it in half a year? Even if my progress had been quicker, I doubt whether it would have made much difference, for he did not like me. I had no chance in Breslau. Where should I go?

I have found that questions of this kind are in most cases answered not by deliberate choice and planning but by a series of accidental events. And so it was in this case.

One day Reiche asked me whether I knew a paper by a man named Einstein on the principle of relativity. He said Planck considered it most important. I had not heard of it, but when I learned that it had something to do with the fundamental principles of electrodynamics and optics which years ago had fascinated me in Hilbert's and Minkowski's seminar, I agreed at once to join Reiche in studying it. From that moment relativity became our principal interest, and both Loria and Ladenburg were infected by our enthusiasm. I began to think and work on some relativistic problems, and when I got stuck over some difficulties I wrote a letter to Minkowski asking his advice.

His reply was a great surprise. He did not answer my questions but said that he was himself working on the same subject and would like to have a young collaborator who knew something of physics, and of optics in particular. Would I like to come to Göttingen once again? And he added that this might lead to my future entrance into an academic career. Then he proposed that I should attend the annual meeting of the German Society of Scientists and Physicians, which was to be held in Cologne in September.

There he could answer my questions and substantiate his proposition.

I have related that when I left Göttingen after my graduation I was rather fed up with the place and decided never to return, and that Toeplitz used to tease me by prophesying I would return not once but twice, namely first as lecturer and later as professor. I did not like the idea of Toeplitz's forecast proving right, but that was no argument against an opportunity of such importance. I went to Cologne, met Minkowski and heard his celebrated lecture 'Space and Time', delivered on 21 September 1908. Outside the circle of physicists and mathematicians, Minkowski's contribution to relativity is hardly known. Yet it is upon his work that the imposing structures of modern field theories have been built. He discovered the formal equivalence of the three space coordinates and the time variable, and developed the transformation theory in this four-dimensional universe. He told me later that it came to him as a great shock when Einstein published his paper in which the equivalence of the different local times of observers moving relative to each other was pronounced; for he had reached the same conclusions independently but did not publish them because he wished first to work out the mathematical structure in all its splendour. He never made a priority claim and always gave Einstein his full share in the great discovery.

After having heard Minkowski speak about his ideas, my mind was made up at once. I would go to Göttingen and to help him in his work.

XI.   Göttingen Again.
'Habilitation'

I arrived in Göttingen rather late, on 2 December 1908, long after the semester had started; I know this from an old letter but I cannot remember what caused the delay. Toeplitz had taken my old room for me in Kirchweg 4, where he still lived, and although I disliked the idea of being so close to him again, I moved in as nothing else was available. The following day I called on Minkowski and was received very kindly. He explained to me his ideas on electrodynamics and relativity, and he listened patiently to my own suggestions. Once again I was fortunate in meeting a leading man in my branch of science and in being allowed to watch him working on a subject which fascinated me. But alas, it lasted only a few weeks. When I returned from a short Christmas holiday at home, I learned that Minkowski had been taken to hospital dangerously ill, and that an appendicitis operation had been performed. I did not see him again. He died on 12 January 1909. And with him went all my hopes and plans. The following months were one of the critical periods of my life. I had to decide whether to give up my academic ambitions or to fight hard. For I had lost my patron and I knew full well that Felix Klein would oppose any attempt of mine to join the teaching staff of the Philosophical Faculty.

The day after Minkowski's death I was sitting dejectedly in my study when a representative of the mathematical students appeared and asked me to speak in their name at the funeral. I had never done anything of this kind before and was frightened by the idea of addressing that solemn assembly. But I accepted and worked hard day and night, writing and memorizing. It helped me to overcome my grief and depression, and my words, as I discovered later, won me friendly feelings in many hearts, even in that of Klein.

Minkowski's numerous papers and unfinished manuscripts were given to Speiser, a young Swiss mathematician, and myself for sifting and possibly finishing. We first of all separated the pure mathematics from mathematical physics; the latter was entrusted to me. It was a formidable bundle, but most of it turned out to be either manuscripts of paper already published, or indecipherable sketches. Only one investigation which I knew in outline from my last interview with Minkowski was far enough advanced for me to try to finish it for publication. This I succeeded in doing. The article appeared in May 1910, together with a reprint of Minkowski's celebrated paper *Die Grundgleichungen für die Elektromagnetischen Vorgänge in Bewegten Körpern* (The basic equations for electromagnetic processes in moving bodies) as No 1 of a series of monographs edited by O. Blumenthal. It contains a derivation of the electrodynamic field equations for moving bodies from the hypothesis introduced by Lorentz, that all electric and magnetic properties of matter depend on the existence and the movement of electrons; but it differed from Lorentz's own derivation in two ways: the electrons are introduced not as particles but as a continuous distribution varying in density and velocity from that of matter; and secondly this variation is treated by Minkowski's new relativistic methods. I have just read the paper again and I think it represents Minkowski's ideas fairly well.

Apart from this work on Minkowski's posthumous papers, I continued to develop my own ideas which I had discussed with Minkowski. He had not been enthusiastic about them but had raised no objections. These ideas have to do with the problem of the self-energy of an electron and have accompanied me throughout my scientific life. Twenty-five years later I found a solution from the standpoint of classical electrodynamics, by generalizing Maxwell's linear field equations in such a way that they became non-linear; but even at that time classical field theory was obsolete and had been replaced by quantum theory of the field, and here my solution did not work. So the old problem remains unsolved. It can be circumvented by a method given by Dirac which the younger generation, amongst them Heitler, Bhabha and also my own son-in-law Maurice Pryce, take as *the* solution; but I do not believe that a fundamental problem can be discarded in this way, and I expect it to turn up again when the time is ripe.

In my young days the situation was this. J. J. Thomson had remarked that the charge on an electron would produce an additional mass since acceleration generates a magnetic field, and he directed attention to the possibility that the whole mass of the electron might be of this electromagnetic nature. This idea was eagerly taken up by others, fascinated by the suggestion that one of the fundamental concepts of Newtonian mechanics, mass, might be not a primary but a derived quantity, and that electromagnetism lay behind mechanics. Abraham, who I have mentioned before in connection with the 'Schwarzer Bär' circle, and who at that time was a lecturer at Göttingen, was a keen advocate of this theory. He had worked out the properties of a fast moving rigid electron in detail and found that the electromagnetic mass should depend on the velocity. This strange result was experimentally confirmed by Kaufmann (and later by many others). At that time relativity was already in the air; the contraction of moving bodies in the direction of velocity had been proposed by Fitzgerald and by Lorentz (before Einstein's theory was published) as a convenient explanation of the negative result of Michelson's experiment. Lorentz calculated the self-energy for a contracting electron and found a much simpler formula for the dependence of mass on velocity than that of Abraham; experiments have proved that Lorentz's formula is right.

All that was going on at the time of my return to Göttingen. I could easily look up the correct order of events in the literature, but that does not matter much for these recollections. Anyhow, these were exciting days and, as I frequently came into contact with Abraham, I was well informed about his controversy with Lorentz. Abraham was anti-relativist and objected to Lorentz's derivation. I also doubted it, but doubted Abraham's derivation as well. Both proceeded by calculating the self-energy of a charged rigid body in uniform motion (Lorentz with contraction, Abraham without it) as a function of velocity and using this energy as Hamilton's function for obtaining the equations of motion. This procedure assumes that the energy calculated for constant velocity also holds for accelerated motion. My doubts were concerned with this point, and I decided to derive the equations of motion for an accelerated electron in strict accord with the principle of relativity.

This led at once to a great difficulty. For if a body is accelerated, different points of it have different velocities, hence different contractions: the idea of rigidity breaks down. My first problem was therefore: how far can the concept of a rigid body be preserved in relativity? Rigidity means lack of deformation. I worked out the mathematical expression for the deformation of a moving body on relativistic principles, the so-called strain-components, which are differential expressions containing the derivatives of the coordinates as functions of their initial values and of time. Then I postulated that these strain-components should be zero; I obtained in this way differential equations for the possible strain-free motions and I found a solution of them which represents a uniformly accelerated rigid motion in a straight line, uniform in the sense that the acceleration in the instantaneous rest-system is constant, and rigid relative to the same system. This 'hyperbolic motion', as I called it, is in fact nothing more than a 'rotation' in the four-dimensional space of Minkowski. In this space a rotation has not a one-dimensional axis (line of points at rest), but a two dimensional one, a whole plane. Now the four coordinate axes x, y, z, t determine six planes yz, zx, xy, xt, yt, zt, and to each of these belongs a rotation; the first three are ordinary 'cyclic' (timeless) rotations, the last three are my 'hyperbolic' motions. This interpretation leads to a simple method for deriving the electromagnetic field of a hyperbolically moving electron, as was shown later by Sommerfeld and others. I obtained the field by straightforward, rather cumbersome calculations. The main result was a confirmation of Lorentz's formula for the electromagnetic mass as a function of velocity, and its dependence on acceleration. I worked on this investigation all through the winter and spring of 1909, and when it was finished I showed it to Hellinger who, of all my friends, was the one most interested in mathematical physics. He approved of it; backed by his opinion I ventured to apply to the Mathematical Society for permission to give a report on it and was admitted.

This lecture was a complete catastrophe, much more serious than anything I had encountered before in my scientific career. The whole atmosphere of that learned Society was neither pleasant nor encouraging. At a long table parallel to the blackboard were seated the most formidable mathematicians, mathematical physicists and astronomers of Germany: Klein, Hilbert, Landau (Minkowski's successor), Runge, Viogt, Wiechert, Prandtl, Schwarzschild, often strengthened by guests, German or European celebrities. The younger members and less important guests were seated at two long tables at right angles to the 'high table'. This younger crowd, if not as famous as the 'Mandarins', was yet no less critical and perhaps more conceited: Zermelo, Abraham, the Müllers, Toeplitz, Hellinger—all known to you from previous mention in these pages, and many newcomers: Gustav Herglotz, Alfred Haar, Hermann Weyl, Paul Köbe and others. Books were piled on the green cloth of the tables; at the beginning of the meeting Klein gave a short account of his impressions of some of these new publications and then circulated them. So everybody soon had a book in his hand and paid very little attention to the speaker, and what attention he gave was mostly in the way of objection and criticism. There was no friendly listening nor a vote of thanks at the end. It was extremely difficult to catch the attention of this audience, to create a spell of interest, and scarcely possible to arouse enthusiasm.

No wonder that the first time I had to face these people, I failed dismally. I had no experience in lecturing and I am sure I did it badly enough. But the main obstacle was Klein's preoccupation with some aspects of the theory of relativity which, together with his animosity against me, prevented him from grasping the point of my work. He was accustomed to interpret the Lorentz transformation as relations of a rigid frame in Minkowski's four-dimensional space, meaning by rotation a timeless change of orientation, while I used the expression rigid for a four-dimensional structure moving in time. He at once objected to my terminology and advised me in harsh terms to study mathematical literature before addressing the Society. I defended my standpoint by saying that it was a straightforward generalization of physical concepts—whereupon Abraham joined in the debate to tell me that my knowledge of physics seemed to be just as scanty as that of mathematics. He was annoyed because my theory led to Lorentz's formula for the electromagnetic mass and not to his; yet it was not fair to stab me in the back and side with the 'Great Felix' whom he used to ridicule on every other occasion. The continual interruptions and attacks confused me, I got hopelessly entangled in my formulae and hardly opened my mouth again. Suddenly Klein declared that that was enough; he had never listened to such a bad lecture in all his life.

I retired to my seat completely broken and expected to see grinning, contemptuous faces around me. But nothing of the kind. I do not think many had listened, and those who had did not seem quite convinced that Klein was right. When I left the room after the meeting, making my way as unobtrusively as possible, I met Runge, who had waited for me. He said: 'I am interested in your work. Klein seems to have missed the point. Call on me tomorrow morning and explain it to me.'

This was a great consolation. I do not know how I should have survived the night without it. Next morning I went to Runge, who listened patiently to my explanations and said when I had finished: 'Seems all right to me. I shall speak to Hilbert about it.' So he did. Meanwhile Hellinger, who was then

135

Hilbert's private assistant, had already discussed the scandal of the previous night with Hilbert and severely reproached him for not having interfered in my favour. Then Runge appeared and backed Hellinger, until it was decided that I should explain my work to Hilbert also. This I did with good results, and at the next 'Bonzen-Spaziergang' Hilbert and Runge jointly attacked Klein, insisted that he had been mistaken and unjust, and urged him to make amends.

All this I learned much later. For Hellinger wisely did not speak until he knew that Klein had given in. Meanwhile, I was in a miserable, dejected mood. The first day after the catastrophic lecture I prepared to leave Göttingen for good, and my friends had the greatest difficulty in persuading me to stay. I seriously meant to give up science and go to one of the technical colleges in order to become an engineer, according to my original intentions. Never since have I spent such a wretched time, in spite of the wars and revolutions and exile I have been through. For all these were general misfortunes coming from outside, while my failure in the Mathematical Society seemed to me a clear proof of my own inadequacy and presumption.

One day I got a letter from the secretary of the Society, a Dr Timpe, who informed me that the council had decided my lecture had not been received in the right way and should be repeated, and asked me whether I was prepared to do so.

At first I was not at all prepared to do so. I considered the sentence to be final and just. For if I could not explain my ideas in a comprehensible way, I had no right to become an academic teacher. My friends tried in vain to persuade me. Finally Runge had a serious talk with me and said that my obstinacy was silly. He made it clear to me that I was simply afraid to undergo the ordeal again, and this was sufficient to reverse my decision.

The following weeks were not pleasant, for I was indeed afraid. But I prepared myself carefully, going through my lecture with Hellinger point by point. When the day came the meeting was even more crowded than usual. It was an outstanding event that Felix the Great had had to retreat. Everybody knew that this had happened, and the books which were circulated did not absorb the attention of the audience to the same degree as usual. Runge introduced me with some friendly words, saying that Klein has misunderstood my intentions and himself had suggested that he should listen to me a second time. So he did, at first with a somewhat sullen expression, but soon with full attention. There were no interruptions, and when I had finished I sat down among friendly smiles—applause being unknown in this society.

After the meeting Voigt said to me that Minkowski had once mentioned to him that I might become a lecturer and asked me whether I still had this intention? The present paper seemed to him a very suitable thesis, and I could count on his support in the faculty.

This meant complete victory. But I was too exhausted to enjoy it. We celebrated it in Mütze's Weinstube, an old wine restaurant in mediaeval style where they had an excellent but expensive Rüdesheimer. This I ordered for my faithful friends and we emptied quite a number of bottles, enjoying the exquisite wine and commenting on its qualities with the air of connoisseurs. I expected a considerable bill, but when it came it was amazingly moderate. What had happened? They had two kinds of Rüdesheimer, one the old, celebrated and expensive one, the other a new, cheap, ordinary one, which the waiter had served to us. This shows not only my absentmindedness—for I had chosen the wrong number on the list—but also the powers of the imagination and the spell of a name. For we had thoroughly enjoyed our 'expensive' Rüdesheimer.

My paper was published in the *Annalen der Physik*, in 1909. My definition of rigidity produced a considerable discussion and it turned out that it was not fully analogous to the classical concept. For while a rigid body in non-relativistic mechanics has six degrees of freedom, three coordinates of one point and three angles of rotation, it was found by Herglotz (*Annalen der Physik*, 1910) that my relativistic rigid body had only three degrees of freedom, namely the three coordinates of one of his points. When I read this brilliant paper by Herglotz, I felt admiration and a sting of envy, for his presentation is so clear and convincing—how much wretchedness I would have been spared if I had had this power of analysis. Later (1911) Herglotz published another excellent paper on the relativistic mechanics of deformable bodies, where he used my strain definition, quoting Planck, Laue and others with much emphasis, but me only in an inconspicuous footnote. I remember that I felt hurt, for I had a high regard for Herglotz's abilities. Later, when I knew him better, I found out that he was quite unconscious of having committed an offence. His own mathematics was so much better than mine that it had not even occurred to him. I committed similar misdeeds myself later, in more than one case. It is difficult to be just to one's predecessors in a work in which one is enthusiastically involved. For the progress of science such questions are of course indifferent. Yet we scientists are human beings, and no human being likes to be pillaged. Therefore I cannot promise that questions of this kind will never appear again in these recollections.

I got my admission as a lecturer, called 'Habilitation', in the summer 1909. At the beginning of the winter semester (23 October 1909) I gave my inaugural lecture 'On Thomson's "plum pudding" model of the atom', a subject not in line with classical tradition, which at that time was still prevalent at Göttingen, but indicating the direction which my interest had taken and which was to absorb me more and more. The lecture was published in *Physikalische Zeitschrift* in 1909 (Vol. 10, p. 1031) and I have read it again with interest. Regularities in the line spectra of gases were known at that time, but not understood. I described in poetic language how a physicist, looking into the spectroscope, had the sensation of getting a glimpse into the mysterious depths of matter. Then I compared his impressions with that of a musician who listens to a new symphony through the closed doors of a concert hall and tries to understand the structure of the score and the distribution of the parts of the various instruments. Unable to reconstruct the whole score, he may succeed in a summary for the piano. Physics also has such gifted artists: Thomson's atomic model is the first sketch of the atomic symphony. After this inspired introduction, I gave a reasonable account of Thomson's suggestion; I was particularly impressed by the weak analogy between the stable arrangements of electrons inside a sphere filled with a positive charge and the periodic table of elements. It seems to me today that I had quite a sound judgment about the relevant features of physics and chemistry. However, Thomson's model did not survive very long and was soon replaced by Rutherford's nuclear theory of the atom, which was based on direct observations, or, to use my poetical simile, made by one who had pushed open the doors of the concert hall. Only five years later the mystery of the laws of line spectra and of the structure of the atom was solved by Niels Bohr.

XII.   *Lecturer in Göttingen*

My first lectures were concerned with rather special departments of classical physics, e.g. on vibrations of elastic bodies, on electromagnetic waves, etc. I was certainly a very bad lecturer. One of my pupils in this first year of my academic career was Richard Becker, who was to become my successor in the Göttingen chair of theoretical physics when I was driven out by the Nazis—much against his inclination, I am sure. His academic promotion had been slow, and once, after a disappointment about a vacant chair, he told me in a fit of frankness: 'That is all your fault, because you were such a scandalous lecturer.' Though meant as a joke, it certainly contained a good deal of truth.

Another incident which shows that my lecturing was abominable happened a few years after my 'Habilitation'. I was asked by a group of chemists, medical men, pharmacists, etc., to give them a short private course on thermodynamics. One of them was Rudolf Ehrenberg, lecturer in physiology, who told me years later, when he had become my brother-in-law, that this lecture had been an ordeal for all of them; they did not understand a word but dared neither to interrupt me or nor to stay away as I was one of their colleagues and had volunteered for the job. I think they did stay away eventually and the course came to an inglorious end. My mistake was to suppose that everyone approached a subject in the same abstract way as I did. I disliked compromising in the question of logical rigour, as I was still under the spell of pure mathematics. How this spell was broken, I shall have to relate presently.

Soon after my 'Habilitation' I moved from Kirchweg and Toeplitz to a house, Nicolausberger Weg 49, which was run as a boarding-house by a widow, Frau Hampe, and was traditionally inhabited by young scholars and university teachers. At that time Abraham and Zermelo lived there, and also the philosopher Grelling, whom I have mentioned as one of Nelson's disciples. But this proximity did not produce a greater intimacy between us. My immediate room neighbour, however, was a man of quite a different type, and to him I became very attached. Joseph Partsch was the youngest professor at Göttingen University, a member of the law faculty. He, like myself, came from Breslau, where his father had been professor of geography. He had been a very brilliant student; after his graduation he fell ill with tuberculosis and was sent to a sanatorium in the Swiss mountains. He spent several years lying in his bed on an open verandah, working all the time. He discovered the existence of a number of Egyptian papyri, which contained most valuable material for the history of ancient law. He has described to me how he wrote his book on this subject lying in the open air, sometimes in winter at such low temperatures that he could hardly hold his pen and that the ink froze in the pot. But this book made him famous overnight, and he was offered a chair in Göttingen when he was not yet thirty. He was a blond giant, of fresh complexion and stout, full of life and sparkling wit. I liked him immensely, and I think he liked me, too. He later married a very beautiful girl from Breslau; but he died very young from a fresh attack of his lung trouble.

Most of the men in our house took their meals in their rooms, singly or in groups. I had my lunch with Grelling and a few others who lived elsewhere. Two of these must be mentioned as they became great friends of mine. Both were Hungarian Jews and mathematicians.

Alfred Haar was the son of a big landowner who was on good terms with his aristocratic neighbours and shared their opinions. Alfred was a small, delicately built and neatly dressed fellow, pleasant to look at, friendly, gentle, sharp-witted, sometimes with a good sense of humour, yet often melancholy and depressed. After lunch he and I used to join Partsch in his room for a cup of coffee, and the debates which ensued have remained in my memory. After his graduation, Haar left Göttingen; before disappearing he introduced us to his friend Theodor von Kármán, also a Hungarian Jew, no less clever than Haar, but lacking his elegance and gentleness. Haar later became professor in Budapest; I saw him but rarely, and he died very young. Kármán's early scientific career was very closely connected with my own. Later we separated; he attached himself to Prandtl and became one of the leading men in aerodynamics. He was a professor in Aachen but emigrated to America before Hitler came to power, and became one of the top authorities on his subject in the U.S.A., perhaps in the world. During the second world war he was scientific adviser to General Arnold, the chief of the U.S. Air Force. I met him a few months ago (June 1945) in Moscow where we were attending the celebration of the 220th anniversary of the Academy of Sciences of the U.S.S.R. He was still the same as in the old days in Göttingen: a strange mixture of kindliness in his relations to his friends, and of bitter cynicism in regard to human affairs in general.

Both Haar and Kármán greatly influenced my scientific development, pulling me in opposite directions. Haar was a pure mathematician, a devoted disciple of Hilbert, of the same type as Erhard Schmidt, whom I have described previously. He loved the beauty of mathematical construction, the purity and rigour of demonstration. I enjoyed—and still enjoy—this element of mathematical thinking, although I had already discovered that I lacked creativeness in this line. Kármán, on the other hand, was the best type of applied mathematician. He could well appreciate the artistic value of a deduction, but was not deeply interested in things which had no practical application in natural science or technology. I think that it was during this period of our meals together in Frau Hampe's house that he worked out his well-known theory of the double row of linear vorticles in a liquid flowing around a cylindrical obstacle. This theory was beautiful, but still more impressive was the first photograph of the phenomenon predicted by it: the simple trick of dispersing a powder on the surface of water made it possible to see and photograph the lines of motion and to observe the formation and propagation of the vorticles. I liked this kind of thing even more than the most brilliant theorem of pure mathematics. At that time I was working on consequences of the theory of relativity which had little connection with observable facts. But through our daily discussions I was slowly converted to the problems of actual physics, and in our last year together we joined forces in a problem of atomic physics, the application of quantum theory to the theory of crystals.

I do not think that Kármán's contribution to our common effort was much greater than mine, but he was the driving force, the directing mind from whom I learned the essentials of mathematical physics. These are: to regard the problem in its right perspective, to estimate by rough and ready methods the order of magnitude of the result expected before going into detailed calculations, to use approximations adapted to the accuracy needed, even if no rigorous proof can be given, and to be constantly aware of all the facts.

That our attention was drawn to crystals and their lattice structure was connected with some work by Madelung, who was an assistant in the physics department. Lattice structure of crystals was at that time (1911, 1912) much under discussion, but still hypothetical; Madelung's work came just a little before the discovery of X-ray diffraction by von Laue, Friedrich and Knipping (1912). He obtained a numerical relation between the elastic constants and the wavelength of infra-red absorption (Rubens's 'Reststrahlen') by studying the vibrations of a model lattice for rock salt. (This model had in fact already been guessed by other people long before, e.g. by Crum Brown in Edinburgh, and was later confirmed by the Braggs with X-rays.) I have already mentioned in Chapter VI how Abraham let loose his caustic wit on poor Madelung when he congratulated him on having 'cheated our learned Academy' by making them accept his paper. Kármán did not share this opinion, for he saw clearly that the fundamental idea and the order of magnitude were right; but he was too great an expert on mechanics not to be critical of Madelung's primitive handling of the dynamic problem of lattice vibrations. He suggested to me that we should reconsider the problem from the standpoint of the mechanics of a system of coupled particles. Which of us first connected this line of thought with Einstein's paper on specific heat of solids, I cannot remember. We were well satisfied with our results but a great disappointment was to follow.

Our first paper appeared in April 1912 in *Physikalische Zeitschrift* (vol. 13, p. 297). But in March 1912 Peter Debye, professor in Zurich, had given a lecture on the same subject to the Swiss Physical Society and published a short paper in *Archive de Genève* (March 1912, p. 256). He used a very different and much cruder method, replacing the crystal by a continuous substance, but he obtained simpler formulae than those in our first paper, in particular the well-known T^3-law for the specific heat at very low temperatures. Without knowing this, we had meanwhile derived the same law from our general formulae. While we were writing our second paper (*Physikalische Zeitschrift,* vol. 14, 1913, p. 15), we learned from a lecture given by Sommerfeld in Göttingen that Debye had obtained the same results. So we had to acknowledge Debye's priority in the whole matter. I found later that our papers are hardly known in the English-speaking world. Yet they are more satisfactory than Debye's, which applies only to quasi-isotropic substances. The superiority of our method was definitely established by my pupil Blackman (about 1932).

The work on crystals was for me not a transitory phase, as for Kármán, but was to prove of lasting interest. I discovered that hardly anything was known about the conditions which the atomic forces have to satisfy in order to produce a periodic arrangement or lattice. The dynamics of crystal lattices became my central interest, and although it has at times been replaced by other problems, I have again and again returned to it and am still today (December 1945) occupied with new aspects of this subject.

But now I have talked enough about science and have to give you an account of these last years of real peace, of that bourgeois civilization which broke down in 1914 and will perhaps never return. I enjoyed it and can assure you that it was not as bad as it is depicted today. I shall tell you how I lived, and you may well be envious; for instance, if you hear how easy it was to travel: no passport required, no exchange problems, as you could get foreign money to an almost fixed value, very little customs formalities, excellent hotels everywhere, fast trains, not crowded and not expensive, and so on. So let me tell you about a few of my journeys.

I went to Italy several times. When my brother Wolfgang had passed some examination, I invited him on a journey to Naples. It was not a complete success, for Wolfgang had even at this youthful age an overwhelming enthusiasm for art, to the exclusion of everything else, and liked to use artists' jargon in a way which was apt to drive me out of any museum. However, my sufferings ended when we met Hans Mühlestein, whom I have mentioned before as a member of the Friesian school of philosophy and a friend of Leonard Nelson. That was very convenient for me, for the two art fanatics neutralized each other, and I could enjoy myself in my own fashion. I had to pay for three instead of two, but that did not worry me at all.

Later I preferred to travel with Ladenburg. Easter 1909 we spent in Bozen (Bolzano), staying at the Hotel Austria in Gries. No place in the world have I loved more than the South Tyrol, and I hope I may return there, to see the Schlern and Rosengarten in the purplish evening light, to sit in the Stadtcafé or the Greif, to stroll under the 'Lauben', to buy grapes and apples in the fruit market, to be carried by the funicular to the Virgl, or higher up to the Vigil-Joch, to stroll up the Oswald Promenade and to look down on the garden-like valleys of the Eisack and the Etsch. I still remember that first visit with Ladenburg when the world was one of peace and plenty, and nowhere more so than in Bozen. There one also met interesting people. The celebrated chemist Adolf von Baeyer was staying in the same hotel with his whole family. Ladenburg knew them well from his Munich days; one of Baeyer's sons, Otto, a physicist, was his friend. I remember the patriarchal figure of old Baeyer, a noble example of the ancient type of dignified German scholar. He was one of the founders of modern organic chemistry and contributed much to the development of the German chemical industry.

The following spring (1910) we went to Italy and made Florence our headquarters, from which we travelled all over the country. We saw Pisa, Siena, San Gimignano and a number of smaller places, such as the monastery of Vallombrosa. We were young and carefree, and we enjoyed it thoroughly. We believed ourselves to be experts in Italian art and were extremely proud when we guessed correctly the painter of a picture without looking at the catalogue. Two years ago (1943) when the war swept over this beautiful country, all these old impressions came vividly back to my mind. But what a change: battles where we had been roaming, and the destruction, filth and stench of modern war sweeping over the lyrical hills of Tuscany. It seems the height of madness, and yet I am not sure that the human race is not preparing to beat even this record.

For the summer vacations I developed a programme which I followed for several years, with one interruption, when in 1911 I went with Mama and Wolfgang to the Dutch seaside resort of Zandvoorst. That programme consisted of Bayreuth and the Engadine.

The Neissers were great Wagnerians. They were among the first subscribers to the erection of the Festspiel-Haus and never missed a performance in the years following its opening. They had often tried to persuade me to join them, but I had refused. Not that I disliked Wagner's music—I think I told you that I drove my house-mate Toeplitz almost mad by playing the piano scores of the Wagner operas—but I disliked spending twenty-five or thirty marks (or whatever it was) for a single ticket, and I had a horror of the international snobbish crowd which assembled in Bayreuth every year. But one day Toni wrote to me that they would have a car at their disposal, belonging to a former pupil of Albert Neisser, who would be joining their party, and that they intended touring the countryside of Franconia with its lovely cities, villages and castles. That made me change my mind, and I have never regretted it.

It was a pleasant party; apart from the owner of the automobile, there were the Erlers with their pretty wives and one or two young girls, from amongst the former 'protégés' of the Neisser house. The Neissers knew many of the Bayreuth musicians, whom they invited to supper after the opera. One of them I remember particularly well, Siegfried Ochs, the conductor of the Berlin Philharmonic Choir, who was a charming man and most entertaining, especially when he performed musical parodies at the piano. His eldest daughter later married my friend Fritz Reiche. The standard of the opera at that time was extremely high, and the performances were so perfect that a genuine feeling of reverence and awe was produced even in that spoiled and snobbish audience. I heard in those years all the Wagner operas, some of them, like *The Ring* and *Parsifal*, even more than once. But I never became a complete Wagnerian like Albert Neisser. I even once had a little clash with him, following a casual remark of mine—something about Bizet, whose opera *Carmen* I loved very much. This I should not have said in Albert's presence. He became quite furious: 'If you prefer that coffee-house musician to the great "Master", you should not have come with us to Bayreuth.' I was in disgrace for a whole day.

But this devotion to Wagner and his work was not without a tragic note. For when the Wagners' Haus Wahnfried gave a garden-party and many of our friends were received by Frau Cosima and young Siegfried, the Neissers were not invited—for they were Jews. All his fame as a scientist, all his devotion to Wagner's work, all the money he spent in aid of the Bayreuth theatre made no difference: Jews were excluded from Wagner's house. On such a day we used to go on an excursion by car, and the Wahnfried party was not to be mentioned, but I felt that an old wound was smarting. Yet the Neissers never wavered in their devotion to Wagner and his cause. I am glad they died long before this world of theirs was shattered. Siegfried Wagner's widow was one of the first persons of rank to open her house and purse to Hitler. No place has contributed more to the fateful slogan of the 'German master race' than Wagner's Bayreuth. His heroes became the symbols and prototypes of the Hitler Youth and the anti-semitism of 'Haus Wahnfried' spread over the whole country. Yet who can deny the genius of Wagner as a musician? I loved *Siegfried* and *Die Meistersinger*, and I still know the score so well that I can recognize any scene heard on the radio from a few bars. I gave up playing these operas after my marriage since Hedi did not like them. My taste has also changed. Now I play almost nothing but Bach. Yet when I hear a good performance of one of Wagner's works I still feel a kind of thrill. Is this the magic of the tunes, or just the reminder of happy and harmless days?

From Bayreuth I used to go to Switzerland and spend a month in Pontresina or Sils Maria where I met Ladenburg. He was a mountaineer of sorts and climbed, with a guide, Piz Bernina, Piz Palu and other giants. I was content with the smaller tours, Piz Languard, Piz Corvatsch and such like. The Neissers, too, visited the Engadine nearly every year and had their regular evening parties where one met interesting people. One of these was Fritz Haber, the chemist, later well known for his method of producing ammonia directly from the nitrogen of the air, and ill-famed as the originator of chemical warfare in the first world war. He used to take me for walks in order to discuss problems of thermodynamics which occurred in his physico-chemical research. One day I found him in the company of a small, sturdy man who wore not mountain dress like the rest of us but looked like a businessman from Berlin or London, in a black suit, grey striped trousers and a bowler hat. He turned out to be Gilbert Lewis, the American chemist, who had travelled from California to Berlin in order to discuss problems with Haber, and when he did not find him there, followed him straight to Pontresina without even changing his attire. When I was introduced to him he at once forgot all about his chemical problems and began to talk relativity with me. He was an enthusiastic relativist and admirer of Einstein, and he had read some of my papers. Later I visited him in Berkeley, California, and we have met at many a scientific congress, for instance in Russia (1928) when we travelled down the Volga with a group of physicists (see Part 2 Chapter II).

Back in Göttingen, most of my time was of course occupied by my work, lecturing and research, which I have already described, but there was also plenty of entertainment. For recreation and exercise I continued riding and playing tennis. At that time we had a real tennis club with excellent courts on the slopes of the Hainberg. I became quite a good player, a fact which had some influence on my life, as you will hear later.

As a member of the university staff, the houses of the professors were open to me, and I soon felt much more at home in Göttingen than as a student. There was for instance the lovely house of the mathematician Landau, Minkowski's successor, whose wife was a daughter of my father's friend Paul Ehrlich, the discoverer of salvarsan. The Runge family were very hospitable, and I became a good friend of some of their numerous daughters. It was at this period that I was first introduced into the household of my future father-in-law, Viktor Ehrenberg. Some families were connected by friendship, and if a young man was introduced to one, the others followed automatically. To this group also belonged the Merkel family; a professor's widow with two daughters and two sons. These boys were very musical, one an excellent 'cellist, the other a violinist. I had regular trio evenings with them for many years, sometimes enlarged to a quartet or quintet.

In winter there were parties and dances, in the private houses and in the hotels, and in summer, excursions to some inn in the woods where we played and danced in the open air. I remember vividly a big party given by the Ehrenbergs in the style of a rural 'Kirmess'. There was an enormous tent in the garden where father Ehrenberg, dressed as a Bavarian village innkeeper, served beer from a big barrel while mother Ehrenberg, in the corresponding female costume, dispensed sausages and other simple refreshments. An accordion played for the dancers, and a big pole was erected from which you could collect prizes, if you were not afraid of spoiling your evening dress by climbing to the top. Partsch, my house-mate, was at this party, and I enjoyed watching his enormous bulk waltzing around with cheerful devotion. But suddenly I saw a kind of white flag appearing at the base of his back: his trousers had been split by his exertions and his shirt had slipped out. I succeeded in manoeuvring him into a corner before his misfortune was discovered. But never did I see such despondency as his: the lovely evening spoiled, how could he, with his girth, hope to find another pair of trousers? We lived not very far away, so I took him home and sewed the tear in his trousers as well as I could. It did not look neat, but who would inspect with critical eyes the bottom of a learned professor? Nobody did. He danced the whole night unsuspected, and when we went home together at dawn he swore to me eternal gratitude and friendship for my life-saving work.

During this period a guest professor was invited every year by the mathematical section of the Göttingen Academy. The money came from the

Wolfskehl Fund, and I think that the use made of it to pay the fees of guest lecturers, though very laudable, was not strictly legal. For this man Wolfskehl, from a wealthy Darmstadt family, had bequeathed a considerable sum, a hundred thousand marks, to the Academy as a prize to be awarded to the solution of the celebrated problem of Fermat. This problem is easy to understand. You know that $3^2 + 4^2 = 5^2$ (namely $9 + 16 = 25$); the equation $x^2 + y^2 = z^2$ has therefore a solution in integers, $x = 3$, $y = 4$, $z = 5$; it has in fact more than one. The problem is: has the equation $x^n + y^n = z^n$, where n is any integer, 3, 4, . . . , a solution where x, y, z are integers? Fermat once casually remarked that he had a simple proof that this is not the case; but he never published it. All mathematicians believe that he is right, but none has ever succeeded in demonstrating the non-existence of a solution. I suppose Herr Wolfskehl had also made some effort. Anyhow, he offered the prize, with the effect that hundreds of manuscripts arrived at the office of the Academy claiming that they contained a solution to Fermat's problem. Some of them were voluminous, others just letters or even postcards. But altogether they were a nuisance. For in 90 per cent of the cases the solution was based on a misunderstanding and in the rest on an obvious error. Time and energy was wasted by reading and answering this flood of communications. In the end the Academy found a way out of this difficulty: they would consider papers only if published in a periodical. Meanwhile the interest from the Wolfskehl Fund was to be used to bring guest lecturers to speak on mathematical problems of the day. Some of these I wish to mention as they had some influence on my life.

The first guest professor was Henri Poincaré from Paris; he lectured on pure mathematics and I do not remember anything except his strange appearance: a dwarfish man with a slightly hunched back, a rough, short beard and very sad eyes. In 1910 the guest lecturer was H. A. Lorentz, the celebrated Dutch mathematical physicist. I was given the task of taking notes and working out these lectures for publication. In this way I came into close contact with a man who at that time was generally considered the top authority in my field. It was a thrilling experience to discuss with him all the controversial problems of physics. These lectures have appeared in *Physikalische Zeitschrift* (vol. 11, p. 1237, 1910) under the title 'Alte und neue Fragen der Physik' (Old and new questions in physics) and offer a splendid cross-section of the physics of that period culminating in a derivation of Planck's radiation formula. From that time on, I had a good friend and adviser in Lorentz. I visited him in his house in Haarlem, where I was received with the utmost friendliness.

The next visiting professor (1911) was Albert Michelson from Chicago, whose celebrated experiments were the main foundation of relativity. He came with his second wife, Edna, and three little girls, aged between six and ten. His lectures were not very inspiring and did not bring me into close contact with him. Our friendship instead developed on the tennis court where I happened to be his regular partner. He was a man of over sixty, incredibly vigorous and an excellent, hard-hitting player. In the final tournament he beat us all and won the first prize. On the tennis court I also met Mrs Michelson and became much attached to her and the sweet little girls, particularly the second one, Beatrice, whom I still remember as a lovely child. I was sorry when at the end of the summer semester they had to return to America.

But that was not the last I heard of them. A few months later I got an invitation from the University of Chicago to give a course of lectures on relativity during the summer term of 1912. It was accompanied by a very nice letter from Mrs Michelson, expressing her and the children's wish to see me again. I accepted at once, secured a leave of absence from my university and sailed for America.

XIII. Chicago.
'El Bokarebo.' Marriage

Almost every British scientist of today has at some time been in the United States, but in Germany before the first world war this was not the case. Very few of my colleagues had crossed the Atlantic. Among my circle, only Max Abraham had been there, spending some years at the State University of Illinois at Urbana, but he did not like the American way of life and had returned. So it was quite an adventure for me to go on this trip in April 1912. I sailed from Bremen on one of the fast German steamers, had a colossal flirtation with a lovely American girl named Mabel but arrived safely in New York.

Though I enjoyed the grandeur of Manhattan's skyscrapers and the turbulent life of the city, I was disgusted by the shocking social contrast between the rich and the poor. I watched the latter in the crowded quarters of the Eastside and Harlem. On the other hand I called on some wealthy families in their beautiful houses near Central Park. These families were mostly those of great Jewish physicians, to whom I had introductions from my father's friends, for instance from old Ehrlich. They were very kind to me and it will seem ungrateful when I tell you that I disliked their gorgeous dinner parties. I remember one in particular where, amongst many courses there came a dish consisting of a whole chicken for every guest, not a tiny one, but one big enough for a family of four. Every guest took a little piece and the rest was taken away, probably to the wastebin. That very same morning I had seen in some Jewish quarters near the East river poor starving people digging in refuse heaps for something to eat.

From New York I went straight to Chicago and stayed a few weeks in the Michelsons' lovely house. Then I moved into a room in one of the students' dormitories. It was one of the most miserable places I have ever lived in, comparable only with German military barracks: filthy, bleak, depressing. But there was nothing else to be had, for the nearest decent hotel was some miles from the University. When in July I travelled north to visit Mrs Michelson and her children in their summer resort on Lake Michigan, I found the sleeper berth in the train luxurious compared with the dirty and uncomfortable bed in my room at the dormitory, a room which the terrific heat of the American summer transformed into a miniature hell.

Another trip from Chicago took me to Niagara Falls and over to Toronto in Canada.

My work in Chicago consisted of lecturing on the theory of relativity to a group of research students and of younger members of the staff. At that time my lecturing technique had already improved, and I think that I was fairly successful. I also worked a little in the laboratory. Michelson gave me one of his wonderful concave gratings and showed me how to use it. So I spent pleasant hours in focusing, observing and photographing spectra of many substances, and I was very happy when I got a plate which Michelson approved of. But he was only interested in the technique of producing faultless photographs and hardly at all in the meaning of all the lines and bands seen on the plates. I could not fail to observe numerous regularities, in particular in the spectrum of the carbon arc, and I asked Michelson whether he knew an explanation. His reaction was very curious. In one of the bands which seemed to me obviously to follow a simple law (I suppose it was one of the CN-bands) he showed me the existence of irregularities, consisting of lines which were displaced from the position where they should have been, and said: 'Do you really think that there is a simple law behind it if horrible things like that happen?' I do not know whether he thought that nature acts in a haphazard way; but he was certainly not interested in the secret behind these phenomena. In fact the first step towards lifting the curtain had at that time already been made, as I soon learned, in the establishment of Deslandre's formula for simple bands which led finally to the explanation of the complicated band system, including those apparent irregularities.

Shortly before I left Chicago I witnessed the Convention of the Republican Party for the nomination of the next president of the United States. The strongest candidate was Theodore Roosevelt, the first of this name, the 'rough-rider' from the Spanish war. I attended the meetings, which were held in a colossal hall. The crazy procedure of these conventions has been described often enough, but I do not think anybody who has not seen it can imagine the pandemonium. The walls were plastered with large pictures of the candidates, mainly Roosevelt, and with heads of the elk, his emblem. A number of bands were playing, sometimes different tunes simultaneously. Each candidate was greeted with cheering, the duration of which was a measure of his popularity. Roosevelt's cheering lasted over fifty minutes. I bore this ordeal heroically in order to hear him speak. He looked and spoke like a man of powerful mind and will. But altogether I found this convention a poor advertisement for democracy.

After the end of the summer term I started on a long trip through the States, planned with the help of Mrs Michelson. I went to Minneapolis and there took the Canadian Pacific transcontinental train. After a pleasant journey through the endless wheat plains of southern Canada, I reached the Rocky Mountains and stayed a few days in all the celebrated mountain resorts such as Banff, Lake Louise and Glacier Point, enjoying the alpine scenery and air. Lake Louise is certainly one of the loveliest spots I have seen, comparable with the Engadine.

149

The train journey down the canyon of the Fraser river to the Pacific coast was remarkable because our pullman cars became a kind of timber market. There were men coming in at little stations who would offer you a piece of forest which they pointed out to you from the windows of the carriage, very cheap indeed. I almost yielded to a chap who wanted to sell me some acres of primeval forest for twenty or thirty dollars. Later the train passed through regions where wide stretches of woodland had been destroyed by fire, and then I was quite glad not to have exchanged my dollars for such a perishable commodity. I spent a day in Vancouver, visited Victoria, British Columbia, and then went to Portland, Oregon, where I took the Shasta Express to California. The colossal snow-covered cone of Mount Shasta could be seen for half a day. At the first Californian station we were standing alongside a freight-train loaded with some dark material which looked like coal but turned out to be dark grapes, ripe and sweet. All the passengers on our train ate as many as they could. I joined in, with catastrophic consequences for my bowels. But when I arrived in Berkeley, my friend Professor Gilbert Lewis, the chemist, cured me by taking me across the bay to San Francisco to a little Italian restaurant where he ordered tea and sandwiches, with a twinkle in his eye. The tea was Scotch whisky. Later we had Chianti under another disguise. At that time prohibition was not universal in the States; I met it here for the first time and found its practice greatly amusing. However, the stuff cured me.

From San Francisco I went to the Yosemite Valley for a week's touring. It was before the days of the motor car, and there was still something left of the original loneliness and majesty of this wild and lovely country. Transport was by horse-drawn coaches. I stayed a few days in the Sentinel Hotel in the centre of the National Park and went climbing in the surrounding mountains. Then I travelled by coach through endless forests to the Mariposa grove of giant trees, the biggest and oldest in the world (*Sequoia gigantea*). The people in my coach were all very well dressed, almost elegant, but from their language and the topics of their conversation I found out that they were simple farmers from the Middle West who had become rich and were now taking their wives and daughters for a holiday trip. One of them became quite attached to me. He was incredibly enthusiastic about everything. When we stopped to see the tallest tree of all he took out a ball of red thread, gave me one end to hold and walked around the tree measuring its circumference. I could have told him the result from my Baedeker, but I did not and so preserved his friendship and esteem. Half a year later I received in Göttingen a letter containing an exact copy of this red thread with a few lines of greeting. I kept it for many years as a reminder of American kindliness.

I took a train to Los Angeles and stayed there for a fortnight. This visit has become completely mixed up in my memory with my second one, fifteen years later. I know that I visited Pasadena and Mount Wilson and met Hale, the founder of the observatory, and I remember some excursions to the coast, which at that time was still wild and lonely, in particular one to Santa Catalina Island. It was reached from San Pedro by a small steamer; a lovely mountainous island, with a fine rocky coast, covered with wild woods and scarcely inhabited. There one took a glass-bottomed boat and went out into the bay to see the 'Marine Gardens', the wonderful world of tropical sea plants and animals. It was like a fairy tale. When fifteen years later I visited the place again, the steamer from San Pedro was an enormous vessel with eight rows of benches, for some thousand passengers. The beach around the bay on the island was lined with big hotels and a casino from which a pier in the Blackpool style stretched into the water. And there were all types of modern noise-making 'amusements', merry-go-rounds and dancing tents and shooting ranges. Instead of the small glass-bottomed boats rowed by fishermen, there was a motor launch of considerable size, also with a glass bottom, but not open to the sky and therefore filled with exhalations from the hundred or so passengers who were sitting on long benches on both sides, expressing their admiration at the wonders of deep-sea life by noisy shouting, laughing and screaming.

On my journey back East, I took the southern route to see the Grand Canyon of the Colorado in Arizona. This was undoubtedly the greatest impression I got in America. The first glimpse over that stupendous gorge with its marvellous colours and fantastic shapes is overwhelming. The El Tovar Hotel at the rim of the abyss was at that time still small and comfortable. The air of the high Arizona plain is incredibly clear and refreshing, and so I stayed longer than I had intended. I was able to make the descent to the river, 4000 feet below, on muleback with Red Indian guides, a strenuous but fascinating trip. One day climbing in the rocks I made the acquaintance of an elderly gentleman with a big camera who at once made me help him in taking pictures. He told me that he was the 'first' amateur photographer in the States and had made a special study of the Grand Canyon. Indeed he knew a tremendous number of spots from which one had surprising views.

A year later I was reminded of that photographer by a postcard signed by him and Lindemann (now Lord Cherwell). The latter was my 'successor' at the summer school of Chicago University and on my advice visited the Grand Canyon, where he met this camera-man in the same way. So things repeat themselves.

On my return to New York I received a cable from my friend Theodor von Kármán, reading something like this: 'Will you share house with me and three other lunatics?' To this I replied at once 'With pleasure'; for I knew what it meant. We were not satisfied with our house in Nikolausberger Weg, Göttingen, where I had lived for several years and where Kármán used to share my meals. We had often made plans about renting a new and comfortable house where we, together with some friends, could live under the care of a well-trained housekeeper and a maid. But this idea had never come to anything because of the difficulty of finding a suitable housekeeper. Now Kármán's cable must mean that he had found that treasure and a house as well. But when I arrived in Göttingen (October 1912) I found that there was more behind it than I thought. Kármán had met an elderly woman who had given up nursing with the idea of establishing a home for old and decrepit people. Kármán at once asked whether we would not do, but she answered that we were not decrepit enough, and that she greatly preferred patients who were a little feeble-minded. Then Kármán took great trouble to convince her that we were just the kind she wanted—though physically quite fit yet mentally decidedly abnormal. Well, he must have been convincing for eventually she accepted.

So I found them already installed in the new home, Dahlmannstrasse 8. Sister Annie, the nurse-housekeeper, a stout, plump woman with a kind, silly face, showed me my two spacious rooms on the first floor of the modern villa. Kármán had the other two rooms of this floor. Downstairs was a common dining room.

The names of our house-mates were Hans Bolza, Bep (Robert?) Renner and Sergei Boguslavski, all of them students of mathematics and physics; the last one a young Russian of noble descent, very gifted and attractive, but of delicate health. He soon became my favourite pupil and produced some very good papers; when the first world war broke out, he returned to Russia and became professor in Charkov, but died very young of tuberculosis during the Russian revolution.

When I arrived they were all in shirt-sleeves working hard getting the place ship-shape, painting doors, moving furniture, hanging pictures. I took off my jacket too and joined them in their work. Suddenly a taxi drew up, a young and pretty girl with black hair stepped out and rang the bell. We hurriedly slipped into our jackets and opened the door. She asked to see the room which had been advertised in the paper, was shown by Kármán to the attic and declared at once that she would take it. Out came her luggage from the taxi; and there she was, a member of our new community. I was rather stunned by this quick development—I did not even know that there was an attic room available, and I did not expect to have a female co-lunatic. But the others explained. They had already been living in the house a few days and found that Sister Annie presided at the dinner table as a matter of course. This was not as they had expected and it was embarrassing indeed, for she kept a strict discipline, not allowing any freedom of expression or little harmless indiscretions, and even corrected their manners. The situation was unbearable, and a council of war was summoned. Many plans were proposed and rejected until one of the group had a brilliant idea. There was still an attic room free, which was intended as a spare room for guests. Let it be advertised for a single lady; from the answers they could choose a person capable of 'neutralizing' Sister Annie at dinner, by involving her in female chat of absorbing interest. This was accepted, a carefully worded advertisement was published, and now we had the result; not an elderly spinster, as expected, but a pretty young girl student with extremely decided manners. None of us had the nerve to refuse her.

It turned out to be a good choice. Her name was Ella Philipson. She did not 'neutralize' Sister Annie but she was not in the least shocked by strong expressions, she had a good sense of humour and laughed away Sister Annie's attempts to correct us. Our dinner table consisted not only of the residents but of a number of people from outside. One of them was Paul Ewald, a young theoretical physicist who had just returned from Cambridge. He and Ella discovered at their first meeting that they were distant cousins and became great friends. If one looked for Paul, one only had to knock at Ella's door. Sister Annie was deeply shocked, but could not object after she was told that the two young people were engaged to be married. It all went very quickly, and it gave the signal for all of us; at the end of one year most of us were engaged or married. But that is to anticipate events. First came this crazy but enjoyable year in the house El Bokarebo. That was the name we gave it, composed of the first letters of our names. Years later when we were scattered over the earth, it was still readable over the entrance gate.

Due to the ever-changing dinner-guests, our community was not restricted to the six 'residents'. We often had fascinating discussions and debates, or parodies of such. We gave parties and performances. Hilbert's fiftieth birthday fell in this period. We concocted a poem called the 'Love-Alphabet'. Hilbert had a great weakness for the female sex, and he was always flirting with new love in a grotesque way, whether students or wives of colleagues, or actresses. But the dominating figures in his life remained Käthe, his wife, and Clärchen, the maid. Our composition aimed at a couple of rhymed lines for each letter of the alphabet, and I think we almost succeeded. I remember only one, letter M:

Malvinen gab er nie ein Busserl,
Dafür sorgt schon ihr Mann, der Husserl.
(To Malvina he never gave a kiss,
Her husband Husserl looked after this.)

Referring to the stiff and immaculate wife of the philosopher whom I have described earlier.

Of course, our life was not all fun and games. We had our work and we did it with earnest devotion. It was during this time that Kármán and I published our papers on the vibrational spectra and the specific heat of crystals which I have mentioned in a previous chapter. I wrote a series of articles for Berliner's *Handwörterbuch der Naturwissenschaften* on the propagation of light, on principles of physics, and on space. In collaboration with Kármán and Bolza, I published a paper entitled 'Molekularströmung und Temperatursprung' which was an application of Hilbert's general kinetic theory of gases to a special case, the diffusion of a gas through a fixed set of equal obstacles, when the collisions of the gas molecules with each other are neglected, but those with the obstacles taken into account, with an application to Knudsen's observations of a discontinuity of temperature at the wall of the vessel containing the gas.

My chief, Professor Woldemar Voigt, in no way interfered with my activities. I was a lecturer but not his assistant. He allowed me to do some experimenting in his laboratory. On his suggestion I tried to determine the dielectric properties of water in the region of short electric waves—or what was called 'short' at that time: several decimetres—by using a Lecher system and determining the resonance of a condenser filled with water or a comparison liquid. I obtained smooth and reproducible curves with many maxima and minima but could not explain them. Other people had found similar but different curves, and as I had no reason to trust my experience more than theirs, I never published my results. Later I ventured into the domain of X-rays. At that time serious attempts were being made to settle the question of 'waves of particles' by producing X-ray diffraction patterns. Experiments by Walter and Pohl (1908/9) with wedge-like slits seemed to give positive results. From these Sommerfeld estimated a wavelength of $0·4$ Å during this year (1912); I began to ponder about small natural objects which could be used instead of artificial slits for producing diffraction, and I hit on colloid particles which were much under discussion in our circle, since we had a new department of colloid chemistry under Zsigmondy. I was just going to plan an apparatus to study the passage of X-rays through colloid solutions of gold particles, when the celebrated paper by Laue, Friedrich and Knipping appeared (1912), describing their use of crystal lattices as gratings with atomic distances. Kármán and I were enraptured by this marvellous idea and at the same time depressed by our failure to spot it ourselves, although our crystal work had led us so near to it.

Professor Voigt was a great music-lover and musician, specializing in J. S. Bach. He conducted his own Bach choir which gave annual concerts of cantatas, and motets, and he had a good organ and a complete collection of Bach's works in his spacious house. When he invited guests he often played to them on the piano or the organ for hours on end. There was a peculiar atmosphere at these parties; everybody felt the friendliness of host and hostess and their eagerness to entertain at the highest level, but not everybody was musical enough to listen to Bach for a whole evening, and not everybody was well-bred enough to conceal that he was bored. Sometimes Voigt condescended to have a performance of Mozart or Beethoven. I remember that one day I was asked at very short notice to play a Beethoven sonata for two pianos, together with some female celebrity. I practised day and night, and it went off quite well. I had heard people playing on two pianos often enough in Neisser's house, but this party at Voigt's was the first time I did it myself.

Of other people outside the El Bokarebo set, I wish to mention one man who became a friend of mine, Heinrich Rausch von Traubenberg. He was a genuine Baltic Baron from Riga not, like Erhard Schmidt, an 'aristocratic bourgeois'. But he made little use of his noble birth and ancient family. He was not rich and had to earn his living, which he did first as an engineer, later as a physicist. And what an enthusiastic physicist he was! The art of experimenting, which he called 'tricking nature', was for him an artistic creation. He had the most charming way of speaking of these things, quietly and jokingly, but with a thrilling vibration in his voice which betrayed his enthusiasm. He also liked to embark on long discussions on social, philosophical and religious subjects. At that time he was a lecturer like myself and assistant to Professor Wiechert, the geophysicist, and he lived in rooms belonging to the Rohns Inn on the Hainberg near the seismological institute where he worked. There I visited him often after a stroll through the woods and talked with him over a bottle of Burgundy late into the night. He distinguished very carefully between 'small talk' and 'great talk', but I doubt whether I ever succeeded in being a partner in this highest grade of conversation, from which I was barred by my lack of theological interest. My future brother-in-law Rudolf Ehrenberg was better suited to it because of his religious mind as well as his superior powers in consuming Burgundy. Later when I was married we had some memorable convivial evenings with all three of us together.

Towards the end of this El Bokarebo period, a tragedy happened. One fine spring morning our house-mate Bep Renner was brought home dead. He had gone, the night before, for a bicycle ride in the hills. Nobody noticed when he did not return. He was found by woodcutters in a narrow steep passage, with his neck broken, lying near his damaged bicycle. He was a strong fellow and a perfectly competent cyclist; the mystery of this accident has never been cleared up. I had seen death before, my father's and grandfather's and others; but here it was the first time I had ever stood beside the body of a young man, one of my friends, of my own age. It was a new and terrible experience. We had not to wait long before death gathered a richer harvest.

In the same spring, 1913, Hedi and I became engaged, and I spent many a weekend that summer visiting her at Leipzig, where her father was professor, and later at Bad Pyrmont, where she was sent to strengthen herself for matrimony.

We visited Breslau together, where Hedi had to undergo the ordeal of meeting my innumerable relatives. I remember a party where her lovely hair was much admired; she and my cousin Claire Kauffmann had a competition to see whose plaits were longer, and Hedi won. I introduced Hedi to the Neissers; it was the last time I saw Toni and her lovely house. She was already very ill and died soon after; Albert lived a few years longer and died during the war.

Our wedding was in Grünau. The reason was a little divergence of opinion between my mother-in-law and myself. In fact we were very great friends, and had already been so for years before my engagement. She had an old 'oracle' consisting of a set of small cards on which passages from the Bible were printed. If she was in doubt, she drew one of the cards and got advice provided she could interpret the sentence. Once she asked this oracle about her daughter's future husband and got the reply: *Über eine Zeit wird das Haus David einen freien offenen Born haben wider alle Unreinigkeit.* ('In that day there shall be a fountain opened to the House of David . . . for uncleanness.') When she heard of our betrothal she was not astonished but very pleased. Yet there was a discord in our relationship, owing to the fact that I was a Jew, and objected to being baptized. This made it difficult to have a proper church wedding, which she regarded as essential and desirable. But I could not overcome my distaste for such a ceremony. Eventually a compromise was found. My sister offered to lend her house for the wedding.

There was also a semi-religious ceremony performed by a Pastor Luther, a friend of the Ehrenberg family and perhaps a distant relative. (The second wife of Hedi's grandfather, the great jurist Rudolf von Jhering, was a direct descendant of Martin Luther.) Many members of both families were invited; a big marquee was erected in the garden, where we had an exquisite dinner in the best Ehrenberg tradition. It was a wedding in the grand old style.

XIV.   World War One

On our honeymoon we went first to Copenhagen, then to Sweden, and finally to Vienna, to attend a meeting of the *Naturforscher-Gesellschaft* (corresponding to the British Association). I had been in Vienna before, as a guest of my friend Ernst Philip Goldschmidt, who was a relative of the Rothschild family and lived at that time with his mother in the Heugasse, an aristocratic street in the inner city. He knew everybody of importance, and had taken me to theatres, clubs and ballrooms, and I had had a glimpse of that old gay Vienna which the present generation only knows from operettas. Now I saw it again with Hedi, and we enjoyed it thoroughly: Burg-Theater and Opera, Schönbrunn and Prater, palaces and museums.

There was a big reception in the Burg, the Imperial Palace; the old Emperor Franz Joseph did not appear, but was represented by his grandson Otto, the last of the Hapsburg emperors. There I saw the oldest and proudest court of Europe still in all its pomp and splendour; a few years later it had gone for ever.

Our first address in Göttingen was 'Am Weissen Stein 4'. We had a flat on the first floor of an apartment house. It was not in the fashionable part of the town where the professors used to live. Our back windows looked on to the barracks of the infantry regiment stationed in Göttingen, and we heard the reveille in the early morning and the noise of the parade-ground all day long. Yet this situation was a great asset in the eyes of our maids. The tenant of the upper flat was a Captain Lodemann of this regiment, with whom we had a permanent feud. He disliked my piano-playing; but instead of complaining or asking me to restrict my time of practising, he answered noise with noise by letting his little boy beat a drum or blow a trumpet. So our relationship was tense. But a year later he was dead, one of the first victims of the first world war.

At that time Professor Voigt decided to retire, in order to finish his book on crystal physics, which is still the standard work on this subject. Peter Debye was nominated to be his substitute, later his successor. I have mentioned him before in connection with the theory of specific heat where he forestalled Kármán and myself by a few weeks. It was very fortunate to have this brilliant

Max and Hedi – a honeymoon snapshot

man as a colleague. We held a colloquium together and had many interesting discussions.

The year 1913 is memorable in science because of the publication of Niels Bohr's first papers on the quantum theory of the atom (in the *Philosophical Magazine*, July, September, November, 1913; March 1914). They made a deep impression on us and were thoroughly discussed. Yet the whole atmosphere of the physics department in Göttingen was, in spite of Debye, not favourable to such revolutionary ideas. Rutherford's model of the atom on which Bohr's theory was based seemed to me at that time an alternative hypothesis to Thomson's model, which some years earlier I had taken as the subject for my inaugural lecture. But we had a man in the department, Dr Bestelmeyer, who had studied radioactivity in depth, and one day I asked him about the details of Rutherford's work. He explained to me the deflection experiments of α-rays and gave me the literature, which I devoured. Suddenly I saw that Rutherford had not 'invented' a model but 'interpreted' experimental facts in the most obvious and natural way. It was much later that I learned that Lenard had come to similar conclusions about the nuclear structure of atoms (he called the nucleus 'dynamide') by experiments on the scattering of electrons. After having overcome this difficulty, I very soon became an ardent follower of Bohr. Yet it was several years before I made any contribution of my own to the problem of atomic structure.

At that time (1913) I still thought in terms of classical physics. A series of papers by Gustav Mie (*Annalen der Physik*, vols. 37, 40, 41) appeared in which

157

he tried to generalize Maxwell's electromagnetic equations in such a way that they could explain the existence of electrons with finite self-energy. It was the first attempt at a non-linear field theory, and it fascinated me by its natural conception and elegance. I wrote a little paper on the energy-momentum theory of Mie's electrodynamics, which appeared in *Göttinger Nachrichten* (1913). It contains a simple derivation of the differential conservation laws which has become the standard method. Mie did not in fact succeed in deriving the existence of electrons as a consequence of his field equations. Twenty years later I took up this question again and found a solution which was quite satisfactory from the standpoint of 'classical' (i.e. non-quantum) physics. But meanwhile quantum theory had become well established, and all attempts to adapt my non-linear electrodynamics to the quantum concept failed. Today I see clearly the reason for this failure; I shall speak about it when I reach that time in my tale—the period of Hitler's ascendancy and my emigration. Yet I still think that the fundamental difficulties of modern physics can only be solved by some general non-linear set of equations.

At about the same time that I was working on this classical theory, I also made, in collaboration with my friend Courant, an attempt at a modest contribution to quantum theory, namely a derivation of the so-called law of Eötvös, which expresses the temperature-dependence of the capillary constant. Our method consisted in applying Planck's formula for the energy of an oscillator to the vibrations of the surface, in the same way as in the theory of specific heat it is applied to the elastic vibrations. We omitted, however, the zero-point term in Planck's formula which at that time was uncertain. Therefore the quantum effect which we believed we had found is spurious. Nevertheless, I think it quite satisfying that we obtained the temperature coefficient of the Eötvös constant fairly accurately—it is of course nothing more than an application of the classical equipartition law. I never took up this question again; it seemed to me that the physical chemists with their mysterious concept of 'parachor' had made a mess of it.

Meanwhile I continued my investigations on atomic vibrations in crystals and studied in particular the diamond lattice. I demonstrated that one could express many physical properties with the help of only two atomic constants, provided that the interaction of all but the nearest neighbour atoms can be neglected. In this way I obtained a quadratic relation between the three elastic constants, but no measurements were available to check it. A few months ago, thirty-three years after establishing the theory, I eventually obtained these data, and in such an amazing way that I must write it down. I had invited one of my pupils, Miss Smith, to re-develop the old theory of the diamond lattice with modern means, and she had confirmed my old formula. Then I wrote to my friend Professor F. Simon in Oxford to ask whether he could let one of his men make measurements of the elastic constants of diamond with the help of modern methods (ultrasonics). As I was taking this letter downstairs to be posted, the bell rang and the postman delivered a big registered letter from the Royal Society, most likely, I thought, one of those manuscripts for reviewing, a job which I dislike intensely. Cursing, I opened the envelope and read the title:

'A Measurement of the Elastic Constants of Diamond', by the Indian physicist Bhaghavantam, a reliable man. I at once introduced his values into my old formula and lo, all was perfect. I ascribe this miraculous confirmation of my old theory after a third of a century to the fact that Hedi had helped me by calculating numerical tables of thousands of figures, a miracle which has never happened since!

All this time I was under slight pressure from my parents-in-law to embrace the Christian faith. In the end I yielded. I had never practised the Jewish religion and was quite ignorant of its rites and traditions. My main argument for sticking to it was that what was good enough for my father and his father was also good enough for me. I had also a strong bias against all Christian churches from the time of my childhood, when my schoolmates laughed at me because I used a somewhat different wording of the Lord's Prayer (as I described in Chapter I). The figure and the teaching of Jesus could not but appeal to me, but my father's upbringing had taught me not to rely on revelation and tradition, still less on miracles, promise of eternal bliss or fear of punishment, but on my own conscience and on an understanding of human life within the framework of natural laws. If one is not brought up to revere Christian symbols and confession, the history of Christianity, or rather of the churches, is nothing to be proud of. It is quite possible—and the events of the last decade have made it appear even more probable—that Western or Nordic man is so savage by nature that without Christianity his history would have been even worse; but it is still bad enough, and much slaughter and persecution has been performed in the name of Christ.

Such arguments were strong in my mind. But there were other forces pulling in the opposite direction. The strongest of these was the necessity of defending my position again and again, and the feeling of futility produced by these discussions. In the end I made up my mind that a rational being, as I wished to be, ought to regard religious professions and churches as a matter of no importance. So I allowed Pastor Luther, who had married us in Grünau, to give me a few lessons in the Christian faith until he considered me prepared to receive baptism. It has not changed me, yet I have never regretted it. I did not want to live in a Jewish world, and one cannot live in a Christian world as an outsider. However I made up my mind never to conceal my Jewish origin. There were others who tried to, but Hitler found them out anyhow. When the Nazi persecutions began, I felt again a member of the Jewish people, in spite of Pastor Luther, and I suffered with them.

In May 1914 our daughter Irene was born. We chose the name not only because we liked the sound of it but also because of its symbolic significance. There was a rumbling in the world like distant thunder—Agadir and the Panther affair had just occurred; people began to speak of war as if it were imminent, yet nobody took it seriously. This summer of 1914 was lovely, blue skies and blazing sun, gardens full of fruit and fields promising a rich harvest, everything made for the happiness of mankind. The newspapers played their usual game, intrigue and strife, rebellion and murder, guns and battleships. There was a more than ordinary murder in Sarajevo, followed by a more than ordinary display of threats and ultimatums. But who cared? Things like that had happened before. And then one morning big headlines: general mobilization in Russia, in France, in Germany. That meant war. It was the 2 August, the first anniversary of our wedding. The same day an important letter arrived which was to prove fateful for me.

That terrible day when peace ended and war began was very different that first time from the next time twenty-five years later. There was no real cause, either in the economic or the ideological sphere. Europe was flourishing and getting rich so quickly that an automatic solution of the social problem was in sight. Fascist or communist ideas had no followers. I do not know whether anybody understands today what really happened. To me it seemed utter folly, just an outbreak of insanity. That is of course an over-simplification. But it seems to be true that every civilization carries the germ of self-destruction, of suicide, in itself.

The second war did not come like lightning out of a clear blue sky. It was not the first step from sanity to madness but the end of an accumulation of stupidity and brutality. Once Hitler had been allowed to assume power there was no escape. The world was accustomed to calamity and it took this last and greatest one with quiet, sad determination. And this is the other great difference.

In 1914 there was a patriotic outburst of foolish enthusiasm in all countries. We had it in Göttingen in full measure: flags and marching and singing. Troops marched through streets lined by people throwing flowers. Flags everywhere—in the streets and on the trains carrying soldiers to the front. The barracks behind our house was the centre of excitement, for the regiment had to march in a few days' time. Many of our young friends had to join up, among them my colleague Richard Courant. He was Sergeant-Major in the reserve army and at once got a commission on the outbreak of war, like many other Jews who would never have become officers in peace-time. I had an excellent pair of binoculars of unusual construction, a gift from grandfather Kauffmann, which produced an image with magnified stereoscopic effect. These I gave to Courant before he marched. They were smashed a few months later by a French shell which hit his kitbag—which was fortunately not on his back.

The patriotic frenzy was coupled with wild rumours and a spy hunt: wells were said to be poisoned, the horses of the regiment paralysed, bridges blown up. All foreigners were rounded up and put into custody, amongst them Boguslavski and other Russians. We had great trouble in getting them out of prison and out of the country.

The newspapers were full of pep articles. I hated the war, but I could not escape the influence of the propaganda. I believed, like all the others, that Germany was being attacked, that it was fighting for a worthy cause and that her existence was at stake. I felt terribly useless and restless. Men were needed for harvesting. So I and one of my students, Alfred Landé, set out on bicycles to a farm near the little town of Northeim, some twenty miles north of Göttingen. There we worked in the fields from morning to nightfall, and we saw no newspaper or army bulletin. That was a good thing. But harvesting is not easy for a bookworm, unaccustomed to hard physical labour in the sun and dust, and in my case the dust produced asthma which disturbed my night's sleep. At the end of the week I was exhausted and had to give up while Landé continued.

When I cycled back and passed through Northeim I saw posters announcing the fall of Liège. It was a terrible shock to me. I knew of course that a violation of Belgian neutrality was one of the features of German strategy and that the refusal to give up this plan had brought England into the war. But I still believed that the German army would wait until the French occupied Belgium; that was what we had been told. Now there was open aggression. I rode on in a state of deep despondency. When I arrived home I learned that Captain Lodemann, our neighbour, had been killed in action at Liège. From that day on the black dress and tearful face of Frau Lodemann were a daily reminder of what war really means.

But now I must speak about that fateful letter which arrived on the day the war broke out. It was from Max Planck, the great physicist, discoverer of quantum theory and professor in Berlin. He wrote that he was feeling the burden of his teaching and had asked the Prussian Minister of Education to found a second chair for his subject. This had been granted, and he was thinking of proposing my name as the first occupant; would I be willing to accept? You can imagine how glad and proud I would have been in normal times. But this offer exactly coincided with the declaration of war, and though I was well aware of the honour and of my private luck, no jubilant feeling could develop. I knew Planck only superficially from scientific meetings and had never discussed problems of physics with him. He must have read and liked my papers. That was a great thing for me. But what did personal matters mean in those days? I answered Planck at once, thanking him for the honour, and accepting his offer, provided that my service was not needed elsewhere in the present emergency. During the next weeks Hedi and I frequently discussed the question of my future position and our removal to Berlin.

But no official letter came, and slowly the whole matter was drowned in the flood of exciting and tragic events in the world. First the victorious advance of the German armies in the West; every day a victory, the town permanently flagged, the newspapers full of boastful military articles—but the back pages also full of endless lists of casualties, amongst them many young men we knew.

One of the first was Steiger, the husband of Felix Klein's daughter; theirs had been a war wedding, and after a few weeks she was a widow. Two brothers named Stechow, great friends of ours, went to the East with a cavalry regiment; in a skirmish with the Russians one was killed and the other disappeared (he returned after the war from a Siberian prisoner-of-war camp). There was news like that every day. Those autumn days of 1914 stand out in my memory as a nightmare of conflicting emotions, the wish for a quick end to the war with a German victory and fear of the consequences—domination of Prussian militarism—anxiety about numerous friends at the front, grief and

161

pity for those who suffered, and as a contrast our peaceful little home. Hedi and I took refuge in music during these days; she took to singing Lieder by Schumann and Brahms. One of our favourites was Brahms's

> Warum denn warten von Tag zu Tag?
> Es blüht im Garten, was blühen mag.
> (Why, then, wait from day to day
> In the garden bloom the flowers that may.)

I cannot hear it even now after thirty-two years without seeing before my eyes the scene of those tragic days. Why wait from day to day? Waiting for an end to strife and hatred, that has been our lot ever since, and it seems that even now when I am writing this the time has not come to sit carefree in the garden.

I cannot deny that during that time I felt very much against the English, the French and above all the Russians. We were told every day about the abominable atrocities committed by the Cossacks in East Prussia. The idea of these 'Asiatic hordes' destroying the nice tidy German villages, torturing women and children and so on, infuriated me. I wrote a stupid letter of this sort to my friend Ehrenfest in Leiden and duly received a sharp reply, telling me just what the Dutch thought about the 'German hordes' destroying the nice tidy Belgian villages, torturing women and children, and so on. This greatly helped me to regain my mental equilibrium. Then the battle of Tannenberg took place, the Russians were driven from German soil and this sore spot in my mind began to heal.

But now I shall not attempt to give you an account of all the events of the war and my reactions to them, but return to my private affairs, which are the subject of these pages. As most of the students were called up, there was very little teaching, and I had a lot of spare time. This was difficult to bear and I decided to write a long paper on the dynamics of crystal lattices in which I intended to collect the results of my previous papers, together with some new investigations on crystal optics and elasticity. At the same time as Kármán and I were developing the thermodynamics of crystals based on the theory of infra-red lattice vibrations, my friend Paul Ewald had laid the foundations of modern crystal optics (including the X-ray region) in his well-known thesis. He showed among other things that the bi-refringence of a tetragonal crystal could be derived from its lattice structure. Hilbert was interested in this work and gave a new presentation of it in a series of lectures which I attended. It seemed to me necessary to unite these two things—infra-red and optical vibrations of a lattice—into a consistent theory. It was also clear from Madelung's work, mentioned before, and my own investigations that the atomic forces producing the infra-red vibrations are the same as those producing ordinary elastic stress.

Moreover, my interest in the problem of elasticity was roused by a remark of Debye's. One day he asked me the following question: Take a straight row of equidistant equal particles, a model of 'one-dimensional crystal', and assume that each of them acts with some force of finite range on its neighbour. Since each particle is surrounded perfectly symmetrically by its neighbours, the resultant force must be zero—except of course for the particles near the ends. If you change the distance of neighbours, the so-called lattice constant, the forces on every anterior particle still cancel each other. Hence it appears as if the lattice constant were determined by the ends of the chain, and the same would hold for the forces resisting a change of the lattice constant. In three-dimensional space this would mean that equilibrium and elastic stress would depend on things going on near the surface—they would be surface effects, while it is quite evident from the ordinary theory of elasticity that elastic constants are a bulky property of matter itself. Now I do not doubt that Debye did know the answer to this question even if he had not read the classical papers of Cauchy on the molecular theory of elasticity and the continuation of this work by Voigt. Anyhow, he made me think about a simple presentation of this theory in modern terms. Voigt's work in particular was difficult to follow. Cauchy's main result was that a system of equal particles acting on one another with central forces forms an elastic body with not more than fifteen constants (in the case of lowest symmetry), while the phenomenological theory which treats the crystal as an anisotropic continuum—also according to Cauchy—leads to twenty-one constants. Voigt had shown that this reduction of the number of constants from twenty-one to fifteen is due to the assumption that the particles are points of equal mass; if you take instead small rigid bodies of finite extension, you obtain the twenty-one constants of the descriptive theory.

I had never quite understood Voigt's reasoning and decided to tackle the whole problem independently. I took the particles to be extensionless mass points but I rejected Cauchy's assumption that they were all equal. In Cauchy's time nothing was known about the real structures of solids; but now we had the experimental structures determined by the Braggs with the help of X-rays. They showed clearly that a lattice was a periodic repetition of a group of atoms, or in other words that it consisted of several simple lattices, each of a different kind of particle. I succeeded in showing that even with central forces between all particles one gets the multi-constant theory (twenty-one constants) except in very singular cases. This was the beginning of my investigations which soon grew to such volume that it seemed impossible to publish them as a paper in a periodical.

At that time the publishing company of Teubner in Leipzig had a kind of monopoly on mathematical books. I approached them as to whether they would like to publish a monograph on crystal dynamics. After lengthy negotiations they agreed but under conditions not very favourable for me. For instance, I had to take half the small royalties in the form of books published by them. Well, I was completely in their hands and as my scientific library was rather scanty I had no great objection to replenishing it from Teubner's stock. Hedi also took some books and some nice big prints well suited for our nursery. Under these colourful pictures my children were brought up.

To have an absorbing task to do was a great relief in wartime. Winter came and the fighting settled into the trenches. I remember very little of how we lived in those last months of 1914. But shortly before Christmas we had a sad surprise, in the form of a letter from Planck in which he told me that the Berlin appointment had come to nothing. The gist of it was that Laue had heard of this new position and he had written to Planck saying that he was very keen to get it. Laue was at that time professor in Würzburg but did not feel happy there—why I cannot imagine. He was Planck's pupil and friend, and already a celebrity in consequence of his discovery of X-ray diffraction. Laue was willing to sacrifice a full professorship in Würzburg, an 'ordinary' one, for an 'extraordinary' one in Berlin because the latter would bring him near to his revered teacher. Planck had replied to Laue's letter saying that he had made a personal offer to me and could not withdraw it; but he advised him to call on the director of the department for universities in the Prussian Ministry of Education and apply for the post. He, Planck, had little doubt that they would accept this application and he deeply regretted having raised hopes which might not be fulfilled.

Indeed, we were disappointed, Hedi and I, and we drafted an answer which expressed what we felt. But we did not send it off and revised it the following day. And this procedure was repeated, until a few days later a letter was posted which showed natural regret but no trace of resentment, acknowledging Laue's superiority and expressing the feeling that in these days of war personal affairs did not matter anyhow. It must have been the right kind of letter, for years later Planck told me how relieved he had felt and how much he appreciated my quiet resignation.

Meanwhile we celebrated our first war-Christmas quite cheerfully; it was also the first with our baby, gazing with her big eyes into the candles of the Christmas tree.

Then came another surprise, a letter from the Ministry of Education offering me the chair in Berlin.

I heard later from Planck how things developed to this quick conclusion. The appointment had been delayed by the Ministry due to the usual red tape. One day Laue appeared and offered himself as a candidate for the new chair. But he did not find the reception he expected; the director of the department for the universities seems to have regarded Laue's visit as interference with his own competence, and instead of accepting his offer, he at once despatched that letter to me. So poor Laue had to stay in Würzburg, although not for long. In these war years several new universities were founded: Frankfurt am Main, Cologne, Hamburg. In all of them there were already medical, scientific and law schools of good academic standard, and considerable funds were available, given by wealthy merchants. Now the high patriotic mood of the first war years was used by the municipal councils to carry through their plans to give these schools full university status. Frankfurt, if I remember rightly, was the first one; there existed the Physical Society, the Senckenberg Foundation— which had given their gold medal to my father in the last year of his life—big modern hospitals and other such departments. These were now all combined into a university, and Laue became its first professor of theoretical physics. But there again he was not happy and four years later he asked me to exchange chairs, as I shall report later.

In 1915 my connection began with the *Physikalische Zeitschrift*. This journal had been run for a long period by E. Riecke and H. Th. Simon. When Debye came to Göttingen, he replaced Simon (who I think was ill); the actual work, however, was done by assistant editors, H. Busch (later well known as a pioneer in electronic optics) and myself. To this period I owe a considerable knowledge of physics authors and also some experience in editorial work. This work was made rather difficult when I went to Berlin (March 1915), and after a year of exhausting correspondence with Busch in Göttingen, I gave it up. But six years later I became editor again and ran the journal for several years.

In spring 1915 we moved to Berlin.

XV. Professor Extra-Ordinarius in Berlin. The A.P.K.

We had chosen our new home in the Grunewald suburb, Teplitzer Strasse 5. At that time Berlin still consisted of half-independent boroughs with separate administrations, councils, mayors and budgets—although there was also a central administration with an *Ober-Bürgermeister*. Along our street ran the boundary between the wealthy 'Colonie Grunewald' where rich and influential people had their villas, and the district of Schmargendorf, with a large middle-class population. Consequently rates and taxes were quite different on the two sides of the street—much lower on the Grunewald side. In this way we could afford to live in a very nice house in pretty surroundings. We had our own little garden in the midst of other gardens and parks. Less than a mile away the forest was still untouched and offered pleasant walks through fir woods, so characteristic of the Mark Brandenburg, to quiet little lakes.

The house itself had four apartments and we felt the only ordinary family among rather queer people. Below us lived a widow, Frau Gregory, with two daughters, one of whom was an artist of local fame, a singer who accompanied herself on the lute. Hedi took singing lessons with her and practised scales. Next door to us lived Rita Sacchetto, a dancer of more than local fame who had introduced a 'Greek style' into dancing. She was, though middle-aged, still a very beautiful woman; some of her pupils, whom we met on the stairs, were extravagantly dressed creatures of dazzling loveliness and strange behaviour. Below Rita Sacchetto, a very strong-minded former nurse opened a private convalescent home. One of her first inmates was a prince from a very old German family who slept most of the day but spent half his nights in town, after which he would return and make a terrible racket in his room. Then the caretaker woman below would knock furiously on the ceiling with a broomstick in order to silence his Highness. All these domestic events helped to distract our minds from the bleakness of the war.

Every morning I travelled by tramcar or bus the long way to the University, Unter den Linden. There was a staff-room where we assembled before the lectures. I have an unpleasant recollection of the short periods spent in this room. I knew nobody and nobody cared to introduce himself to the newcomer. They discussed academic and political questions; most of them were apparently philologists, historians or lawyers, and I felt an outsider.

Slowly I began to distinguish a few prominent scholars, for instance the Germanist Roethe, who was well-known as an *All-Deutscher* (Pan-German), and proclaimed in this room of learning his nationalistic and annexationist ideas. Another was the historian Delbrück, a very different type, small and delicate, with a fine expression and kindly face. I was much attracted by him and glad when he addressed me. He was an example of the best type of liberal German scholar, and I enjoyed the short talks we often had before or after the lectures. Later his son became my pupil; he is now living in the United States and well known for his work on the application of modern physical methods to biology.

One of the first people I visited was Albert Einstein. I knew him superficially from scientific conferences. He was not a professor at the University but had a special position at the Berlin Academy. He soon came to our house with his violin to play sonatas with me and he astonished Hedi by his delightfully unconventional manners. He at once took off his coat and the loose cuffs of his shirt and fiddled away with the greatest ease and enthusiasm—he played well, with a fine understanding of music. We had many talks on physics. I had studied his first papers on general relativity (theory of gravitation); in fact I had taken the two red-covered papers, from the Proceedings of the Berlin Academy, with me on our honeymoon and studied them, much to Hedi's disgust. So I could discuss these problems with him.

The most striking feature of his way of thinking was his faith in the simplicity of fundamental laws. But he was not an 'a-priorist'. All his theories were directly based on experience. He had the gift of seeing a meaning behind inconspicuous, well-known facts which had escaped everyone else. The most important example is the equivalent of gravity and acceleration, known since Newton's time, but not previously recognized as a clue to the understanding of the cosmos. There are many other cases, as for instance the interpretation of the law of photoelectricity in terms of Planck's quantum of energy. It was this uncanny insight into the working of nature which distinguished him from all of us, not his mathematical skill. At that time he was struggling hard to find the correct analytical expression for his physical ideas, studying Riemann's papers on differential geometry and their generalization by Ricci and Levi-Civita. Gradually he became a real mathematician but moved away from actual physics. A great deal of quantum theory is due to him, in particular the notion of the light-quantum or photon, but when, ten years later, quantum mechanics was discovered he did not accept it because he objected to its statistical interpretation. He has often expressed his dislike of indeterminism and stubbornly continued his attempts to derive quantum laws from classical field theories. In a letter I got from him about two years ago (7 September 1944) he writes: 'We have become Antipodean in our scientific expectations. You believe in the God who plays dice, and I in complete law and order in a world which objectively exists, and which I, in a wildly speculative way, am trying to capture.' These wild speculations are the feature which distinguish the Einstein of today from that of 1915. But if any man has the right to speculate it is he, whose classic work stands like a solid rock.

Hedi too became a great personal friend of Einstein. At that time he had just recovered from a severe illness during which he was near death's door, saved mainly through the efforts of a cousin of his, who had been divorced and had two daughters. He himself was divorced and had two sons who lived with their mother in Switzerland. He now married this cousin. But though she regarded it as her duty to manage his affairs and to give him the comfort he needed for his work, he never changed his almost ascetic way of life. His own room never contained more than a bed, a table, a deck-chair and a primitive bookshelf on which he kept bundles of reprints. He once said to Hedi that there was absolutely nothing, either person or thing, without which he could not do at a moment's notice.

My audience at the University dwindled from week to week as the students were called up. In the middle of the summer semester it seemed to be hardly worth continuing to lecture. Because of the terrible losses in battle, the standard of health required by the recruits became much lower, and even an asthmatic might be conscripted. My book on crystals was finished and had been sent to the printer. There was nothing to keep me from doing my duty to my country. In June 1915 I learned that under the auspices of Professor Max Wien a unit of physicists and technicians was being formed to work on the development of wireless in aeroplanes. After consultation with Planck, I decided to join this group. They were stationed at a big military camp, Döberitz, near Spandau, and there I went with a small suitcase, once again to become a soldier in the German army, the third time in my life. After having been a dragoon and a cuirassier I was now to become an airman. But that was not the end of it; soon I was to wear the uniform of the field-artillery.

About the same time my friend James Franck also joined the army. It is said that the first day he was given the command of a platoon he shouted 'Pla-toon, turn right—please', a courtesy which, though quite in accordance with his nature, considerably retarded his promotion. Soon he joined a group of physicists who, under Fritz Haber's leadership, prepared and later directed the gas-war. I disliked this chemical warfare and refused all offers to take part in it. I stuck to my wireless men who were less brilliant than the gas people, but did a much more harmless job, if war jobs can ever be considered harmless. Yet I still think that the airman, especially the fighter, is the only remnant of the times when warfare had a halo of romance, when personal courage and skill were decisive; and it seemed to me not a bad job to help these knights of the air by giving them an instrument of communication with the ground. We were strictly ground staff and I was never in the air during my war service.

We lived in large wooden huts in the usual military way and were instructed in operating ordinary field telegraphs and later small wireless sets. We also had technical lessons which were not without amusement, since our teacher, a sergeant, had only just learned the little he was passing on, while the instruments remained a complete mystery to him. Poor fellow—for quite a while we purposely concealed the fact that we were all physicists, some even wireless experts, and we listened in deadly earnest to his explanations, swallowing the most grotesque slips. But his authority was restored when it came to learning and using the Morse alphabet. He was a master in taking down an extremely fast message through his earphones, and even the best of us could not attain half his speed after four weeks of training.

We also had some military drill, including marching. All went well with me for five or six weeks, as long as the weather was pleasant and not too hot. My comrades were pleasant fellows, a few of them young professors like myself, such as Valentiner and Kalähne, who were brothers-in-law; the rest mostly members of the staff of the National Physical Laboratory. We had our own mess, where we spent the evenings in at times quite interesting discussions. I was greatly teased because I used every free moment to read the galley proofs of my book, which now began to arrive. The weekends we were often allowed to spend in Berlin and I could go home. One Sunday night Hedi and I invited the whole lot to our house in Grunewald and had a jolly party.

But the weather broke, we had to do a long march in pouring rain and could not change afterwards. I got a bad cold and an asthmatic attack at night. The next morning was a Saturday, and knowing the ways of military doctors, I concealed my coughing as well as I could and did not report to the sick-room. Instead I went with the others to Berlin for the weekend and arrived home in a pretty bad state, with quite a temperature. It was an awkward situation, for if I did not report on Monday morning at Döberitz Camp I was a deserter, yet I did not feel able to make the long journey. The solution was found by our friend Arnold Berliner, the editor of *Naturwissenschaften* whom Hedi rang up. He knew some famous physician who served as a high-ranking medical officer during the war. This man came to our house on Sunday, examined me and sent a certificate to my unit that I was ill and unable to report.

There I was in bed, most comfortable apart from asthma and coughing, but uneasy in my mind about my military future. Just at that moment my friend Rudolf Ladenburg arrived and saved me from all my troubles. He had been at the front from the very beginning; he was a cavalry officer of the Reserve with the rank of captain and had led his squadron during the days of the initial advance, far ahead of the bulk of the army, reconnoitring and fighting in the fashion of old classical warfare. Later, however, they had to give up their horses and dig in. Ladenburg found this trench life very dull and began to think about technical problems in warfare. One of his ideas was what was later called sound-ranging, a method to determine the position of guns by measuring at several observation posts the relative times of arrival of the sound report emitted from a firing gun. Many other people had had the same idea, as we soon found out. But none of them had the rank of cavalry captain or his considerable war experience. So he was the first to convince the military authorities that there was something in it, and he was sent to Berlin to take part in the organization of a sound-ranging department.

There was a technical institution in the Prussian army called 'Artillerie-Prüfungs-Kommission', or A.P.K. (Artillery Testing Commission), which investigated all inventions concerning artillery. At the beginning of the war it had almost completely disintegrated, when the technical officers were sent to the front—so sure was the General Staff of having the best possible technical equipment. But it soon turned out that the French artillery was considerably better, and the Commission was re-assembled in a hurry. However, many experts had been killed, and outsiders had to be taken in. One of these was Ladenburg who, together with an infantry captain, von Jagwitz, was to organize a department for all methods of 'scientific ranging', optical, acoustical, seismometric, electromagnetic, etc.

When Ladenburg heard of my plight he at once declared that I must join his department. A few days later I had an order to report at the A.P.K. As soon as I was well enough, there I went. The department of Captain von Jagwitz was not in the main building, Kaiser-Allee, but in a flat in a private apartment house in Spichern Strasse, a dull street near the so-called Bavarian Quarter where all the streets were named after Bavarian towns. Einstein's house was quite close by, which was a great blessing as I could easily visit him.

Captain von Jagwitz received me with a strange mixture of military authority meant for the 'private' and deference meant for the university professor. But when he found I was no high-brow he dropped all his mannerisms and became friendly and human, explaining his plans and asking my advice concerning the physical problems and the choice of his staff. He was a typical 'Junker' and Prussian officer, lean and tall, fair, clean-shaven, with blue, somewhat expressionless eyes; ambitious and energetic where his own career was concerned, but tending to laziness and neglecting his duties when they could not help his promotion.

He was kind and friendly to his staff. Of course he could shout and abuse in the correct Prussian style, but a moment later he would grin and pat his victim on the back. This method was particularly evident with his orderly, a clever little Jew named Rund, who acted as his secretary and actuary. Like most common soldiers this fellow hated more than anything the thought of being sent to the front, and in this respect he was in perfect agreement with his chief. The method they used to keep themselves at home was also similar, namely to be indispensable to the A.P.K. Jagwitz did this by gathering together all the experts he could lay his hands on, and making them collaborate for the glory of his department; Rund by keeping the files in perfect order, his system, however, being secret and known to him alone, so that nothing could be found and nothing done when he was away for a few days' leave. Two years ago (1944) when I was in Dublin lecturing, my name was mentioned in the papers and good old Rund appeared at my hotel and told me that he had escaped the Nazis and was living with his family quite comfortably in Eire, earning his livelihood by making and selling lampshades. We spoke of olden times and of von Jagwitz in particular. I had heard that Herr von Jagwitz had gone to South America after the defeat in 1918, but had returned when Hitler came into power, and that he had risen to a high position in a German ministry—hence he was bound to be a Nazi. But Rund refused to believe this and persistently said: 'But he was such a kind, decent gentleman.' So he was—human nature is complicated and difficult to classify.

As a member of the A.P.K. my life was very comfortable. Our work began at 9 a.m. and ended at 4 p.m., with a short break for lunch. First I had to help

Ladenburg in working out numerical and graphical methods for the use of sound-ranging at the front, and then to transform them into military regulations. The latter job was supervised by von Jagwitz himself; we had to learn to translate clear directions from our good German into 'military language', which is, as in all countries, an ugly and involved jargon. Later I had to work out the influence of the change of wind with altitude on the propagation of sound. I did this by assuming a simple analytical law for the increase of wind with height, for which the equations of the motion of sound could be rigorously solved; two different laws were found to be useful, each depending on two constants only, an exponential and a power law, and the propagation of sound was determined with the help of the methods of Hamiltonian dynamics. From my formulae, tables were calculated by my staff. The results were checked by experiments on a shooting range and quite well confirmed. I think it was a good piece of applied mathematics, and when the second world war broke out, I offered this method, together with other material which I had preserved, to the British authorities. They returned it with the remark that similar methods had been worked out by themselves. I have no right to doubt it, but I suspect that they did not want to waste time studying my manuscripts since sound-ranging was only important in stationary warfare.

My 'staff', already mentioned, consisted of students of mathematics and physics whom I claimed from other army units. The first was my pupil Landé, with whom I had harvested near Northeim at the beginning of the war. I began to regard this chance of saving gifted young men from being wasted at the front as a major task and pursued it with vigour. In several cases I succeeded, in others I had grievous disappointments. The worst was the case of young Herbert Herkner, who had studied mathematics during my last year in Göttingen and showed the greatest talent to appear for many years. He was now an infantryman in the fighting lines. I had a long struggle in getting my application through for his transfer to our department. At last I succeeded. The order reached Herkner a day before a great battle, where he was killed (Battle of Cambrai, 22 November 1917). I was terribly upset by this news, accusing myself of not having pressed my application for his release more strongly. I was convinced that mathematics had lost in Herkner a genius of the first order, and I expressed this view in an obituary which I offered Arnold Berliner for his *Naturwissenschaften*. He accepted it, and you can read it in Volume 6, page 174, 1918. It is perhaps the only case where a young student has been honoured by an article like those usually devoted to great scholars. But I believed indeed that 'in this noble youth a part of the spiritual future of Germany has been destroyed'. And looking back today I think I was right. Many of the best men perished during these years.

After our method of sound-ranging had been established, we often had to visit the front to inspect sound-ranging units. My first journey West I made with Jagwitz. We arrived at some place near the front and heard a thundering fire. Something was going on, we were told. It was difficult to find a motor car; eventually an infantry major gave us a lift and promised to take us near the sound-ranging station we wanted to inspect. I was still a sergeant and had to sit beside the driver. When we approached the front we were warned not to pass through villages as the enemy was sending over heavy shells into every inhabited area. The two officers declared they had no time for a roundabout route. But the driver grinned and took us carefully over country lanes. Our heroes, deeply involved in a strategic discussion, did not protest, and so we arrived safely at our destination.

The dug-out of our unit was on top of a flat hill. When we had finished our work there, we intended to proceed to the next unit, but it was impossible as all communications were under heavy fire. So we had to stay there the whole day. I could not stand the bad air in the dug-out and preferred to lie in a flat ditch on the hill from which I could see miles of the front, a terrible wall of smoke and fire, and a hellish noise. By a lucky chance no shell fell near us, but I saw several exciting air battles. In this way I had a glimpse of a modern battle, the great Battle of the Somme. I found it quite enough. Although I did not really belong to the military machine I was sufficiently under the spell of discipline to feel nothing of the natural reactions of a human mind; only later I realized that I had witnessed that abominable mass-slaughter to which war has degenerated and that only a mile or two in front of me men were killing and dying. Up there on the hill it seemed to me just a spectacle, something to be taken in with all my senses so that I might never again forget it. And, indeed, I have never forgotten it. At nightfall we returned to one of the staff quarters where we got transport back.

I later visited several sections of the Western front, in Alsace and in Flanders, but always in quiet periods. As I never stayed longer than a few days in one place, I did not become familiar with the local trench system and other conditions; there was always some sporadic firing and I felt much more uncomfortable in these so-called peaceful sectors than at the Somme. The A.P.K. had an experimental station in a little chateau near the Ypres front; there I went once with Ladenburg to try out some new gadget. It was a cold autumn, and I suffered badly from bronchitis and asthma.

I never saw the Eastern front, and had only a glimpse of the South. At the beginning of my career, I had to accompany Jagwitz and a number of officers to Austria and Hungary in order to compare our sound-ranging methods with those of the Austrians. In Oderberg, the frontier station, we were received by a delegation of Austrian officers, headed by a General Austerlitz, a rather ominous name; in fact he was a very decent, cultured man. Jagwitz introduced me, the simple sergeant, adding my professor's title, whereupon the general, who was a mathematician and knew some of my work, took me alone to his reserved compartment and talked all the way to Vienna about current scientific problems. Jagwitz regarded this preferential treatment as an honour to himself and was nicer than ever to me. So were the others, and so the whole journey was very pleasant. It took us to Budapest, and from there to a big military camp in the Carpathian mountains where our methods were demonstrated. The only black spot was the last night when the Austrians gave a sumptuous dinner to their Prussian comrades – to which I, as a mere N.C.O., was not invited—and made them drink so much Hungarian wine that they had to be carried away to their beds.

172

Meanwhile our scientific staff had increased. Apart from Ladenburg and myself there was Wätzmann from Breslau, a specialist in acoustics, old Kurlbaum from Charlottenburg, Madelung and Reich from Göttingen and others. Soon it became necessary to attach some psychologists to our department. The reason was this. The proper physical way of measuring short time intervals, such as occur in sound-ranging, is by electric recording, with the help of an oscillograph. But the firms which could supply such apparatus, Siemens and Halske, General Electric, etc., were overworked and could not accept a big new order. Therefore we bought a great number of excellent Swiss stop watches which could measure a fiftieth of a second. Now we needed psychologists who could work out methods to select and train specialists with a fairly constant reaction-time. This rough procedure worked better than one would expect, although the British automatic instruments used in later phases of the war were definitely superior. A few years after the war I met William L. Bragg (the present Sir Lawrence Bragg), who was one of the British sound-ranging experts, and compared experiences with him. He did not think highly of our method, but I doubt whether the greater accuracy of the British instruments was of much use, because of numerous sources of error, such as wind and weather, all kinds of noise and, most of all, the head-waves of the shells.

We got the services of two leading psychologists, Wertheimer and von Hornbostel, both Austrian Jews and members of the group who, under Köhler's leadership, had founded the 'Gestalt' psychology. Wertheimer became a great friend of ours. He was a deep thinker, but of a different type from any I had known before: sceptical in the extreme, inclined to take nothing for granted and to regard any observation as a deception of the senses or the mind, until its truth was shown by correct psychological experiment. We spoke a rather different language; when I said 'space' I meant the geometric space of the physicist, while he understood the spatial order of sensory impressions. It was most instructive for me to see the other side of the picture, to learn something about the psychological phenomena, the 'Gestalten', which are the raw material of all observation. On one point we could never agree. He did not even accept logic and arithmetic, and he tried to construct a meta-logic and a meta-arithmetic, based on different axioms. He said: 'Look at primitive people, savages, who cannot count past five: They have a different arithmetic'. But that did not convince me. I think today that to some extent he was right although the material available at that time was insufficient to show the fertility of such ideas. Meanwhile Brouwer's attacks on the 'law of the excluded third' and Neumann's investigations on the foundations of quantum mechanics have established a kind of non-Aristotelian logic.

Our psychologists were extremely valuable in our dealings with 'inventors'. I must tell you of one of these cases. We received a letter from the War Ministry that the German Ambassador in Stockholm had sent two Swedish inventors to Berlin, who claimed that they had an apparatus for detecting ammunition over large distances, and we were asked to investigate the matter. It appeared that these two men were lodged at the Adlon, the best hotel in

Berlin, where they were very comfortable; of course they did everything possible to prolong their stay. First they declared they needed ten kilogrammes of platinum, and it proved very difficult to reduce this amount to a more reasonable quantity which could be procured from the firm of Heraeus in Hanau. Then they refused to be tested on an experimental range and said that they could demonstrate their apparatus at the front—where of course no control was possible. Thus weeks went by. At last they were persuaded to do a test on our experimental station. Now the psychologists went to work and devised a fool-proof scheme. 100 identical boxes were made, 98 of them filled with sand and only two with some ammunition of equal weight. These boxes were arranged in a big circle and the two men with their instrument posed in the centre. When the day of the test came our general and a lot of other high-ranking officers attended. After long mysterious preparations the men began to turn their instrument. The box where they stopped was opened, and lo, was found to be filled with ammunition! The general shouted: 'Well done', and to us scientists: 'Your scepticism, gentlemen, is for once disproved. I think the test is over.' We had the greatest difficulty in persuading him to allow at least one more test. Here von Jagwitz proved of great help, feeling that his belief in science was at stake. He finally succeeded, and the next test of course yielded sand. The thing went on, and the two poor fellows had extremely bad luck and in another 99 tests they hit on nothing but sand. In the end we had to open all the boxes to show them that two really contained ammunition. They tried all kinds of mystical explanations. The general had disappeared, and von Jagwitz disposed of the two impostors.

Hornbostel and Wertheimer constructed an interesting instrument, a direction-finder by sound. They had discovered that the ability to estimate the direction from which a sound comes is based on the time difference with which the wave front reaches the two ears. The accuracy is limited by the base (distance between the ears); but this can be easily increased by mounting two microphones at a greater distance and by listening with earphones. With such an instrument the direction of a firing gun can be determined to a fraction of a degree, and it has been used successfully at the front.

It was suggested (I think by myself, but I am not sure) that this method could also be applied under water for discovering submarines. This and another affair brought us into contact with the German Navy.

One day when von Jagwitz returned from a conference at the War Ministry, he summoned the senior members of his staff; we found him pale, excited and vexed. 'Gentlemen,' he said, 'I believe I have commited a major blunder. There was a complaint from the front about unsatisfactory searchlights, and as nobody was prepared to produce better ones, I said I would. I have never seen a searchlight in my life. Have any of you?' No, we had not. His spirits fell, but he begged us to get to work, to find out everything about searchlights, how they were constructed, which firms made them, etc. One of us suggested that the battleships would have strong searchlights. Hence a group of us were at once sent to Kiel to get in touch with the Navy, others visited factories, and so on. The job was not easy but it was done, and von Jagwitz's prestige saved.

Once I had to go with von Jagwitz, Ladenburg and some other officers to the Baltic for an inspection. We were told to wait at a small port, Neustadt, on the East coast of Schleswig-Holstein, for a destroyer which would take us to our destination. It was a cold, foggy November day, and after having waited in vain for a while at the quayside we went to a nice-looking inn, called the Voss-House, after the poet Voss, the translator of Homer, who had lived there. We were soon joined by a group of naval officers stationed in Neustadt, and settled down comfortably at a big round table, soon covered with numerous bottles. Such an Army–Navy fraternization was not an everyday event and had to be properly celebrated. So the amount of alcohol consumed was considerable, and I had my share of it. After a while I felt an urge to retire, and looking round, I discovered a small door which seemed to lead to the place of loneliness. There I went, the door snapped to behind me and I found myself in pitch darkness. I turned back to open the door again to get some light but, to my amazement, knocked my head against some hard obstacle. This was absolutely beyond the powers of my intoxicated brain to comprehend. Then I tried another direction, but my stomach landed against a similar obstacle; in a third direction I hit my feet against something, causing a great deal of pain. I sat down and tried to think hard. It was clear I was immured, enclosed by walls on all sides. How this had happened I could not find out. After a considerable time light began to pour in. They had missed me and sent out a search party. They found me sitting on the floor in the narrow space inside a spiral staircase.

This was the comical beginning of a series of incidents which became more and more unpleasant until they ended in a complete breakdown. About two o'clock at night the destroyer arrived, and we embarked. It was a cold, rough journey. We had to sit in the tiny officers' mess, and I had the choice between the smoky atmosphere there or the icy fog outside. The result was an asthmatic attack. We were taken to a big liner used as an auxiliary cruiser and an experimental station. There the drinking continued, and my asthma became worse. A visit from a submarine did not improve it. The captain offered me his card-room on the bridge 'for some fresh air'; but the cold drove me away, and finally I found myself in the ship's hospital. The doctor decided that I should not remain on board, and as we were near the port of Warnemünde, where mail was to be fetched, I was taken ashore in a launch. That was the end of my military expeditions.

The war drew to its end, and a feeling of impending catastrophe was in the air. Even our professional officers became nervous. They had often discussed with me the chance of a victory and had found out that I did not share their optimism and their faith in the invincibility of the Prussian army. Now when things were not going according to plan, when the big summer offensive of 1918 petered out and the front line began to retreat, they came to ask my opinion. They were all afraid of their own troops, for there was a lot of grumbling amongst the men, of which I was well informed, as Landé and others of my scientific staff were privates or lance-corporals and lived with the men. I could not resist the temptation of having a little fun by scaring the officers, who had once been so proud. I said that if we were beaten there would be a socialist revolution and every single man wearing an officers uniform would be killed. After that many of them kept civilian dress in a drawer at the office so that they could easily disappear if the evil day should ever arrive.

During these last months of the war I worked on a new sound-ranging problem. The conical head-wave carried by the shell is much louder than the report from the gun itself and is the main source of error in sound-ranging. I had the idea of using this violent noise itself for sound-ranging. The mathematical problem was formidable, but I solved it by a semi-graphical method. The trajectory of the shell itself was of course an unknown quantity and all depended on having a formula for it which was not too complicated. Such formulae were contained in the textbook of ballistics by Crantz, a professor at the Military Academy, who was also chief of the ballistics department of the A.P.K. So I got in touch with Crantz, who was the greatest authority on this subject in Germany. Although he did not wear uniform, he was of very Prussian appearance, clean-shaven, with hard features and cold, ruthless eyes. I was quite glad when he gave me one of his men as assistant, a Lieutenant Rückle. He often came to our department and was very useful and entertaining. For it turned out that he was an arithmetical prodigy, who could do incredible calculations in his head. We had great fun making him compete with our computing machines directed by our best men, and he often won. In addition to his extraordinary computing talent, he knew a lot about mathematics, and I think he had a doctor's degree, or got one later. I always thought him unique, until I met A. C. Aitken, now Professor of Mathematics here in Edinburgh, who has the same gift, perhaps even to a higher degree. Rückle's end was tragic. He appeared in variety shows in order to make money, and in the company of the artistes began to drink more than was good for him. Eventually he became a hopeless drunkard and died very young.

I remember very little of the last weeks before the defeat, and I was not present at the closing of our department, as I was in bed with one of my usual attacks of bronchitis. Landé reported to me that on 11 November, the day of the revolution and mutiny, the officers quietly disappeared in their civilian clothes, while the soldiers looted our depots and then dispersed. I later visited our precincts and found all the apparatus gone, including the stock of Swiss stop-watches; only some of my theoretical papers were left. I still regret that I did not secure one of the stop-watches for myself.

XVI. End of the War. Revolution

Now I shall try to give you an idea of our private life in Berlin during the first great war. There was a general breakdown involving a shortage of food and other necessities, rationing, queuing. The first indication of what was to come was an order from the Government that a large percentage of pigs should be killed because there was not enough food available for them. The meat was to be preserved and people encouraged to lay in a store. We accordingly purchased a considerable quantity of salami and cured ham from a pork butcher in Göttingen which we asked him to send to Berlin. One day we received a note that a big box was lying at the goods depot of one of the Berlin stations, and we asked a big transporting firm to collect and deliver it. But week after week passed and no box appeared. We telephoned, we called at the firm's office, all in vain. When finally the box arrived, the summer had come and the heat, and what we found in the box was indescribable: it was all 'alive'. Hedi's brother Rudi, when he learned of the sad affair, reminded her of that famous book by her grandfather, Rudolf von Jhering, entitled *Der Kampf um's Recht* (The Fight for one's Rights), in which he insisted that it was everybody's duty to defend his legal rights before a court of law. So we did. It was the only lawsuit in which I was ever involved—and I lost it of course.

So we had to stand the 'turnip winters', as they were known, without the little additional protein contained in our lost salamis and hams. This second winter of the war was bleak indeed. Turnips served for everything; not only as vegetable, but also as substitute for jam, a supplement to flour in bread and cakes and I do not know what else. In November 1915 our daughter Gritli was born, in our house in Teplitzer Strasse, and Hedi became very ill after the confinement. Her nerves were exhausted from all the excitements of the time and the food shortage. Luckily we had excellent domestic help, two maids, Minna and Lina Klinge, who were sisters and came from Schlarpe, a village near Göttingen. Irene and the baby Gritli could be safely left in their charge. I took Hedi to a sanatorium at Bad Nassau on the river Lahn, where she recovered after two months. I well remember those gloomy days in March 1916 when I stayed with her for a few days in Nassau and went for lonely walks on the snow-covered hills. Strolling through the enchanted forest in its winter dress of glittering snow one might have forgotten all the worries and sorrows at home and in the world, but for the constant thunder of big guns from the Western Front, faint but distinct, a reminder of what was really going on. It was the murderous battle of Verdun.

The following summer when Hedi was well again, we made the acquaintance of the Blaschko family, who became our greatest friends during this period. We had already heard of Dr Alfred Blaschko, a celebrated dermatologist and friend of Arnold Berliner, and we had seen the little man on the road in the tramcar. But although he attracted us in a mysterious way right from the beginning, we had never dared to address him. At that time Irene was suffering from a bad furunculosis, and on Berliner's advice we called Dr Blaschko in. He came at once, and after the consultation, involved us in a fascinating talk about medicine, children and politics—mainly politics. He was an active politician, a deeply committed socialist. We soon accepted his invitation to visit him in his home in Taunus Strasse just around the corner. It was a lovely house, spacious and in exquisite taste. His wife and children received us kindly and soon there was hardly a day when we did not see each other.

Blaschko was a self-made man. Coming from a humble Jewish family, he had worked his way up until he became one of the leading physicians of the city. Being a dermatologist, he of course knew the Neissers well. Yet he was a type very different from Albert Neisser, not wholly absorbed by either research or practice, but deeply interested in social conditions, specially as they affected public health, and the menace of venereal disease. While Neisser, by his discovery of the gonococcus and his research into syphilis, had done much to conquer the disease itself, Blaschko studied the human and social background of the question. He was a great authority on venereal disease legislature and played a prominent part in the drafting of a bill which just then was about to come before the *Reichstag*. This activity involved him in practical politics and brought him in contact with many politicians. He was a socialist of the moderate type, which in Britain would be called 'Fabian', whose leading figure in Germany was Eduard Bernstein, whom Blaschko counted among his friends. Several times I listened to their discussions, which were mostly concerned with political problems of the day, but sometimes touched the principles of socialist doctrine.

If I try to remember the effect on my mind of those disputes of two leading socialists, I find the impression of their personalities in the foreground, and of their theories in a vague and nebulous background. They were both enthusiasts, full of spirit and wit, clever and learned, but more than that I felt their love for suffering humanity. Blaschko's socialism was certainly more one of the heart than of the brain. Yet he tried to educate me and made me read a lot of socialist literature. At that time I read considerable parts of Marx's *Captial,* but found it extremely tedious and dull. I have never understood how such a dry and voluminous book could have such an appeal to wide circles of educated as well as simple people. Bernstein's writings appealed to me much more. But altogether I have remained sceptical with regard to economic theories. My own socialism is certainly not based on rational arguments but, like Blaschko's, on ethical principles. In the sphere of ethical and religious ideas I have been strongly influenced by Blaschko. He shared my scepticism of revealed religion, yet he was not without deep convictions which might be called religious. I loved listening to him when he talked to me about these things in his quiet, clear way. In less serious matters there was often a twinkle in his eyes and a joke on his lips. Since my father's death I have loved no man more than him.

Frau Blaschko had many brothers and sisters, and one of the latter was married to Hermann Ullstein, one of the five brothers who owned the big publishing firm. Their house was opposite Blaschko's, bigger and more splendid. We were introduced there too and met a number of interesting people, writers and artists. Ullstein invited me to visit his publishing and printing works where, among other daily and weekly papers, the *Berliner Illustrierte* was produced. I had never thought before of what it means to produce about a million (or was it two?) copies of an illustrated periodical every week, not only to print it but to fold and bind and pack and distribute it. It was in fact the biggest printing factory then existing in Europe—there may have been a few bigger ones in the U.S.A.—and it was equipped with the most modern monsters of fast automatic printing presses. It was worth seeing. A considerable part of the publishing trade was created by and in the hands of a few inter-related Jewish families, the Mosses (owners of the leading newspaper, *Berliner Tageblatt*) and Ullsteins (owners of the equally important *Vossische Zeitung* and the popular *B.Z. am Mittag*). What has become of these people? I know only the fate of our friend Hermann Ullstein, whose property was confiscated by the Nazis and who died as a night-porter of a cinema in New York.

As soon as I had settled down in the A.P.K. (as described in the previous chapter) I took up some scientific work. My book on crystals appeared under the title *Dynamik der Kristallgitter* (Dynamics of Crystal Lattices), but owing to the war it did not attract much attention. I started another line of research, on the theory of the so-called optical activity or natural rotatory power. This well-known phenomenon consists in the ability of certain crystallized or dissolved substances, to turn the plane of polarization of a linearly polarized beam of light passing through, and it is of great importance to the chemist as it allows one to distinguish asymmetric molecules which differ, like the right hand from the left. Theories of this effect had been developed, the latest by Drude, based on the electromagnetic theory of light. But Drude used a rather artificial model of the molecule, assuming that an electron can move either in a right- or a left-hand spiral. Debye had once remarked that the effect should be explicable without such an improbable mechanism, and I took up this idea. I assumed that a molecule corresponds, in its optical behaviour, to a coupled system of electronic oscillators and found that optical activity resulted automatically if the relative distances of the oscillators was not neglected in comparison with the wavelength of the light. I published this theory in several papers, the most detailed of which appeared in the *Annalen der Physik* (vol. 55, 1917, p. 177). It contains not only the theory of rotary power but of all optical consequences of molecular anisotropy in a medium which is isotropic in bulk by nature but can be made anisotropic by the action of external fields. The latter effects I studied in more detail in several special papers (*Berliner Berichte*, 1916). My object was to explain the optical behaviour of liquid crystals which at that period were a topic of great interest. I assumed the cause of the parallel orientation of the molecules to be electric dipoles. Later investigations have shown that the orientating forces are of a different kind. Yet the optical part of my work is quite independent of this mechanism and is well confirmed by experiment.

The preoccupation with anisotropic molecules then led me to an investigation into the scattering of light (*Berichte der Deutschen Physikalischen Gesellschaft*, 1917). I discovered the depolarizing effect of anisotropy and gave formulae for it—without being aware that the same effect had been discovered long before by Lord Rayleigh and studied by Gans and others. The only excuse for this is the fact that there was a war on, which made it difficult for me to study the literature. Yet I think my derivation has some advantages over those of my predecessors. My second paper on this subject contained new ideas and results; it was a detailed calculation of the scattering of light by the permanent gases hydrogen, oxygen and nitrogen, based on Sommerfeld's quantum models (*Berichte der Deutschen Physikalischen Gesellschaft*, 1918). Sommerfeld's papers on the application of Bohr's theory to higher atoms were the great event in theoretical physics of this period (*Annalen der Physik*, vol. 51, 1916, p. 1125). He generalized Bohr's quantum conditions for systems of several electrons and applied them to the explanations of the laws of line spectra with extraordinary success. In those papers, the fine-structure constant $2\pi e^2/hc = 1/137$ appears for the first time in physics literature. I think that it ought to be called Sommerfeld's constant, just as the names of Avogadro, Boltzmann and Planck are attached to the fundamental constants of atomic theory, statistical mechanics and quantum theory. Sommerfeld's constant is perhaps even more fundamental than these, as it is a dimensionless number, independent of the units chosen; it touches one of the outstanding unsolved problems of theoretical physics. I think that Sommerfeld's merits have never been properly appreciated; he ought to have received a Nobel Prize. These papers of 1916 contain not only the main laws of ordinary line spectra, but also those of the X-ray spectra. I studied them with great enthusiasm and became convinced that his atomic 'ring-models' must contain a considerable amount of truth. I was particularly interested in a subsequent paper by Sommerfeld (*Annalen der Physik*, vol. 53, 1917, p. 497) where he applied his ideas to the derivation of the dispersion formula of the permanent gases hydrogen, oxygen and nitrogen. The ring-models of these molecules are essentially anisotropic and should therefore produce the depolarization effect which I had deduced previously. On the other hand experiments of Smoluchowski (Cracow Academy, 1916) had shown no indication of this effect. To explain this I made a detailed calculation of the scattering of Sommerfeld's molecular models and obtained a depolarization so small that it might well have escaped observation.

Another paper of this period contains a derivation of the infra-red dispersion formula for simple ionic crystals and its application to the determination of the ionic charge (*Berliner Berichte*, 1918, p. 604). It gave a complete confirmation of the assumption made by many previous investigators (Madelung for instance) that the particles in the alkali halides are monovalent ions, and corresponding results for other crystals, such as calcite.

I also wish to mention a paper on the electromagnetic mass in crystals (*Berliner Berichte*, 1918, p. 712) in which the contribution of the mutual electrostatic energy of the particles in crystals to the mass density is deduced. I think that this result is a strong argument for the view that all mass is the effect of interacting fields of forces.

All this work I carried out in the late afternoons and evenings when I had returned from eight hours' work at the A.P.K. At that time I must have had a solid mental constitution. Blaschko used to tease me about my ability to attend to several things simultaneously. Once on entering my study he found me at my desk writing a scientific paper, with my left hand rocking the pram containing baby Gritli while Irene was playing with her doll between my legs under the desk. That, however, was exceptional and happened only during Hedi's illness. After her return there was plenty of time for work and music and many other things. The rush in the A.P.K. began to subside and we had hours of leisure which we first used to squander with chatting, but after a while we began to discuss scientific questions and even to work them out. At that time Madelung used to sit opposite me at a big desk with many drawers. Each of us emptied a drawer of all military papers and filled it with scientific books and notes. We were both interested in crystals and began a kind of collaboration.

I was convinced that Sommerfeld's ring-models of atoms were essentially correct, i.e. with respect to their geometrical dimensions, charge distribution and energy. The idea struck me to use these models for building up a crystal. The simplest crystals whose structures were well known were the alkali halides, such as rock salt (sodium chloride). It has been suggested that these crystals were built not from neutral atoms, but from ions. A general physical theory of inorganic chemistry had just been published by Kossel using the assumption that a chemical bond is formed in two stages: first the reacting atoms exchange electrons and so become ions; then the ions attract one another by electrostatic (Coulomb) forces. It was a revival of the old Berzelius theory of the electric nature of the chemical bonds; it turned out to be very successful for a wide range of chemical compounds, which he called 'heterpolar'. In these molecules, the ordinary idea of 'valency', i.e. of a saturated force between two atoms, is of no importance. Instead, one has to use the notion of 'coordination number', introduced by the chemist Werner in his investigations on complex molecules; that is the number of ions of one sign of charge which surround an ion of the opposite sign.

I decided to calculate the simplest properties of a lattice of the sodium chloride type, assuming that each lattice point is occupied by a Sommerfeld ring-model of an ion, in proper orientation. I asked my pupil Landé to collaborate with me. We showed that the electrostatic potential of a nucleus plus a ring of equal charge is, at some distance from the centre, inversely proportional to the fifth power of the distance. But in our case the charges of nucleus and rings are not equal; in the metal atoms (Na) one electron in the rings is lacking, while in the halogen atoms (Cl) the rings contain one electron more than in the neutral state. It was quite easy to sum up the contributions from all lattice points due to the fifth power part, as the series converges rapidly enough. But the excess charges were troublesome; a direct summation was a hopeless task because of the extremely slow convergence. Here Madelung stepped in. He followed our work with great interest and at once grasped this difficulty. It is strange that an elegant solution of this purely mathematical problem was found by a man who was by training an experimental physicist and not at all well equipped for such an analytical problem. But he had a natural gift and a great love for mathematics and tried hard to improve his knowledge. A fruit of his studies is a book containing a collection of mathematical formulae useful for the theoretical physicist (published by Springer). The constant representing the electrostatic energy of an ionic lattice per unit distance of neighbour atoms is rightly called Madelung's constant. Other methods of determining it have been devised by several people, the most general and powerful one by Ewald.

This result of Madelung's made it possible for Landé and me to finish our calculation and to determine the lattice constant and the compressibility of the crystal from no other knowledge than the number of atoms (or ions) in the periodic table of elements. The results were amazingly good. We gave the manuscript of our paper to Einstein who submitted it to the Berlin Academy for publication. After a short period we got the proof sheets and were happy and proud to see our tables and graphs in print. Then a little catastrophe happened. One morning, on entering our room in the A.P.K., I found Landé pale and desperate, and his first words were 'Our paper is wrong.' It was a nasty shock. He showed me our mistake. I am truly ashamed to confess it. For it is one of those elementary errors which students might commit in their first course on statics. If you have to calculate the energy of a homogeneous distribution of particles in equilibrium, you can assume that every particle is surrounded by its neighbours in exactly the same way; therefore you calculate first the energy between one particle and all its neighbours and—now comes the point: multiply the result not by the number of particles, but by half that total. For otherwise you would obviously have taken the contribution of each pair of particles twice. Well, that is what we had done. Now fortunately it did not spoil our whole work, but only the last section concerning the compressibility. Our results about the lattice constants were not affected; for they are obtained by determining the atomic distance for which the total energy is a minimum, and there a constant factor of the energy does not matter. But our calculated compressibilities turned out to be all twice as large as the observed ones. Well, the question was whether the latter negative result was so serious that it invalidated our whole work. I at once went to Einstein and discussed the question with him. His opinion was that he had definitely shown that the cohesion forces in ionic crystals were of electric origin, and he regarded the discrepancy about the compressibilities as a matter of secondary importance which should be tackled separately. We followed his advice to omit the last section of our paper, and so it was published (Berlin Academy 1918, p. 1048).

But we at once took up the challenge of that discrepancy, and we did it in a manner which today still has my full approval. Instead of changing and doctoring the atomic model until agreement was reached, we reversed the whole question. We noticed that the magnitude of the compressibility (the electrostatic part of the energy being known) depends only on the exponent n of the leading term ar^{-n} of the remaining repulsive forces and of the equilibrium value of the lattice constant. Taking the latter from observations we could determine the value of n from the measured compressibilities. We found instead of the value $n=5$, characteristic for the Sommerfeld ring-models, a value about twice as large. This result was of considerable importance. It made it possible to obtain reliable values for the total lattice energy, and this enabled me, a short time later, to make an essential contribution to thermochemistry. And it produced in my mind a state of doubt with regard to Sommerfeld's models and Bohr's quantum theory of the atom in general. The high exponent $n=10$ showed clearly that the ions of the type Na^+ or Cl^-, i.e. of structures like those of the rare gases ('closed shells', in Bohr's later terminology), had a much higher spatial symmetry than ring systems. I became sceptical about these models. They are of course not the essential part of Bohr's theory. As he has frequently stressed, the main point of this theory is the postulate that atomic systems exist in stationary states and that transitions between these states are accompanied by absorption or emission of finite quanta of energy hv (h=Planck's constant, v=frequency). The representation of these stationary states as orbits of electrons obeying classical mechanics was regarded by Bohr himself as only an approximation, according to his principle of correspondence. But others, including myself for a short time, took the orbital theory more seriously; but this confidence was badly shaken by our discovery that the exponent of the repulsive force was not five but ten. From that moment on, my endeavour was not to find confirmations of the orbital theory but to provide arguments for its insufficiency. That led ten years later to the discovery of quantum mechanics.

But now to return to Berlin during the final period of the war. There were one or two occasions when a negotiated peace seemed possible. On one such occasion the point at issue was, whether 'unrestricted' submarine warfare should be embarked on. The moral aspect of this type of mass murder, though very important to people like Einstein and myself, was hardly mentioned in the press. The question discussed in numerous articles and private conversations was whether it would bring the United States into the war against us, and if so, whether this would make much difference. There was a strong group, mainly of intellectuals, who believed that it would be disastrous. One day, in the winter of 1917, I received an invitation to attend a conference on political questions in the house of a young colleague of mine, a lecturer in the history of art. I went there one evening. It was a villa in Margareten-Strasse, one of the elegant streets near the zoo, and I had a glimpse of splendid rooms filled with works of art. I found myself in an assembly of about twenty men, amongst them Einstein and a few others I knew, and I was informed that the purpose of the gathering was to discuss the submarine problem with a few high officials from the Foreign Office. We heard a report by a historian and a lively debate.

At the end we were asked by the chairman to meet again in a week's time and to keep the proceedings secret. This created a problem for me. For as an officer I was bound by a regulation not to take part in any secret political activity. I decided to risk it, and I took part in several such meetings. It was the first and only time that I had the feeling of being in direct contact with and possibly an influence on the powers that be. However, our attempt at a moderating influence came to nothing. One morning we read in the papers that unrestricted submarine war had been declared. Ludendorff had his way again. The disastrous consequences emerged half a year later, as we had predicted.

As I said before, I was down with one of my bronchitis troubles when the mutiny in the navy broke out. So I cannot tell anything from my own experience of how things developed in Berlin. It has been described by others: the appearance of sailors from Kiel with red flags who spread the revolt over Northern Germany; the complete absence of resistance from the Imperial Government and of the ruling classes, the shocking news of the Kaiser's flight to Holland, Ludendorff's sudden demand for an immediate armistice, and so on.

When I was up again, the last Imperial Chancellor, Prince Max of Baden, had resigned and a people's government under the socialist leader Ebert had been formed. That same day Einstein rang me up and told me that the University had been occupied by revolutionary students and soldiers and that several professors, including the rector, had been interned. He, Einstein, was afraid that their lives might be in danger. As he believed he might have some influence on the students, he intended to intervene; would I join him? I of course agreed and set out at once. I had to walk the long way to Einstein's house in the Bavarian Quarter through streets full of wild looking and shouting youths with red badges. I found with Einstein our friend Max Wertheimer, the psychologist, whom he had also summoned. We started at once on our journey to the inner city, partly by tram. When we approached the Reichstag building we found it surrounded by an immense, excited crowd. We forced our way through but were prohibited by 'red' soldiers from entering. We stood there for some time, watching cars coming and going, carrying important-looking people, some of whom were cheered by the crowd, and we felt we were in the thick of it. Eventually Einstein was recognized by a socialist journalist, Hollitscher, who helped us to get through the cordon and into the building.

After many inquiries and walking through endless corridors we were shown into a small conference room where the newly established Students' Council was holding its first meeting. Again Einstein was identified and we were asked to sit down at the end of the conference table until their deliberations were finished. So we sat down and listened. It was not very easy to keep serious. They discussed a general programme on 'Students' Rights', in the same ceremonial manner as the Constitutional Assembly of the French Revolution would have discussed the 'Proclamation of the Rights of Man'. Now German students had always enjoyed great freedom, much more so than British students, and therefore it was not easy to proclaim new freedoms and new rights, and what it all came down to was just the opposite, namely restrictions on freedom: only socialist doctrines should be taught, only socialist professors and students allowed, and so on. Einstein was well known to be politically left-wing, if not 'red'. But this was too much even for him. When eventually the chairman asked his opinion, he made some critical remarks. I can still see before me the astonished faces of these eager youths when the great Einstein, whom they believed wholeheartedly on their side, did not follow them blindly in their fanaticism.

Yet Einstein was too kind and wise to disappoint them, and quickly turned the discussion to the actual question which had brought us there: the occupation of the buildings and the arrest of the rector and other professors. They declared that they had given over buildings and prisoners into the hands of the new socialist government, and that we ought to negotiate with them. We learned that the government was assembled in the Chancellor's Palace in the Wilhelm Strasse—the same building, by the way, from which Hitler later ruled and where he died. There we proceeded and were admitted again with the help of our journalist friend. On the wide imposing staircase footmen were standing in the old imperial livery, as if diplomats and generals in splendid uniforms were to be ushered in. What was now running up and down those stairs were men in ordinary clothes, socialist parliamentarians, journalists, trade-union men. The big reception hall was full of them. Einstein was soon recognized and welcomed. There was much cheering and elation, for the news of the rising in Munich had just come in.

We tried to find out who was responsible for university matters, and learned that the only man who could give us power to settle them was Ebert himself, the new President of the Council of the new government. This Council was just in session and we had to wait a long time. But it was not dull—things were happening every minute. Eventually Eduard Bernstein came and took us into a small room where Ebert was waiting. He shook hands with us, listened to Einstein's short report and answered: 'You will understand that this matter is not important enough for my personal attention when I tell you that I have just received the terms of the armistice from Versailles. Terrible terms, indeed. But I shall give you a paper which will help you.' Whereupon he scribbled on a visiting card some words of recommendation, signed with his name. This ended the audience, and we began our retreat. After a day crowded with impressions, it was a long exhausting journey home for one who had only a few days ago recovered from illness. But I would not have missed that day for anything. We then believed in a free, democratic, socialist Germany, and we had seen its birth.

Now I shall quote again some lines from Einstein's letter, parts of which I have cited elsewhere. He wrote (7 September 1944):

185

Do you still remember the occasion some twenty-five years ago when we went together by train to the Reichstag building, convinced that we could effectively help to turn the people there into honest democrats? How naive we were, for all our forty years. I have to laugh when I think of it. We neither of us realized that the spinal cord plays a far more important role than the brain itself, and how much stronger its hold is.

I have to recall this now to prevent me from repeating the tragic mistakes of those days.

XVII. Frankfurt am Main

When I first heard of Laue's offer to exchange positions (see Chapter XIV), it appeared fantastic to me. Candidates for professorial chairs at German universities were proposed by a Faculty and appointed by the Minister of Education—a private exchange of professorial chairs at different universities did not fit into this scheme. But in these revolutionary times everything was possible. Planck himself was willing to recommend it if I would agree. After careful consideration I did agree. It meant promotion to a full professorship (*Ordinariat*). Hedi and I welcomed also the opportunity to get out of the turmoil of revolutionary Berlin.

Life was indeed not very pleasant during this period (winter 1918/19). There was sporadic fighting, flaring up here and there. When Hedi had to go to town to shop, she had to avoid streets where there was fighting. There were rumours in the air that the left-wing 'Spartacists' would loot the homes of the well-to-do. One night our neighbours told us that 'Spartacus' was marching up the Kurfürstendamm, the long road leading to our suburb, to attack the wealthy 'Colonie Grunewald', and they began to pack their precious property in boxes to remove them to safer places. We refused to be disturbed and went to bed. Nothing happened. It is altogether remarkable how little violence to the life and property of ordinary citizens was done during these days of revolution. But the noise and excitement were considerable. One focus of fighting was the quarter where the University Department of Physics was situated; the river Spree in front of the building separated the northern parts of the city occupied by 'Spartacus' from the centre which was in the hands of the moderate socialists. There was permanent rifle fire across the river, and Professor Heinrich Rubens, the physicist, had to evacuate all the front rooms. But when the medical students working at the Charité, the big hospital beyond the river, had to attend their physics class, a white flag was hoisted and the firing ceased until they had safely crossed the bridge. 'Ordnung muss sein' in Germany, even in a revolution. I had no lectures during this period because of the lack of students, but we opened a colloquium very soon. Numerous physicists came to Berlin at that time to be demobilized. Many of them had not seen a scientific book or periodical for years and were very keen to hear what had happened in physics during the war, particularly in other countries. So they approached Rubens to arrange a colloquium. But he was tired and worried by the damage done to his institute; so he asked me to take the matter in hand. In this way I became the head of the biggest and most distinguished colloquium I have ever attended; there were indeed almost all the German physicists who had already a name or were to make one in the future—it would be impossible to give a list. There was also my old friend James Franck, and his collaborator Gustav Hertz, with whom, just before the outbreak of the war, he had made the celebrated experiments on the excitation of atoms by electrons, which confirmed directly Bohr's quantum theory of atomic structure. They had heard nothing about the progress of the theory during the war years, and so I arranged reports on this subject, and on other questions as well.

During this winter and spring (1919), I worked out the chemical consequences of the theory of ionic crystals which Landé and I had developed. I calculated the lattice energy, i.e. the work done by the atomic forces to build the crystal from the ions. But there was no experimental material with which to compare it. The only related quantity known was the heat of formation from the neutral atoms, and even this only indirectly, since the experiments were actually not made with monatomic gases, but with molecules or solids. I discussed these questions with Franck, who brought me into contact with Fritz Haber, the physical chemist, whose name I have mentioned before on several occasions.

During the war Haber had done important service. The German General Staff had not reckoned with a protracted war and had failed to accumulate stores of war material. Not even the nitrogen necessary for making explosives was available, and this almost led, in the first war winter, 1915, to a catastrophe. Munitions were extremely scarce. The situation was saved through Haber's invention for fixing nitrogen from the air. The German chemical industry, in particular the 'Badische Anilin und Soda Werke', developed the Haber process on a huge scale, and it was just ready for use when all other sources were exhausted. Haber and Walter Rathenau were then appointed to high positions in the Ministry of Supply to organize the whole war economy of the country, which they did with great success. Then came the period of trench warfare, which seemed to lead to an interminable, indecisive struggle. To overcome this deadlock, Haber suggested the use of poison gas and organized this new weapon. I hated the idea of chemical warfare, and (as I said before) refused to take any part in it. I even broke off all personal relations with Haber.

Now, the war having been lost in spite of Haber's efforts, I met him again. He was then director of the Kaiser Wilhelm Institute for Physical Chemistry in Dahlem, near Berlin, and Franck was head of one of his departments. There I was sitting with Franck discussing problems when Haber came in and joined our dispute. He became at once enthusiastic about my idea, and I could not resist his charm. He was in fact a fascinating personality, full of life and vigour, with perfect yet somewhat old-fashioned manners, clear-minded, fast thinking, interested in all branches of science and expert in many of them. From him

I learned that, apart from rough estimates (like those of Kossel mentioned before, and of Haber himself), no chemical energy had ever been calculated, and that my attempt was the first one which, in his opinion, looked promising. He gave me confidence in my own work and taught me how to find and sort out the numerical data needed for comparison with my theory.

The first method I used consisted in taking second differences of the heat of formation of four crystals, formed of two anions and two cations (e.g. NaF, NaCl, KF, KCl) and to compare them with the corresponding differences in the lattice energies. Although these differences are small compared with the total lattice energies and the results therefore not very decisive, they were good enough to encourage further investigation. I learned from Franck the concepts of ionization potential and electron affinity; the former is the work necessary to remove an electron from an atom to which it is bound, and the latter is the energy liberated if an atom with incomplete electronic structure (like the halogen atoms) picks up an electron. The ionization potentials of the alkali metals could be calculated from optical data with great precision, by applying Bohr's interpretation of the line spectra, which by then had been firmly established by Franck and Hertz. The only unknowns were the electron affinities of the halogen atoms. I calculated these from my lattice energies and showed that the results obtained for one atom, say chlorine, were independent of the crystal used (i.e. whether LiCl, NaCl, KCl, RbCl or CsCl). Years later these values of the electron affinities were experimentally confirmed by Joseph Mayer in Baltimore.

Haber improved my numerical method of dealing with chemical energies by inventing a graphical representation, which is now in general use as the Born–Haber cycle.

These papers caused a minor sensation in the chemical world, as Haber had predicted, and in the following years I was frequently invited to talk to chemical societies and to participate in their discussions. My ignorance in chemical matters was profound, and I felt very uncomfortable on these occasions. I had also some controversy with leading chemists, Fajans and Nernst. The differences of opinion with Fajans were of a minor kind, but Nernst attacked my whole idea of calculating chemical energy, instead of using 'free energy' as the chemists are accustomed to do. In fact, the difference between these two quantities is negligibly small compared with the total value. The chemists had never considered these total values but only temperature effects, for which precisely those small differences are essential. I tried to explain this to Nernst in several discussions, but in vain. A few years later, at a natural science conference in Innsbruck, it came to an open clash after a lecture of mine, in which all the members of the meeting clearly supported my view. Nernst was not convinced however; every time I met him he said in his most friendly high-pitched voice: 'My dear young colleague, you ought to have considered not the energy, but the free energy—you have not read your Helmholtz properly, nor my book on theoretical chemistry.' In spite of this difference he was invariably nice and kind to me, and I have pleasant recollections of hours spent in his hospitable house in Berlin.

This same work on lattice energy brought me also my first honorary degree; many years later (1927) I was awarded the D.Sc. of the University of Bristol, together with some very illustrious scientists: Rutherford, William Bragg, Eddington and Langevin, on the occasion of the opening of the H. H. Wills Physical Laboratory. I owe this honour mainly to Lennard-Jones, who was at that time at Bristol and was working on problems closely connected with my crystal investigations.

Now looking back on this episode of my career, I find it rather strange and puzzling. For after the lattice energies had been determined, the chemical application consisted in some additions and subtractions, which a schoolboy could perform. Yet this obvious application has brought me more recognition than the lattice theory itself or any of my other work, which I regard as much better and less trivial, such as my contribution to quantum mechanics. I am convinced that the new ideas inaugurated by quantum mechanics will have the deepest influence on science and philosophy in general, and I am proud to have a share in it. Yet even the fact that I was the first person ever to write a physical law in terms of non-commuting symbols has not been acknowledged by the scientific world in the same way as those little chemical computations. Perhaps the world is right: it is much more difficult and more important to do some trivial sums at the right moment than to participate in a philosophical revolution.

After this digression into future events, I return to the winter of 1919, when we prepared for our removal to Frankfurt. We succeeded in finding a lovely house, in Cronstetten Strasse, on the northern fringe of the city, with a view towards the Taunus hills. It was the nicest house we ever had, with a big garden full of fruit trees, pears and peaches, which brought us a rich harvest. We let some attic rooms to my friend Ernst Hellinger, professor of mathematics, and his unmarried sister. It was very convenient for me to have a 'tame mathematician' whom I could consult whenever I had a problem beyond my own resources. The physical laboratory was in the building of the Physical Society, near the Senckenberg Institution and other university buildings. My staff consisted of a lecturer, an assistant and a technician. I had the good fortune to find in Otto Stern a lecturer of the highest quality, a good natured, cheerful man with whom we soon became great friends. The assistant was Elisabeth Bormann, who had been trained in Berlin and worked for some time in industry. This period was the only one in my scientific life when I had a workshop and an excellent technician at my disposal; Stern and I made good use of both, as I shall report presently.

The professor of experimental physics, Wachsmuth, was a charming man, but did hardly any research. His first assistant, Walter Gerlach, soon found the atmosphere in my department more inspiring than in his own and became our permanent guest and collaborator. I published several papers jointly with him, including one quite good one on the electron affinities of iodine, oxygen and sulphur, calculated from lattice energies; but his work with Stern was more important. A remarkable figure amongst my other colleagues was the mathematician Schönfliess, well known for his fundamental work on the application of group theory to crystal lattices. I remember also Brendel the astronomer and Lorenz, the physical chemist.

Lorenz was very keen on arousing my interest in the behaviour of ions during electrolysis, and finally he succeeded. The unexplained facts which he insisted on pointing out to me were concerned with the influence of the charge of the ions on the mobility: divalent ions move slower than monovalent ones of the same size, and among the latter the smallest ions are not the fastest (as one would expect) but the largest ones. The usual explanation was based on the assumption of hydration, but was rather vague; I tried to give it a quantitative meaning by using Debye's theory of the dipole character of water molecules. An ion is surrounded by an electric field which has an ordering influence on the water dipoles; if the ion moves the water molecules have to rearrange themselves, dissipating energy. A lengthy calculation led me to a definite formula expressing the mobility in terms of the ionic radius and charge, which agreed reasonably well with the facts.

This theoretical work led also to some quite amusing experiments, done by one of my pupils, Lertes, in my laboratory. A sphere of thin glass was filled with very pure water and suspended between two pairs of condenser plates, which by some electronic device could produce a rotating electric field. The water dipoles tend to follow this motion, but have to overcome the viscosity of the water; hence the whole bulb of water is subject to a moment of force, which produces a rotation. This was indeed observed and could be used to calculate the rotational viscosity which was needed in my theory of ionic mobility. The theory of this effect and its applications to electrolysis was later improved by my pupil H. Schmick in Göttingen (1924) and it has lately (1945–46) been taken up again by other people.

We soon came into contact with circles outside the University. My chair, like many others, was a personal gift by a rich citizen, Herr Oppenheim who, if I remember rightly, dealt in precious stones and lived in a big, old-fashioned house in the Bockenheimer district. His wife was from Vienna and very musical, a good pianist, having been a pupil of Clara Schumann, but so exuberant in her musical enthusiasm that she had difficulties in finding partners to play with her on two pianos. I fell her victim and stood it quite a while, even enjoyed it, as she played well and acquainted me with many compositions which I did not know. The old Oppenheims were good people, very kind and hospitable; we were often invited to their box in the opera house. I saw many excellent performances in those two years in Frankfurt. The Oppenheims had a son, who was a business partner of his father, but addicted to philosophy; he wrote a book concerning which he involved me in many a tedious discussion. His wife Gabrielle, called Gaby, of the well-known Brussels family Errers, was a very pretty and coquettish blonde, who began a flirtation with every male who came within range of her flashing eyes. However, Hedi says that I was silly and did not grasp this opportunity. I do not know what has become of the young Oppenheims. But I have heard that the old people committed suicide under the Nazi régime.

Another grand family with whom we became acquainted were the Weinbergs. This name was well known all over Germany as that of two brothers, who bred the most successful race-horses in the country. They were the founders and owners of one of the great chemical industries in Germany, Casella and Co. whose works were situated on the Main near Frankfurt. One of the brothers, Arthur von Weinberg, was deeply interested in science and frequently attended our colloquia. He had published some papers in which he tried to calculate the chemical energies of organic compounds from an additivity principle, and he wished to interest me in this kind of problem. However I thought that the time was not ripe for it. Arthur von Weinberg had married a Dutch widow, née Huygens, a direct descendant of the great physicist, Newton's contemporary. She had two daughters from her first marriage, lovely girls whom we met when we were invited to their gorgeous house, Buchenrode, in Sachsenhausen, the suburb south of the Main. There they lived in a princely style. Two footmen helped you out of your coat and showed you through the imposing hall to an enormously long drawing room, full of works of art, flowers and lovely furniture. There was also his organ which he liked to play—but unfortunately not Bach or Handel, but transcriptions from Wagner operas. At dinner there was a waiter behind every chair, and the plates and dishes were of gold. We did not like this style of living. He was of Jewish origin, but very proud of his success and his 'von' before his name, and he admired the aristocracy. So he married both his stepdaughters to counts; one of them, Graf Spreti, was, I think, the chief trainer of Weinberg's pedigree horses. I heard later that both marriages were unhappy and led to divorces. Altogether this story, as most of my stories of people I knew, does not have a happy ending. When the Nazis came to power the Weinbergs succeeded in being acknowledged as 'honorary Aryans', presumably on account of their money. But this preference does not seem to have protected them for good, for I have learned that in spite of his sons-in-law, Arthur von Weinberg was deported to the concentration camp at Theresienstadt and died there.

Our real and intimate friends were of a very different kind. My father had a friend from his student days in Breslau, named Lachmann (not the one who was his assistant, when I was a youth of seventeen or eighteen, and whose influence on my development I have mentioned much earlier). This Dr Lachmann was now an elderly practitioner in Frankfurt. He and his lovely and youthful-looking wife received us in a most friendly way, and he became our family doctor. But a real friendship developed between us and two of Dr Lachmann's three blonde and pretty daughters. One of them, Grete, was a schoolmistress and married to a musician, Dr Friedrich Hoff; the other, Susanne or Susi, was a professional musician herself. I had heard of them before, for Hoff was a close friend of the companion of my childhood, Franz Mugdan, who used to tell me remarkable stories about him. Indeed, both the Hoffs and Susi were remarkable people.

As Friedrich and Susi were violinists and could play the viola while Grete played the cello, there was for me the perfect team for playing chamber music, piano trios and quartets, which we did with great enthusiasm, either in our house or in Hoff's little flat. Never in my life have I had a better and more congenial ensemble, and I enjoyed it immensely. They were all of a much higher musical standard than myself but bore my amateurish efforts with great patience and even occasional signs of approval, although our tastes often diverged. For instance Hoff despised Brahms (whom I loved), or at least pretended to, declaring after a sweet melody that it made him want to vomit. He preferred Reger, who seemed (and still seems) to me rather dry, a very learned composer with few ideas. Yet we had enough in common, Bach, for instance. Hoff opened new vistas to me by playing Bach's solo violin sonatas. However, he was not only a musician; he had studied history and obtained a doctor's degree, he painted a little and was widely read in literature. This side of him appealed strongly to Hedi, who had a great understanding for his natural, untamed behaviour and rough language. They had endless serious discussions, but also played practical jokes on each other. One result of their friendship was a trio for strings which Hoff wrote, the Breugnon Suite. Hedi was fascinated by Romain Rolland's novel *Colas Breugnon*; the hero appeared to her a type of man somewhat similar to Friedrich Hoff. In fact, Hoff felt the same and wrote under this spell a lovely piece of music. It is the only work by Hoff of which I have a copy.

Susi became a member of the well-known Bandler Quartet in Hamburg and married a musician, but was divorced. When Hitler came to power she went to Britain with her little daughter and became a music mistress at Kurt Hahn's school, Gordonstoun, where she is still. Grete also went to England and taught, first in Edinburgh, later in Reading. Her marriage with Friedrich ended in divorce. Now, after the war, she is back in Frankfurt helping to reconstruct German education. The third Lachmann sister, and her husband, a Dr Rothschild, perished during the Nazi régime.

Though these friendships and other attractions of Frankfurt made our life very rich and pleasant, there were also many great difficulties and sorrows. The aftermath of a war affects the life of everybody. We had a big house, but our good servants, the sisters Minna and Lina, left us to get married; it was hardly possible to get domestic help, and those we got were unreliable. I do not know how many times we changed. Once we had a brilliant cook, but she stole. Once we had a maid who was pregnant and told Hedi she was in her fourth or fifth month, but one day, when Hedi was away in Kassel, the girl fell sick. I telephoned to all the hospitals and at last got an ambulance, which took her away. When I rang up the hospital I learned that the baby had arrived before the ambulance had turned the next corner—if it had appeared ten minutes earlier I should have had the whole affair in my house. One day Hedi brought a young soldier home, a poor fellow in a ragged uniform. She had picked him up on the street as he looked so miserable, and he had told her that he had just come back from the front and had no home to go to. So we gave him our neat and clean spare room where he stayed a few days, enjoying himself in the house and garden, while I telephoned to all the military authorities to find out what to do with the man. The result made me suspicious and I told him to come with me to the Commandantura. So we took off together, but in the city he seized his opportunity when we were in a dense crowd, and bolted. When I came home I found Hedi in tears, for the spare room was not neat and clean any more but full of bugs, and had to be fumigated at great expense. It also turned out that the fellow's whole story was a lie; he had never been at the front, but only in training, then he had deserted, and his father, an honest craftsman, had turned him out.

Then we had the Kapp putsch and the French occupation. We were ardent supporters of the socialist government and were therefore highly excited when the news of the counter revolution led by Kapp became known. In Frankfurt, like everywhere else, the population answered with a general strike, and the military revolt broke down miserably. The French occupation has left at least one definite trace in my memory. There was much excitement in the town because the French used black colonial troops, which the Germans regarded as a deadly insult. We read in the papers of clashes between civilians and troops, and of casualties. After things had settled a little we went about as if everything were normal. One day our little girls were invited to a children's party in the Bockenheim district, and I took them there by tramcar. We had to change cars at the Hauptwache, the centre of the city, but there was an excited crowd blocking the road and shouting wildly. So I took the children through side streets to another tramway line and delivered them safely. On my way back I learned that there had been serious rioting and firing at the Hauptwache and that quite a number of people had been killed and wounded.

All these were minor incidents and misfortunes. We also had great and grievous losses in this period. Our friend Alfred Blaschko in Berlin fell ill and soon learned that he had cancer and no hope of recovery. When his strength was ebbing he wanted to see me, and so I went to Berlin and spent some hours at his bedside, hours not entirely sad but brightened by the heroism with which he bore his suffering and looked death in the face. He was not religious and did not expect a life after death. Yet he was completely composed. He had done his work and got his share of life. He asked me to keep up my friendship with his wife and children, and this Hedi and I have done. In death as in life he was the example which I wish to follow.

Then my mother-in-law died in our house in Frankfurt. Both parents were our guests when she fell ill with that disastrous influenza which swept the world after the war and is said to have killed more people than the guns. At first it did not look very dangerous, and my father-in-law returned to Leipzig to take up his teaching duties. But a few days later we had to call him back, and he arrived just in time to attend to her in her last hour. She had been a real mother to me, and I felt the loss very heavily.

For Hedi these days were terrible and exhausting. I think it was then that we decided to spend our summer holiday in the South Tyrol. I had always wished to show Hedi my beloved Bolzano; so we chose a mountain resort in this region, Sulden near the Ortler, where we stayed for some weeks, and then went to Merano, Bolzano and even to Venice. Since this first visit, Hedi has shared my delight in this lovely country. I shall never forget the expression on her face when she saw the Piazza San Marco for the first time. Yet there were also disagreeable moments on this journey. It was shortly after the war, and

Germans were still hated and outcast—even more so than they are now after the second great war. In the boarding-house where we stayed in Venice, none of the other guests, mostly Americans, spoke to us; we had to sit at a separate table, which we shared with a lively young man who had a typical American face, yet rather dark skin, like an Italian. We noticed that he was also avoided by the other Americans. The explanation was that he was not a pure-blooded white man; he had dark crescents on his finger-nails which, among Americans, sufficed to exclude him from 'white' society. This was a depressing experience. On the other hand, the Italians were as friendly as could be.

The work in my department was dominated by an idea of Stern's. He wanted to demonstrate and measure the properties of atoms and molecules in gases with the help of molecular rays, which had been first produced by Dunoyer. Stern's first apparatus was designed to produce direct evidence for the velocity distribution law of Maxwell and to measure the mean velocity. I was so fascinated by the idea that I put all the means of my laboratory, workshop and mechanic at his disposal. And I started myself, with the assistance of Fräulein Bormann, on a similar experiment, the direct determination of molecular collision cross sections (measuring the intensity of a beam of silver atoms *in vacuo* and in a gas). Stern's experiments were entirely successful, and Fräulein Bormann's and mine went well enough to meet, many years later, with the approval of the greatest experimentalist of our time. I was then in Cambridge and lecturing in the Cavendish. One morning I met Rutherford, who said to me: 'There must be an experimental physicist who has exactly the same name as you. When I was preparing a lecture of kinetic theory of gases, I found a paper in the *Physikalische Zeitschrift* signed by Max Born and Elisabeth Bormann containing the description of experiments much too good for a mathematician like you.' I assured him that the idea at least was mine, while the actual work was mostly done by my assistant.

Now Stern's plans became more ambitious: he wanted to measure the magnetic moments of atoms by deflecting an atomic ray in an inhomogeneous magnetic field, and he hoped in this way to produce evidence for one of the strangest consequences of quantum theory which Sommerfeld had developed and called 'quantitation of direction'. In view of the extreme difficulties to be expected, Stern joined forces with Gerlach, who had much experience in high vacuum techniques. So they began to construct their apparatus, but that needed money, and there was none.

We were already in the inflation which later became so disastrous; but we were not aware of what was happening. Everything was scarce and expensive. Physical instruments were hardly obtainable. So my funds were quickly exhausted, and I had to look for means.

At that time a wave of interest in Einstein and his theory of relativity was sweeping the world. He had predicted the deflection, by the sun, of light coming from a star. Several expeditions, amongst them a British one under Eddington, had been sent out to tropical regions where a total eclipse of the sun was visible and the deflection could be observed. Now after laborious measurements and tedious calculations the conclusion was arrived at that

Einstein was right, and this was published under sensational headlines in all the newspapers. It caused a tremendous stir in the civilized world, as I have already described in another chapter. There was an Einstein craze, everybody wanted to learn what it was all about, and he became the victim of a publicity racket. I used this for my own purposes. I announced a series of three lectures in the biggest lecture-hall of the University of Einstein's theory of relativity and charged an entrance fee for my Department. It was a colossal success, the hall was crowded and a considerable sum collected. My friends in the Frankfurt business world told me that I would have done even better if I had sent out private invitations to a lecture in the most expensive hotel, in evening dress and with cocktails, and had asked for an assistance fund. But that was not in my line.

The money thus earned helped us for some months, but as inflation got worse, it evaporated quickly and new means had to be found. One day I met a friend of the Ehrenberg family who told me that he had been engaged for years to an American girl from whom he had been separated by the war, and now he was going to New York to be married. I said jokingly: 'If you find a German-American who is still interested in the old country, tell him I need dollars for important experiments in my Department.' I had quite forgotten this remark when a few weeks later a postcard arrived, signed by this man: 'I am happily married and have found your man. Write to Henry Goldman, 998 Fifth Avenue, New York.' At first I took it for another joke, but on reflection I decided that an attempt should be made. With Hedi's help a nice letter was composed and despatched, and soon a most charming reply arrived and a cheque for some hundreds of dollars which helped us out of all our difficulties.

I must tell you a few words about this Mr Goldman. His grandfather had been a poor Jewish pedlar in Hesse, and when the Count of Hesse drove along, he had to kneel down in the dust of the road. He could not stand these iniquities and emigrated to the States, where he continued as a pedlar, with considerable success. His son was already owner of a banking firm and his grandson, our Henry, became the head of one of the greatest private banks, Sachs, Goldman & Co. He told me that it was he who had the idea of what are now known as the Woolworth shops, originally designed to sell nothing for more than twenty-five cents, or sixpence; he financed this company and apparently made a lot of money out of it. He was extremely rich and used his money to foster music and the arts. He was a great friend of Kreisler and gave the money for the education and training of Yehudi Menuhin. He had a marvellous collection of old masters, among them Raphaels, Rubens, Rembrandts, etc. He gave us a copy of the catalogue with wonderful photographic reproductions, but it was unfortunately lost in one of our removals.

I met him for the first time shortly after receiving the cheque; he had come to Berlin to see the state of affairs in Germany, for which he had preserved a deep affection. I visited him at the Hotel Adlon in Berlin and found a rather Jewish-looking, charming old gentleman, who told me at once that he never believed in the guilt of Germany with regard to the war. When a great loan was launched in America in favour of the Allies, he refused to contribute to it, and when his partners insisted, he resigned from his post as head of the firm. Now he had come to Germany to help us as far as he could, and to make good the injustice done to the Germans in the Treaty of Versailles.

Goldman was a great help indeed. He continued to assist our research by giving money, and he did the same with many other German scholars. I brought him into contact with Einstein, and a few years later they both visited us in Göttingen and stayed in our house. One day Hedi received forty large packing cases sent from America on Goldman's order, containing clothing and shoes for the poorer members of the University; she had to work hard with several helpers to distribute these gifts.

Unfortunately Goldman's eyes were bad and became worse every year. When we went to America in 1926 he was already blind, and it was moving to see him amongst his pictures, which he described in every detail as if he could still see them. One day when Hedi and I were with him two experts from Harvard called on him to see his gallery, and he showed them around, with detailed explanations. When they were about to leave he stumbled a little in passing through a doorway, and one of the men asked me, 'Does Mr Goldman not see very well?' When I answered 'He is completely blind', the man looked at me as if I were mad: 'But he certainly sees his pictures.' Well, he did, but with the eyes of a loving memory.

After Goldman's cheque had saved our experiments, the work went on successfully. The determination of magnetic moments of atoms and later of nuclei remained the main object of Stern's efforts, and twenty-five years later

In Baden-Baden with Mr Goldman (centre) and a friend (right), in 1931

197

(1946) he got the Nobel Prize for his results, together with his former pupil Rabi.

The interest shown by the general public in Einstein's theory of relativity induced me to write down my lectures as a book. I attempted to represent the theory without using higher mathematics, a task which is actually impossible. Yet I think I did it as much as is possible, for the book was successful and appeared in three editions. In the first one there was a short biography and a picture of Einstein, but this was removed from the later editions, in consequence of a correspondence with my friend Max von Laue. At that time Einstein was very much in the limelight, and as his fame increased, opposition and hatred became apparent, instigated by anti-semitism and nationalism. The leaders of this anti-Einstein movement were Philip Lenard, professor of physics in Heidelberg, and Johannes Stark, whom I have mentioned in my account of my first period in Göttingen. They declared that Einstein's ideas were all wrong and his celebrity a swindle due to Jewish advertising. Laue was afraid that any reference to Einstein's personality would only strengthen the position of his adversaries and advised me strongly to drop the biography and photo from my book. I followed this advice. But the Einstein controversy went on and reached its culmination at the natural science conference at Bad Nauheim (I think in 1920), where Lenard attacked Einstein at a public meeting, although other prominent physicists, Planck, Willy and Max Wien, tried to appease him. Einstein stayed in our house during this meeting, and we travelled every morning by train to Nauheim. It could have been a pleasant time, but it was spoiled by this conflict, which even led to a temporary estrangement between ourselves and Einstein, as we did not approve of his rather careless attitude and told him so. But this little conflict did not last long, and friendship was restored.

XVIII. Professor Ordinarius in Göttingen

You may remember that when still a student in Göttingen, I used to complain about life in the dull little town, with its philistine inhabitants, and how my friend Otto Toeplitz used to tease me by prophesying that I would inevitably have three periods in Göttingen: as a student, a lecturer and a professor. Now came the day when the last part of his prediction was to come true: I was offered a professorial chair in Göttingen under most favourable conditions.

In my student days there were already two chairs of physics in Göttingen, an experimental and a theoretical one, occupied by Riecke and Voigt. Riecke had died; after an interregnum, when my friend Rausch von Traubenberg ran the department, Robert Pohl, from Rubens's school in Berlin, was appointed, not as full professor but 'Extraordinarius'. Then Voigt retired before reaching his age limit; he wanted to finish his monumental book on crystal physics and be rid of all departmental duties. Yet he remained a professor, also with the rank of extraordinarius, and gave some minor course of lectures. Debye was first invited to deputize for him, and later became his successor, but as full professor and head of the whole laboratory. Towards the end of the war Voigt died, and Debye was offered a chair at Zurich in Switzerland which he accepted, since living conditions in Germany were rather bad. It was his chair I was offered: to be an *'Ordentlicher Professor'* (full professor) and head of the whole laboratory, previously consisting of two independent and well equipped parts. One of these parts had been given over to Pohl and would not be my direct responsibility, but the other part, Voigt's, was to be run by myself.

This was a great opportunity but also a great risk. For I had never directed a big laboratory before, nor given a lecture on experimental physics, and had no intimate knowledge of the technicalities of a laboratory. Moreover I had tried my experimental abilities in Frankfurt and, though my ideas had turned out to be good enough to earn Rutherford's praise, it was quite clear to me that I was no true experimentalist—I lacked training, experience and patience. So I was in a quandary: could I decline such a splendid offer, or should I make the jump into a new job for which I did not feel prepared? My mind was heavy with doubts and fears as I travelled to Berlin to negotiate with the Ministry of Education. I was received by the director of the department for the Prussian

Universities, Wende, and explained my scruples to him quite frankly. In order to become clear about the situation I asked to be shown the documents containing the conditions of appointment of all the professors concerned; and then I made a discovery. There were, as I said before, two positions marked 'Extraordinarius', one for Pohl, the other for Voigt, and one of them bore the condition 'To be cancelled on the death of the occupant'. But this condition was not after Voigt's name, as had certainly been intended, but referred to Pohl's position. Now Pohl was young and very much alive, hence there was no harm done by that condition. On the other hand, there was Voigt's 'Extraordinariat' still existing and vacant; for as I said old Voigt had just died.

I am in general not very quick to grasp a situation, but in this case I did. So I quickly pointed out to Director Wende that there was another vacancy, and that this changed the whole situation. For if they could appoint a second experimentalist to be the head of Voigt's former department I would not hesitate to accept the post, considering the directorship of the two laboratories as purely formal. Wende laughed, saying that this was obviously an error of the copyist; but he very soon saw the possibilities. You must understand we were still living in revolutionary times. Like most of the officials in the Ministry of Education, he was a new man, keen to do new things. One of his aspirations was to stimulate science and in particular to foster the old scientific centre of Göttingen. He therefore accepted my proposal and asked me to nominate a candidate for the second Extraordinariat.

So I returned to Frankfurt much happier than I had left, pondering on the name of the experimentalist to be recommended. There were several men of my age for whom I had a high regard, but one of them appeared to me to be without a doubt the best, and that was my old friend James Franck. The experiments which he, in collaboration with Hertz, had done to demonstrate Bohr's quantum theory of atoms I regarded as most important and fundamental. I had known Franck since my student days and loved him as a most honest, reliable and good-humoured fellow. I also knew that he and Robert Pohl had been at the same school in Hamburg and were great pals, which guaranteed a frictionless collaboration between the two experimental departments.

So I recommended Franck to the Ministry. The details of the further negotiations have vanished from my memory. I only remember a visit to Göttingen, where I had to persuade the Dean of the Philosophical Faculty, the geologist Stille, to recommend Franck to the Faculty as candidate for the second experimental chair. It was not easy, but eventually everything went smoothly. The Faculty saw the advantage of getting two professors for one and made a list with Franck's name on top, and he was duly appointed simultaneously with myself. My faith in him was never disappointed. His brilliant gifts fully expanded in the congenial atmosphere of Göttingen, he produced a splendid series of important investigations, all intended to confirm the quantum theory of atoms, and developed a great school. When he and Hertz received the Nobel Prize, a few years later, I saw my recommendation perfectly justified.

My hope that the collaboration between the two departments would be smooth was also fulfilled. There were some minor disagreements and even a major one during my trip to the United States 1925–26; but even this was settled in a friendly spirit after my return. Pohl too did splendid work in this period. He reorganized the practical courses, as well as his experimental lectures, on a modern basis; these lectures became celebrated all over the world and were published in the form of text-books, and his research school produced valuable results on the photo-electric properties of crystals and other subjects.

Thus the physics school in Göttingen in the twenties was very flourishing and much of this was due to a typist's error.

There were also great changes in the mathematical department. Felix Klein, old and worn out by illness, retired and Richard Courant, one of the Breslau group of my student days, was appointed his successor. He had been a frontline soldier during the whole war but had escaped serious injury. The war over, he became professor of mathematics at the University of Münster, from which he was called to Göttingen. He had married Nina, the youngest daughter of my teacher Karl Runge, and a friend of Hedi; Nina was extremely musical and an excellent violinist. Thus Courant's house became not only a centre of mathematics, but also of music. Hilbert retired a few years later and Hermann Weyl became his successor. Minkowski's former chair was occupied by Edmund Landau. When Runge died he was succeeded by Gustav Herglotz. In this way the old tradition of mathematics in Göttingen was upheld in more than one way, for Courant continued his predecessor's interest in the organization of teaching and in the relations with the neighbouring sciences. Courant's text-books are well known and much used, in particular his book on the partial differential equations of mathematical physics, which he published in collaboration with Hilbert; the latter contributed little to the actual text but the book is written in his spirit and inspired by his ideas. It was published, by chance, at exactly the right moment, when Hilbert's theory of the eigenvalue problems of differential and integral equations found an enormous new field of application in quantum mechanics. Courant took up an old plan of Klein's to establish a great mathematical institute in Göttingen, with modern lecture-rooms and a big library where the excellent collection of books could be properly housed. He succeeded in interesting the Rockefeller Foundation in the project, and in 1927 a splendid building went up next to the physical laboratory, which would have satisfied Klein's most extravagant hopes.

Now to return to the year 1921, when we moved from Frankfurt to Göttingen; I cannot say that we did it without regret. We had come to love Frankfurt and its beautiful surroundings. We had found good friends, we enjoyed concerts, theatres, operas, and we had a pleasant house. Göttingen was a small place, poor in artistic and musical attractions, and the housing problem was extremely difficult. For a long time we could not find anything, until we were offered the ground floor of the house in Planck-Strasse which had belonged to the late Kurator Osterrath (a 'Kurator' is the representative of the government at a university and is in charge of the financial administration); this flat consisted of three enormous rooms and an equally enormous kitchen, used only for official entertainments. We had to divide the rooms into smaller ones by building new walls in order to make them suitable for a family. One room, an annexe, remained as it was and became a combined study, drawing room and music room. My father-in-law gave me his grand piano, an excellent Steinway, which together with my own Steinway were put into a corner of this room. Yet it was still far from being crowded and could easily hold my desk, my whole library and all our drawing room furniture. There was also a big verandah, half covered, half open, with a staircase leading to the large park-like garden, with its lovely old trees and big lawns.

How we got rid of our house in Frankfurt is a sad tale which I shall relate only briefly. Inflation was at that time already in full spate, and even if I had sold the house for a large sum of money, its value would have shrunk to half in a short time. On the other hand, rents were restricted by law to the pre-war level—a negligible amount in current value. Therefore it was common practice to make private and more or less secret arrangements with the would-be tenant to secure a decent rent. This I did, I have forgotten with whom. But whoever it was, he cheated me and as these arrangements were, if not against, at least outside the law, I could not sue him. In this way he compelled me to sell the house at a price which, owing to the growing inflation, meant the loss of a great sum of money.

Our whole life during the next two years was dominated by inflation. It took a long time for people to recognise the meaning of what was going on. We were no exception. When my salary in Göttingen was fixed at about double the usual amount for a full professor (something near 40 000 marks) we believed ourselves very well off, and many envied us. Yet the purchasing power of the mark was at that time already well under half its pre-war level, and sinking rapidly. Even the banks did not understand this. In the spring of 1922 (I believe) we decided to go to Italy again and exchanged a sum of German money—already a large figure—into Italian lire. We had a lovely time in Florence and Rome, and we did not spend all our Italian money. When we returned to Göttingen with 800 lire we had to exchange only about 100 lire in order to repay the bank what they had lent us four weeks previously. The bank had actually paid not only for our whole journey but over and above.

All that was strictly in accordance with the law, for the Supreme Court had decided 'Mark ist Mark'. This principle turned out sometimes favourable, sometimes disastrous. To give examples of both cases: at the beginning of inflation we bought a lovely fur coat for me (which I still have) for a sum scarcely higher than the pre-war price, hence in fact for almost nothing. On the other hand, one part of my capital consisted of a mortgage on a house in my home city of Breslau which became payable at the height of inflation, when the postage for an ordinary letter was already something like 10 000 marks. So one day I received a letter from the debtor containing a few postage stamps as the equivalent of my mortgage of 50 000 marks (£2500), which was originally about a third of the value of the house in question.

In this way the whole middle class lost most of its property. A few clever people bought 'tangible assets' (real estate, jewellery and the like) in time and saved their fortune; some even became monstrously rich by such speculations. The great majority, the backbone of the German bourgeoisie, however, became poor and therefore an easy prey to political agitators. This process, which was one of the causes of Hitler's success, has been described by experts often enough. In everyday life inflation at first seemed a nuisance, but soon became a nightmare, in particular for the housewives. In the end the deterioration of the money became so swift that our salary was paid out every week, or even twice a week. Then Hedi would queue at an early hour at the office of the University bursar for the wretched slips of paper, representing so many millions or billions of marks, and hurry at once to the shops to buy food and other things as available; for in the afternoon the purchasing power might have been halved and the next day dropped to nil. What that meant in senseless effort, fatigue and frustration, it is difficult to describe to people who have not experienced it. Even the present inflation (1948) in Britain is trifling compared with that catastrophe. But suddenly it came to an end, the 'Renten' mark appeared, remained miraculously stable and everything returned to normal—on the surface. In fact all savings had gone, the middle class was destroyed and with it the stability of the German State.

Concerning ourselves, we came out in a somewhat better position than many others. I had inherited a considerable number of shares in our family business, the Meyer Kauffmann Textile Works, and I still possessed them. Some years they would yield a small dividend, but more often nothing. Yet there they were, 'tangible assets' in the full sense of the word, real, not inflated and deflated values. I felt a kind of sentimental attachment to them, for although the business had been transformed into a joint-stock company, the majority of shares were still in the hands of the family, and some of its members were still directors of different factories of the firm, for example my cousin Hans Schäfer and my second cousin Franz Kauffmann. So I kept the shares, although they brought only an irregular and meagre return. But towards the end of inflation (1923) Hedi discovered from the stock exchange figures that the value of the shares was going up suddenly. Just at this time my father-in-law was looking for another flat. He had, a few years earlier, retired from his chair at Leipzig University and taken a small flat in Göttingen, where he was cared for by Fräulein Kati Redlich who became a great friend of ours. This flat was at the other end of Göttingen, so he wished to move to a house nearer to us. Hedi set out to find one for him and she combined this plan with the opportunity of selling the Meyer Kauffmann shares under favourable conditions. She found a man who wished to sell his large, detached house, Wilhelm Weber Strasse 42, at the corner of Dahlmann-Strasse, in the best situation in Göttingen, where he had occupied the top flat, because he wanted to build a small house for himself on the same premises. So after an exchange of telegrams Hedi went to Berlin, negotiated with three different parties who were all interested in the shares, penetrating even to the bank directors, sold our shares very favourably and returned victorious. She bought the house in question and earned the respect of the businessmen amongst our friends.

So we had again become owners of a house. It had three apartments, of which the top one was taken by father Ehrenberg. When in 1928 we had the opportunity of getting a flat in our own house, Wilhelm Weber Strasse 42, we took it at once. But it was a complicated and expensive operation. There was a Frau Robertson on the ground floor, an elderly widow who wanted to marry a young fellow and to follow him to some other town. My father-in-law found climbing the stairs to his top flat rather tiring and wished to take the ground floor flat. We ourselves wanted to have the middle flat, then occupied by a Herr and Frau Lehmann whom we persuaded to move into the top flat, on condition that we paid the expenses. In the end we had to pay for four removals: Frau Robertson, the Lehmanns, my father-in-law and our own. But the result was most satisfactory. Our flat was very comfortable and easy to run. Two little rooms in the attic became Hedi's abode, where she could think and write in perfect solitude, undisturbed by husband and children. We had an excellent housekeeper, Fräulein Schültz, known as 'Schültzli', and a maid. There was a nice little garden behind the house with a lawn and some fruit trees and bushes. I remember in particular the lovely plums. The drawing room was big enough for my two grand pianos, and we had many a pleasant musical evening.

However, the only trouble we had with this house came from those two pianos. One day I was watching Irene and Gritli playing marbles on the parquet floor in the music room when I noticed that the marbles showed a marked inclination to assemble in the middle of the room. Wherever they were thrown by the little girls they rolled slowly to the centre of the floor between the two pianos, as if attracted by a force. This observation set the brain of a physicist working. I got a string, gave the ends to Irene and Gritli to hold at the base of opposite walls and it showed that the string was two or three inches above the floor in the centre. This obviously meant that the floor was not strong enough to bear the two grand pianos. We called in an architect who after examining the floor told us that we were lucky to have found out in time, or we might have broken through during a musical evening and landed in Dr Bensen's dining room below (Dr Bensen, after the death of my father-in-law in March 1929, had become the tenant of the ground floor). A complicated and expensive repair was necessary: an iron joist had to be pushed through a hole in the outer wall and fixed under the ceiling of the Bensens' dining room. Apart from this mishap the home served us well until we were driven out by the Nazis in 1933. It was, together with all my other property, confiscated in 1938 and taken over by the German army administration. And now, in 1948, my claim to it has just been recognized and I hope to get it back soon.

But I must return to the time of our settling down in Göttingen. The first important event was our son Gustav's birth. He was a frail little fellow and gave us much trouble and anxiety during his first two years. But after that he outgrew his initial weakness.

During those happy years in the little town, we went on several excursions in the lovely countryside, to the old castles of Plesse, Hanstein and Hardenberg, into the endless forests of the Göttinger Wald or along the Weser river. We spent our summer holidays at different places, at Ehrwald in the Austrian

Panorama of Göttingen, from the Hainberg

mountains (where Hedi had been so often as a child with her parents), or at the North Sea island of Langeoog, or at Silvaplana in the Engadine, where we shared a beautiful old farmhouse with father Ehrenberg and his Fräulein Redlich. But let me now turn to my profession and my work.

When I made my first call on Felix Klein I was curiously aware of the changed situation. Not so long before I had been a young and shy student who had looked up to the 'Great Felix' as to a god and had been almost crushed by him. In his room, which was crammed with books and manuscripts, I found him looking old and frail but still upright in his majestic attitude, interested in everything that was going on in mathematics and science. He had even read my book on relativity and involved me in a fascinating discussion about my attempt to avoid so-called higher mathematics, i.e. the differential calculus. He expressed the opinion that instead of coming down to the level of the general public one should try to educate it, to raise its scientific taste. Yet he agreed that under the existing conditions of general knowledge my attempt was quite successful. Then he came out with a surprising proposition: would I like to become the editor of the last volume of the complete works of Carl Friedrich Gauss? This publication was one of the monumental undertakings of the Göttingen Academy of Sciences. Only that last volume remained, which was to contain miscellaneous small items by Gauss and some articles by contemporary mathematicians and physicists, with commentaries about Gauss's work on mathematical physics and astronomy. I accepted this offer rather reluctantly as I had never done such editorial work before, and I do not think that I did it very well. The first task was to travel about and visit a number of neighbouring universities, Marburg, Giessen, Tübingen, etc. in order to get in touch with the authors of those essays. They had been chosen by Klein long ago but had never delivered a manuscript. I tried my best and succeeded in obtaining a few articles. But the whole affair never interested me very much, and I later readily agreed to pass it over to somebody else—I have even forgotten to whom.

My relations with Hilbert continued as friendly and intimate as they had been during my previous periods in Göttingen. I had moved away from mathematics, and if I had ever had an inkling of his mathematical greatness, it now became less and less possible for me to follow his ideas, which turned to the most abstract investigations on the logical foundations of mathematics—or the mathematical foundations of logic.

But now I began to appreciate Hilbert's greatness as a thinker in general, a philosopher, though this word can hardly be applied to him, for he had a very low opinion of the recognized philosophical schools and refrained from any attempt to systematize. Yet he was in no way merely critical and negative, but had decided views about all the important questions in science and life. He applied to everything the axiomatic method which he had used with the greatest success in mathematics. But that does not mean in the least that he accepted a set of invariable axioms or dogmas from which he deduced all else—just the opposite. He explained to me more than once how the facts of Euclidean geometry were for him, as for Euclid himself, the given and unshakeable material, and that the problem of axiomatic treatment was to condense these facts into a set of complete and non-contradicting axioms. The possession of these revealed the logical structure of geometry and permitted the contemplation of non-Euclidean deviations. He looked on life at large in the same way. I do not think that he tried to compress the laws of life into axioms—he was too much aware of the complexity of human affairs—but he analysed the roots of beliefs, institutions, laws, customs until he found some short and striking formulation, often so surprisingly sharp and strange that people who did not know him were deeply shocked and inclined to regard him as a revolutionary and even nihilistic mind. But he was nothing of the kind. He simply did not accept traditional rationalizations. I regret that I have never written down one of his characteristic considerations, and I still hope that some of his pupils, Courant or Weyl or Speiser, will have preserved some of them. Here I shall only report one of his remarks which deeply impressed me.

At an evening party, astrology was being discussed and some people argued that, even if science could not justify the influence of the stars on human affairs, neither could it disprove it. Hilbert listened to all that talk quietly for a while, but he kept repeating his usual 'Na, ja, na, ja', so often that everyone fell silent and listened. Then he said in his most acid and piercing voice: 'If you were to appoint a committee of the twenty cleverest people in the world, I say the acknowledged supreme brains of every nation, and give them the task of finding the most stupid thing possible—they would fail to discover astrology. Na, ja, na, ja.' Whereupon nobody had anything to add. But I am sure that one or two left the party deeply offended.

In this way he made enemies, but his superiority was such that they seldom dared to attack him, even in politics, where his behaviour appeared particularly erratic to those who did not know him well. He was very left-wing before the war. He despised the militarism and autocratic methods of the Wilhelminian era in Germany. I did not see him during the war, but when I returned to Göttingen I found that he was just as much opposed to the then socialist government. People commented on this with: 'There, you see how you can rely on "great Hilbert"; he may be a mathematical genius but in practical things he is like a child without deep convictions, just following his impulses.' Nothing could be more wrong than this judgement. Hilbert saw more clearly than most of us the weakness of the Weimar régime, its vacillations between socialist experiments and reactionary decisions (like handing back to the former princes their enormous possessions, which enabled them to destroy the foundations of the republic), but most of all its lack of personalities. He despised many of the leading figures of German politics of that time and expressed his feelings in his usual exaggerated way, comparing the new men with those of the old régime who were at least educated and gentlemen. And so he was regarded as a reactionary.

To my knowledge, Hilbert had no great appreciation of literature, but he himself wrote in an excellent style. Some of his popular articles, for instance his obituary on his friend Hermann Minkowski, are examples of the best German prose. His interest in music was particularly interesting and rather touching. When I met him first he claimed not to be musical at all. He hated amateurish performances at evening parties. But his attitude changed as the gramophone became more and more perfect. I think that he was first attracted to it by its technical ingeniousness. He loved contraptions of this kind and admired men who had the knack of inventing or improving them, a gift totally lacking in him. So he bought a good gramophone and began to enjoy his records. Slowly his taste was formed by listening and comparing. Better instruments and more records were purchased, and finally he had the best possible gramophone and a most complete and well ordered collection of records. He knew them all and had a marked taste, especially with regard to opera. How often did he play to us his newest records of Caruso or Chaliapin, and his joy was genuine and infectious. I shall never forget those evenings in Hilbert's simple and hospitable house.

There were many other remarkable men in Göttingen at that time, and particularly in the Philosophical Faculty, to which mathematics and science belonged, together with languages, history and all the other subjects which at British universities are united in the faculty of arts. To mention only a few of the scientists: Prandtl, one of the founders of modern hydro- and aerodynamics; the chemists Windaus, celebrated for his vitamin research and Zsigmondy, one of the leaders in colloid chemistry—both received the Nobel Prize during this period—and Tammann, the metallurgist; the geologist Stille (mentioned already), the mineralogist Mügge and later his successor, Victor Moritz Goldschmidt, who became a great friend of ours. At my first Faculty meeting I was immediately struck by the difference between its atmosphere and that of the Faculty in Frankfurt to which I had belonged. My last impressions there had been the discussions about my successor; I had strongly, but unsuccessfully, recommended my lecturer, Otto Stern. The Faculty chose instead Erwin Madelung, also a good candidate, who had moreover the advantage of not being Jewish. Yet he never got the Nobel Prize, like Stern. The Faculty in Göttingen showed a far greater objectivity and was always keen to get the best man in a subject, no matter what objections might be raised. When I say 'the Faculty' I mean really the scientific section. Whether the same was true of the historical-philosophical section, I cannot say. There was an unspoken agreement that in questions of appointments the two sections did not interfere with one another. But this changed slowly, and with the emergence of Hitler a rift appeared in our Faculty. Most of the scientists remained as they had been: objective, unbiased with respect to race or religion, and keen to have the best possible scholar. There were of course men of the same attitude in the other section but there were also fanatics, extreme nationalists and anti-semites who became more and more noisy and prominent. But that belongs to a later period.

During the first years all was peaceful, and I remember only one occasion where we had a violent discussion in which I found myself eventually almost alone against the whole Faculty and had the satisfaction that in the end my view turned out to be right. It was the question of honorary degrees. Felix Klein, to whom the growth of science in Göttingen was mainly due, had used the following method: he persuaded some wealthy industrialist to offer a sum of money to the University for founding a new chair or building a new institute, by indicating that the Philosophical Faculty would confer an honorary degree on the benefactor, and the Faculty used to follow his advice. This had worked for years and as many other universities applied the same principle, nobody saw anything objectionable in the procedure. But as usually happens, it was overdone, somewhere some swindler got a degree, and the newspapers made a great scandal out of it. This prompted a group of decent and conscientious members of our Faculty, led if I remember rightly by Stille, to move an amendment to our regulations that honorary degrees should be given only 'for scientific merits' and not 'for merits in regard to science', as was the wording on which the profitable procedure was based. There was a great majority in favour of this change, but a few of us had misgivings. We thought that a method which had been of such great advantage under the wise counsel of Klein, should not be branded as something morally inferior, and we pointed out that the Faculty might judge each case according to its merits. But we were voted down. I however insisted on having my contrary opinion written in the minutes.

A short time later Prandtl was offered a chair in Munich under very favourable conditions. He declared that he would prefer to stay in Göttingen if he could be given the same experimental opportunities as he was offered in Munich, a new wind tunnel and I do not remember what else—a demand which seemed to be beyond our means. But it soon became known that there was some industrialist willing to sacrifice a considerable amount of money for

Prandtl's department, provided he was promised the traditional award. Well, there we were in a nice quandary: to lose Prandtl who was one of our most brilliant lights, or to break the recently established rule and to return to the 'low moral standards' of Klein's régime. The deliberations of the Faculty were a source of secret amusement to the few of us who had expected something like that to happen. But we kept in the background and left the affair to take its natural course, which meant the withdrawal of the amendment and the acceptance of the offer. I remember that I pondered as to whether I should not vote against it, as the man in question was in fact a rather shady character, much below the standard of Klein's donors, but I resisted this temptation, in view of Prandtl's importance to our scientific group.

XIX. Quantum Mechanics

I shall now describe my own scientific activities during this period in Göttingen. My department consisted of one small room in the Physical Institute, and its staff of one assistant and a half-time secretary. There was no workshop and no technician as I had at Frankfurt, since I had entrusted the experimental department to Franck. Nevertheless, I tried to continue some experimental research by having a student working under my direction. But this did not go well. Franck and Pohl were both sceptical about my experimental abilities and showed this in a rather discouraging manner. Pohl was obsessed by a fear of mercury poisoning and took great precautions in his department against spilling this metal in the rooms. Now my student's room was just above Pohl's private study, and one day a drizzle of mercury rained down from the ceiling. A big vessel of a vacuum pump containing mercury had broken and emptied its contents on to the floor. Pohl was mad with anger and anxiety, he ordered the floor of the poisoned room to be removed and made life for my student as unpleasant as possible. I decided to give up experimental research and to restrict myself to theoretical work, always prepared to advise Franck and Pohl and their collaborators if they asked me.

I organized a three-year course in theoretical physics, consisting of six series of lectures corresponding to the six semesters. The students who attended were supposed to know calculus and analytical geometry; therefore most of them began our course in their second year after having taken one year of mathematics—but of course we did not ask them, nor care, where they had learned it. There was, as I have said before, complete freedom of teaching and learning at the German universities, with no class examinations, and no control of the students. The University just offered lectures and the student had to decide for himself which he wished to attend and whether he was able to follow them. Our six lecture series were (1) mechanics of particles and rigid bodies; (2) mechanics of continuous media; (3) thermodynamics; (4) electricity and magnetism; (5) optics; (6) elements of statistical mechanics, atomic structure and quantum theory. Each series consisted of four lectures a week and a tutorial. Apart from the main course, I used to give a special lecture on some problem of modern physics, two hours a week. Later, when my assistant became a lecturer, the main course was given twice with a shift of three semesters; and in the last period, when I had two assistant lecturers, even three times.

All three sections of the physical laboratory, Franck's, Pohl's and mine, held a joint seminar which was a great and exciting affair. Many neighbouring departments took part including those of Reich (applied electricity), Prandtl (applied mechanics), Tammann (physical chemistry), Wiechert, later his successor Augenheister (geophysics), Hartmann, later Kienle (astronomy), etc.; and often the mathematicians attended. Numerous important results were first announced at these informal meetings. Franck, Pohl and I took turns in providing a subject and taking the chair. It was customary to interrupt the speaker and to criticize ruthlessly. We had the most lively and amusing debates, and we encouraged even young students to take part, by establishing the principle that silly questions were not only permitted but even welcomed. There was permanent petty sniping between the extreme experimentalists of Pohl's school and my theoretical men, while Franck maintained an intermediate position: though he was no mathematician, his work was a sound mixture of theory and experiment; he had an uncanny gift for translating abstract ideas into practical demonstrations with relatively simple apparatus.

Franck was an ardent admirer of Bohr and believed in him as the highest authority in physics. I sometimes found this rather exasperating. It happened more than once that we had discussed a problem thoroughly and come to a conclusion. When I asked him after a while: 'Have you started to do that experiment?' he would reply: 'Well, no; I have written first to Bohr, and he has not answered yet.' It was the time before the establishment of quantum mechanics, and I was trying, with my collaborators, to find weak points and contradictions in Bohr's semi-classical theory of atoms. Bohr himself, of course, was doing exactly the same; nobody was more convinced than he himself that his theory was only preliminary. But Franck regarded our endeavours with suspicion and never accepted our conclusions without having confirmation from the prophet in Copenhagen himself. That was at times rather discouraging for me, and even retarded our work to some degree. When finally the solution of the puzzles came not from Copenhagen but from the room next door, Franck was a little shaken. But all this never affected our friendship.

The series of my assistants is rather remarkable, namely Pauli, Heisenberg, Jordan, Hund, Hückel, Nordheim, Heitler and Rosenfeld. The first two got the Nobel Prize, and all the others are well-known leaders in theoretical physics.

Pauli was recommended to me by Sommerfeld in Munich. He was an 'infant prodigy'. In his first semester he took part in a seminar on relativity with such success that Sommerfeld asked him to write the article on this subject for the *Mathematical Encyclopaedia*. Pauli finished this work before graduating. I think it is still, after more than twenty-five years, the best presentation of relativity existing. But when Pauli came to me, he knew all other branches of theoretical physics just as well. We met first during the summer of 1921 in Ehrwald,

Tyrol, where I was spending my holiday (Hedi could not accompany me as Gustav had just been born). I remember that even in the most lovely or majestic mountain scenery, Pauli continued to discuss physical problems. No mental relaxation was possible in the company of this dynamic fellow. Of course he was not a real success as 'assistant'. We worked together, on refined problems of perturbation theory and its application to the quantum theory of atoms, and I learned a great deal from him, certainly more than he from me. But I had no great help from him in my routine work of teaching. I suffered at that time from asthmatic attacks and sometimes had to stay in bed for a day or two. Then Pauli was supposed to give my lecture, which was from 11 a.m. to 12 noon. But he was inclined to forget it, and if our maid was sent to remind him at half past ten, she usually found him still sound asleep. For he worked till late at night, and although a place like Göttingen is accustomed to all kinds of strange people, Pauli's neighbours were worried to watch him sitting at his desk, rocking slowly like a praying Buddha, until the small hours of the morning. But we could assure them that he was quite 'normal', just a genius.

When he left me (I think to become professor in Zurich), he recommended his friend Heisenberg as his successor. He too came from Sommerfeld's school in Munich and was no less an 'infant prodigy'. He was working at that time at his doctoral thesis on a problem of hydrodynamics; Sommerfeld advised him to accept my offer in order to breathe a different scientific atmosphere. When he arrived (it must have been October 1923) he looked like a simple peasant boy, with short, fair hair, clear bright eyes and a charming expression. He took his duties as an assistant more seriously than Pauli and was a great help to me. His incredible quickness and acuteness of apprehension enabled him to do a colossal amount of work without much effort; he finished his hydrodynamic thesis, worked on atomic problems partly alone, partly in collaboration with me and helped me to direct my research students.

At the end of his first semester at Göttingen, Heisenberg returned to Munich for the oral examination, often called 'colloquium' because of its informal and courteous character. In fact, it hardly ever happened that a man whose thesis had been accepted as satisfactory was rejected on account of this oral. Thus Heisenberg left cheerfully and promised to return after the Easter holiday to take up his job again as my assistant.

I was not a little astonished when, one morning long before the appointed time, he suddenly appeared before me with an expression of embarrassment on his face. 'I wonder whether you still want to have me,' he said, and then he explained: he had almost failed in the examination and only got the lowest mark, 'rite' (the order being: summa cum laude, magna cum laude, cum laude, rite). This catastrophe was due to the fact that he was, at that time, not very interested in experimental work and had done his practical course in such a slovenly way that the professor, the great Willy Wien, noticed it. Wien therefore asked him at the colloquium some detailed question about experimental technique, and when he got no satisfactory reply he became very angry and declared that the candidate had failed. As unanimity of the three examiners was necessary, Sommerfeld was in a most awkward position; for Heisenberg was by far his best pupil for many years and his thesis a brilliant piece of work. After a long and animated dispute Sommerfeld procured a pass for the candidate, but only with 'rite'. Sommerfeld had invited a little party to his house to celebrate the graduation of his favourite pupil. But Heisenberg was so depressed by his failure that he excused himself very early, went home, packed his suitcase and took the night train to Göttingen. There he was, putting his fate into my hands. I said: 'Let us see what these formidable questions were which brought you down, and whether I can answer them.' They were certainly rather tricky. So I saw no reason to send him away—in fact I was perfectly confident about Heisenberg's outstanding abilities.

The real joke in this little story became apparent only a few years later when Heisenberg discovered his celebrated principle of indeterminacy (or uncertainty). For one of Wien's questions was concerned with the resolving power of optical instruments, due to the finite wavelength. It is now well known that the formula $\Delta p \Delta p \sim h$ expressing the relation of indeterminacy of coordinate q and momentum p is nothing more than the translation into 'particle language' of the optical formula $\Delta a \sim \lambda$ connecting the geometrical resolution Δa with the wavelength λ (using de Broglie's relation $\lambda \sim h/p$). Heisenberg was conscientious enough to look up all the questions which he had failed to answer in the oral, and among these the problem of resolving power impressed him as fundamental. Thus when the time was ripe he remembered it, and the result was the principle which has made his name famous not only in physics but also in philosophy.

But before I give an account of the birth of quantum mechanics I have to say some words about other investigations. Although my book *Dynamik der Kristallgitter* (Dynamics of Crystal Lattices) which had appeared during the war had certainly not stirred the scientific world, I had received a few very favourable comments, and the most encouraging one from Sommerfeld. The chemical applications of the calculation of lattice energies made in the following years drew the attention of wider circles. Then Sommerfeld wrote me a very nice letter asking me whether I would not write a comprehensive article on the dynamics of crystal lattices for the fifth volume of the *Mathematical Encyclopaedia*. I agreed to this proposal and began to collect material. The actual writing was done mainly during my first two years in Göttingen; the article was published as a small book in 1923 (by Teubner) and appeared a short time later as a chapter of the *Encyclopaedia*. It contains for the first time many of the standard methods used in the dynamic theory of lattices. Today (1948) it is rather out of date, a relic of the pre-quantum-mechanical period, and I am going to replace it by a modern book.

While preparing that article for the *Encyclopaedia*, I came across a great number of unsolved problems which provided suitable subjects for doctoral theses; for it was fairly certain that they could be solved, and they had a certain value for chemistry and crystallography. A considerable number of my pupils graduated Dr Phil. with these on such subjects, which were mostly published in the *Zeitschrift für Physik*. I can mention only a few of these men whose character or fate was unusual.

There was the little Hungarian Jew, E. Brody, perhaps the most gifted of them all, who could solve intricate problems but was unable to write his results in a comprehensible way. We published a few papers together on thermodynamics of crystals (1921), which I had to dictate to him. I have heard that he survived in Budapest all the horrors of the Nazi occupation and the war.

Then there was Carl Hermann, who graduated with a thesis on the optical activity of some cubic crystals (1923) and later became well known as co-editor of the fundamental crystallographic compendium *Struktur Bericht*. He was a pacifist and therefore suffered badly during the Nazi epoch, but he survived prison and concentration camp and visited us after the war.

Another idealist was Gustav Heckmann, who continued the work of my former assistant Fräulein Bormann on elastic constants of ionic crystals and graduated in 1927. He was a pupil of the philosopher Nelson mentioned in Chapter VI, from whom he had derived his liberal ideas, which he combined with a sensitive social conscience. When the Nazis came to power, he left Germany of his own free will—he is a pure Aryan—and joined the staff of a refugee school founded by Minna Specht, first in Copenhagen, afterwards transferred to England.

E. Hückel, who later became known among chemists through his collaboration with Debye on the theory of electrolytic solutions, worked with me on the quantum theory of molecules; we disentangled the interaction of rotations and vibrations. He graduated in 1923, and some years later he became my assistant.

This work on molecules led me to a more fundamental idea; the classification of the harmonic vibrations of molecules and crystals with the help of group theory. At that time a young Dutchman, C. J. Brester, appeared in Göttingen and turned out to be an excellent helper in this work. When his thesis was almost finished, he expressed the wish to use it for graduating in Utrecht, not in Göttingen, and I agreed. This paper has become the theoretical backbone of a wide field of research, the exploration of the infra-red absorption spectra and of the Raman effect. It has been improved and generalized by Wigner and others. But as it appeared in Utrecht, the fact that it came from my school is hardly known. I was surprised when I was invited last April (1948) to the meeting of physicists in Bordeaux to celebrate the twenty-fifth anniversary of Raman's discovery, to receive an honorary degree together with him. I presume that my French colleagues wished to acknowledge my contribution to the theory of Raman spectra and molecular structure.

Meanwhile James Franck pursued, with a very talented school of collaborators and pupils, his experimental investigations into quantum effects in the interaction of atoms and molecules, which I followed with great interest. From our daily discussions there sprang a joint publication on the application of quantum theory to the kinetics of chemical reactions (*Zeitschrift für Physik*, 1925).

My main interest during this period was directed, however, towards the quantum theory of the electronic structure of atoms. The question was how far Bohr's theory did actually account for the facts and to find its limitations.

Bohr assumed that an atom existed in stationary states which were described by special solutions of the equations of ordinary mechanics selected by certain quantum rules. In these states the atom was supposed not to emit radiation; emission and absorption were connected with transitions between two such stationary states, the energy difference of which is equal to that of the light quantum (or photon) $h\nu$ emitted or absorbed, where ν is the frequency and h Planck's constant.

This principle worked exceedingly well in the case of one-electron systems, like the hydrogen atom or helium ion. But could it be generalized to systems with several electrons, such as the neutral helium atom?

To decide this was the first point of our programme. It meant an adaptation of the classical perturbation methods of the astronomers to atomic systems. This problem was tackled in several papers in collaboration first with Pauli (1922) and later with Heisenberg (1923). Applied to helium these methods gave, as we rather expected, results which did not agree with spectroscopic measurements.

On the other hand, the qualitative results showed a fair agreement with the general properties of matter. This was shown in a particularly striking way by Heisenberg and myself in a paper (*Annalen der Physik*, 1924) where we derived the main properties of molecules by our perturbation methods, using the square root of a certain ratio (mass of the electron to mass of a nucleus) as an expansion parameter.

We became more and more convinced that a radical change of the foundations of physics was necessary, and thus a new kind of mechanics, for which we used the term quantum mechanics.

This term appears for the first time in physical literature in a paper of mine (*Zeitschrift für Physik*, 1924) in which an essential step towards the establishment of this new theory was made. As I wish to make my account of this development as objective as possible, I asked Pascual Jordan who was, apart from Heisenberg, my main collaborator during this period, to write down his recollections of the events. I shall record only those things in which Jordan's memories and my own agree. Quantum mechanics was of course not a product of our group in Göttingen alone but had many sources.

We knew that frequencies of vibrations were proportional to energy differences between two stationary states. It slowly became clear that this was the main feature of the new mechanics: each physical quantity depends on two stationary states, not on one orbit as in classical mechanics. To find the laws for these 'transition quantities' was the problem.

An important step was made by my old friend from Breslau, Rudolf Ladenburg (1921). He found that the 'strength' of a spectral line, which measures not only the intensity of absorption and emission but also its power of dispersion, is proportional to the transition probability between the two stationary states involved, a quantity introduced before by Einstein in the theory of heat radiation. A detailed account of this relation was given by Ladenburg and Reiche, my other old friend from Breslau, in *Naturwissenschaften* in 1923.

On the basis of these investigations, Kramers (*Nature*, 1924) succeeded in developing a complete 'dispersion formula' which gave an account of all the essential optical properties of atoms, including the Raman effect, and which contained only 'transition quantities', i.e. such as refer to two stationary states. One can say that Kramers, guided by Bohr's principle of correspondence, guessed the correct expression for the interaction between the electrons in the atom and the electromagnetic field of the light wave. That at least is the way I regarded his results. It was the first step from the bright realm of classical mechanics into the still dark and unexplored underworld of the new quantum mechanics. I made the next step with the question: can one not find, by similar systematic guesswork, the interaction between two electronic systems in terms of 'transition quantities'? Indeed, by a proper re-interpretation of the classical perturbation theory, the corresponding quantum formula could be constructed. It was later fully confirmed by quantum mechanics.

At that time my best pupil was Pascual Jordan, whom I have mentioned already. His strange Christian name, as he told me, was inherited from his great-grandfather, a Spaniard who, as a soldier in Napoleon's army, had come to Germany and settled down there for good. Jordan became my collaborator in the next problem I attacked; it was concerned with Planck's theory of radiation, which led him to the existence of quanta. An inspection of his work showed that he used classical mechanics for the interaction of light and matter, an odd discrepancy. We translated Planck's calculations into the language of quantum theory, introducing 'transition quantities' instead of the corresponding classical quantities. Our paper on aperiodic quantum processes appeared in *Zeitschrift für Physik* in 1925.

We were struck by the fact that the 'transition quantities' appearing in our formulae always corresponded to squares of amplitudes of vibrations in classical theory. So it seemed very likely that the notion 'transition amplitudes' could be formulated. We discussed this idea in our daily meetings in which Heisenberg often took part, and I suggested that these amplitudes might be the central quantities and be handled by some kind of symbolic multiplication. Jordan has confirmed that I spoke to him about this possibility.

Meanwhile Heisenberg pursued some work of his own, keeping its idea and purpose somewhat dark and mysterious. Towards the end of the summer semester, in the first days of July 1925, he came to me with a manuscript and asked me to read it and to decide whether it was worth publishing. At the same time he asked me for leave of absence for the rest of the term (which ended about August), as he had an invitation to lecture at the Cavendish Laboratory in Cambridge. He added that though he had tried hard, he could not make any progress beyond the simple considerations contained in his paper, and he asked me to try myself, which I promised to do.

I remember that I did not read this manuscript at once because I was tired after the term and afraid of hard thinking. But when, after a few days, I did read it, I was fascinated. Heisenberg had taken up the idea of transition amplitudes and developed a calculus for them, by following the connection between the coefficients of the classical expansion of a vibrating quantity and its harmonic components (Fourier series). If two such expansions are multiplied, one obtains a new expansion for the product whose coefficients can be formed from those of the original series by a simple rule. Now Heisenberg suggested that one should forget everything about the series and consider the set of coefficients which represented the physical quantity in question; then one had a multiplication rule for those sets of coefficients. Yet each set was still attached to a single classical orbit or stationary state, while in quantum theory every observable effect depends on the transition between two states. Using our previous experience about the correspondence between the concepts of classical and quantum theory, Heisenberg defined transition amplitudes and their multiplication through the analogy with the sets of classical Fourier coefficients and their multiplication. His most audacious step consisted in the suggestion of introducing the transition amplitudes of the coordinates q and momenta p in the formulae of mechanics, for instance, in the kinetic energy $p^2/2m$, where now p^2 means the symbolic product of the set of transition amplitudes of p with itself. He also succeeded in expressing Bohr's quantum conditions in terms of this symbolic multiplication and applied his formalism to a simple type of anharmonic oscillator.

I was deeply impressed by Heisenberg's considerations, which were a great step forward in the programme which we had pursued. In this introduction he formulated the postulate that the quantum mechanics of the future should get rid of all quantities, such as frequencies and amplitudes of harmonics in classical orbits which could never be observed, and should be built entirely on concepts of its own, which at least in principle are measurable. This philosophy has had a great influence on the development of physics in the following decades. It has often been interpreted as the requirement that all quantities should be eliminated which are not directly observable. Yet I think that in this general and vague formulation the principle is quite useless, even misleading. The question of which are the redundant quantities can only be decided by the intuition of a genius like Heisenberg.

After having sent off Heisenberg's paper to the *Zeitschrift für Physik* for publication, I began to ponder over his symbolic multiplication, and was soon so involved in it that I thought about it for the whole day and could hardly sleep at night. For I felt there was something fundamental behind it, the consummation of our endeavours of many years. And one morning, about the 10 July 1925, I suddenly saw light: Heisenberg's symbolic multiplication was nothing but the matrix calculus, well known to me since my student days from Rosanes' lectures in Breslau.

I found this by just simplifying the notation a little: instead of $q(n, n+t)$, where n is the quantum number of one state and t the integer indicating the transition, I wrote $q(n, m)$, and re-writing Heidenberg's form of Bohr's quantum condition I recognized at once its formal significance. It meant that the two matrix products p and q are not identical. I was familiar with the fact that matrix multiplication is not commutative; therefore I was not too much puzzled by this result. Closer inspection showed that Heisenberg's formula gave only the value of the diagonal elements ($m = n$) of the matrix $pq - qp$; it said that they were all equal and had the value $h/2\pi i$ where h is Planck's constant and $i = \sqrt{(-1)}$. But what were the other elements $(m \neq n)$?

Here my own constructive work began. Repeating Heisenberg's calculation in matrix notation, I soon convinced myself that the only reasonable value of the non-diagonal elements should be zero, and I wrote down the strange equation

$$pq - qp = \frac{h}{2\pi i} \ 1$$

where 1 is the unit matrix. But this was only a guess, and all my attempts to prove it failed.

At that time, the middle of July 1925, a meeting of the Lower Saxony regional branch of the German Physical Society was about to be held in Hanover. A considerable number of the physicists from Göttingen went there by train; it is about an hour's journey on the North-Express. In the train we met several physicists from other universities, among them my former assistant Pauli.

Pauli had meanwhile become famous from many excellent papers, among them that on his celebrated exclusion principle, on which Niels Bohr had built his theory of the periodic system of the elements. He was coming from Zürich (where the summer vacation started earlier than in Germany) to take part in our meeting in Hanover.

I joined him in his compartment, and absorbed by my new discovery, I at once told him about the matrices and my difficulties in finding the value of those non-diagonal elements. I asked him whether he would not like to collaborate with me in this problem.

But instead of the expected interest, I got a cold and sarcastic refusal: 'Yes, I know you are fond of tedious and complicated formalisms. You are only going to spoil Heisenberg's physical ideas by your futile mathematics,' and so on.

I do not remember anything about the meeting in Hanover. The following day I asked my pupil Jordan to help me in my work. I was extremely tired and felt unable to make progress alone. Jordan accepted and after only a few days brought me the solution to the problem: he showed that the canonical equations of motion, applied to the matrices p and q, led to the result that the time derivative of $pq - qp$ must vanish, hence the matrix itself must be diagonal. Then we began to write a joint paper which appeared in *Zeitschrift für Physik*, vol. 34, 1925, p. 858 under the title '*Zur Quantenmechanik*'. We sent a copy of the manuscript to Heisenberg who was at that time on a holiday trip, after having delivered his lecture in Cambridge. We got an enthusiastic reply and decided to complete the work all three together after the holidays.

This paper by Jordan and myself contains the formulation of matrix mechanics, the first printed statement of the commutation law quoted above, some simple applications to the harmonic and anharmonic oscillator, and another fundamental idea; the quantization of the electromagnetic field (by regarding its components as matrices).

Nowadays the text-books speak without exception of Heisenberg's matrices, Heisenberg's commutation law, and Dirac's field quantization.

In fact, Heisenberg knew at that time very little of matrices and had to study them, which he did with his usual incredible quickness and efficiency. Where our suggested application of matrices to fields was concerned, we found not the slightest response; some distinguished physicists (such as the Russian Frenkel) considered our suggestion a mild form of madness. This only changed a few years later when Dirac took up the idea, probably independently, with great success.

Concerning Pauli's refusal to collaborate with me, he explained later to other people that he had not seriously meditated on Heisenberg's ideas and did not wish to interfere with his plans. I cannot remember exactly whether I told Pauli that Heisenberg had asked me to develop his work if I could, but I think I mentioned it; however this may be, I think that even a mind as great as Pauli's is not proof against mistakes of judgement in such difficult questions: he just did not grasp the point.

After Jordan and I had sent off our paper, we separated. I went to Silvaplana in the Engadine to join my family and father Ehrenberg and his Fräulein Redlich in the lovely, big farmhouse which we had rented. It had an enormous tiled stove in the drawing room, surrounded by a kind of wooden cage inside which a ladder led to its top where one could sit, and from there a trap door led to the upper floor. Hedi had told her father, who arrived a little after her, that this was the only access to the bedrooms upstairs; the stout old gentleman was quite despondent about the idea of squeezing his bulk through this hole several times a day and was most relieved when he found a wide and comfortable staircase in the hall.

After my return to Göttingen, about the 1 September, there began a hectic period of collaboration between Heisenberg, Jordan and myself. There were several difficulties: I had received and accepted an invitation to lecture during the winter at the Massachusetts Institute of Technology and I had to leave at the end of October. I wished to have our three-man paper finished by that date, as its contents would give me excellent material for my American lectures. On the other hand, Heisenberg did not return to Göttingen at the same time as myself and Jordan, but stayed in Copenhagen and arrived in Göttingen only to help us in putting the final touches to our paper. Until then, there was an intense correspondence between Göttingen and Copenhagen. The paper was finished in time before my departure (with Hedi) to America and appeared in the spring of the following year (*Zeitschrift für Physik*, vol. 35, 1926, p. 557). It contains almost all the essential features of quantum mechanics: the representation of physical quantities by non-commuting symbols, the generalization of Hamilton's mechanics for these, canonical transformations, perturbation theory, treatment of time-dependent perturbations with application to the optical theory of dispersion (derivation of Kramer's formula), the notion of degeneration, the connection with the eigen value theory of Hermitian forms, an outline of the treatment of continuous spectra, the theorems of momentum and angular momentum (quantization of direction, anomalous Zeeman effect), quantum mechanics of electromagnetic fields with a derivation of Einstein's fluctuation formula.

It is not easy to disentangle the contribution of the three authors. One point sticks in my memory which illustrates the fact that Heisenberg was at that time not familiar with matrix calculus. He sent us a suggestion on how to treat perturbations which I recognized was wrong. The correct method followed easily from the theory of canonical transformation which Jordan and I had meanwhile developed; Heisenberg at once accepted our method. Heisenberg and Jordan contributed, among many other ideas, the treatment of the angular momentum and its applications. My main contribution was the relation of our theory to Hilbert's work on quadratic forms of an infinite number of variables with which I was acquainted from my student days. Here the matrices are used as operators acting on vectors (in a space of infinite number of dimensions, the so-called 'Hilbert space'). These vectors form the bridge from our matrix mechanics to Schrödinger's wave mechanics; there they appear as discontinuous representations of wave functions and are obtained as expansion coefficients. Though we never contemplated waves, and all they imply, in connection with our theory, we introduced in a formal way the concept of the wave function. We used it for an alternative presentation of the perturbation theory which operates not directly with matrices but with these 'wave vectors'. This theory, which I worked out, is completely identical with that later given by Schrödinger and usually called after him.

It was a time of hard but successful and delightful work, and there was never a quarrel between us three, no dispute, no jealousy. This is obvious if one reads our papers. It also appears clearly in a letter which Heisenberg wrote me after he had been awarded the 1933 Nobel Prize, together with Dirac and Schrödinger:

Zurich, 25 November 1933.

Dear Mr. Born,

If I have not written to you for such a long time, and have not thanked you for your congratulations, it was partly because of my rather bad conscience with respect to you. The fact that I am to receive the Nobel Prize alone, for work done in Göttingen in collaboration—you, Jordan and I—this fact depresses me and I hardly know what to write to you. I am, of course, glad that our common efforts are now appreciated, and I enjoy the recollection of the beautiful time of collaboration. I also believe that all good physicists know how great was your and Jordan's contribution to the structure of quantum mechanics—and this remains unchanged by a wrong decision from outside. Yet I myself can do nothing but thank you again for all the fine collaboration, and feel a little ashamed.

With kind regards,

Yours,

W. Heisenberg.

The place and date of this letter are significant: Zürich, November, 1933. Hitler was in power and I was living as a refugee in Cambridge. Heisenberg could not write what he felt from Nazi Germany and had to wait until he was in Switzerland.

I am glad to have this document of his friendship and the testimony of our partnership in the work for which he alone was honoured. Yet the story of these tragic times belongs to a later chapter.

Part 2

Tempestuous Years

I. The 'Heroic Age' of Theoretical Physics

At this point, with the description of the discovery of quantum mechanics and my part in it, the writing of these recollections came to a stop. The second world war was over, the number of students at Edinburgh University increased, and with them my duties as a teacher, examiner and director of a department.

Thus there was little time available for delving into old memories. I postponed the continuation of these recollections to the time of my retirement. When this came in 1953 I actually began writing again and gave a detailed account of our journey to America in 1925/26. But then I lost interest and stopped.

I am now too old (nearing 79) to continue the account of my life in the detailed manner of the preceding pages. You may think this regrettable because the second half of our lives was much more dramatic than the first; in fact, from the usual point of view, the exciting events were just beginning: the rise of Hitler, our expulsion from Germany and emigration to Great Britain, the problem of becoming British and securing a living among Britons, the persecution of the Jews, and other Nazi horrors, the second world war, then the victory of the Allies, peace and finally our return to Germany.

But similar things have happened to many and have been described by others, and better than I ever could. Fate has treated us mildly, and though we shared the anxieties and troubles of this time, we were never exposed to any particular danger or humiliation. These political upheavals will come into my narrative only in so far as our own fate was involved. But I shall not refrain from saying what I think of them.

Taking up the interrupted story, I shall begin by giving an account of the scientific work done during our American journey and later in Göttingen.

My lectures at the Massachusetts Institute of Technology contained in the first part an outline of crystal dynamics, and in the second the elements of quantum mechanics. This was the first systematic presentation of this new field. When I began these lectures, the paper by Jordan and myself, in which matrix calculus was introduced, was still in the press; the big three-man paper by Heisenberg, Jordan and myself appeared just at the end of the lecture course.

225

About the same time, I received reprints of Paul Dirac's first papers, which contain practically the same things in a different form. This was—I remember well—one of the greatest surprises of my scientific life. For the name of Dirac was completely unknown to me, the author appeared to be a youngster, yet everything was perfect in its way and admirable. Later I understood how it happened. In the summer of that year, Heisenberg had given a lecture at Cambridge where Dirac heard of the new ideas; he digested them in an incredibly short time and produced those astonishing papers.

Meanwhile I was not idle in the American Cambridge (Massachusetts). I wrote my lectures down with the intention of publishing them and thus earning some money, as the salary I received from M.I.T. was not large, and was insufficient to cover our expenses. But the President of M.I.T., Mr S. N. Stratton, told me that it was intended to make them the first of a series of M.I.T. publications, a great honour indeed; but it had the disadvantage that it was only an honour, not backed up by a fee or royalties. However, I had my German manuscript (the English translation being done by my temporary assistants) and I sent it at once to the German publishing firm Ferdinand Springer, who accepted it and paid me well. This book has just (1961) been reprinted by an American firm (Fred Ungar) with a new introduction, in which I describe the origin of this work in about the same way as I have done here.

I also continued to think about improving and developing the theory itself. It could be applied only to closed systems in stationary states, not to free particles, nor to collision processes, which are of great importance, as they provide the main method for obtaining empirical information about atomic structures. Discussing these matters with my pupils, I found a lively response from a remarkable young man, Norbert Wiener, now famous as the founder of the new science of cybernetics. He was the son of a very learned Jewish scholar whose interests and accomplishments were of an incredible breadth. If I remember rightly, he had for instance made a new translation of all of Tolstoy's works, had studied and written about the Maya civilization in Central America, and many other things. The son Norbert was a mathematician of repute. He drew my attention to the fact that matrices could be regarded as operators, acting on vectors in multi-dimensional space, and he suggested a generalization of matrix mechanics into a kind of operator mechanics. We worked together hard on this project and published a paper which in some ways is a precursor of Schrödinger's operator calculus in quantum mechanics, but we just missed the most important point in a way which makes me ashamed even to this day. For in our paper we used the differential operator $D = d/dt$ and identified it with $(2\pi i/h)W$, where W denotes the energy, but we failed to see that in the same way d/dq represents $(2\pi i/h)p$, where p is the momentum belonging to the coordinate q. Instead we developed complicated integral representations for q and p which were in the line of Wiener's mathematical thinking, but were hardly assimilated (and now forgotten) by myself. Thus we were quite close to wave mechanics but did not reach it, and Schrödinger is rightly considered its discoverer. Since then I have been a little sceptical concerning Wiener's sagacity and prejudiced against his theories, even where these find world-wide acclamation, as in the case of cybernetics. A few years later he was invited by the mathematicians of Göttingen University to give a course of lectures. These were not very successful, and when the students complained about his teaching there was trouble between him and Courant, which I tried to smooth out. This was the last time I saw him. Once he intended to visit us at Edinburgh but as both Hedi and I were not very well we had to decline. Last year he gave a lecture at the natural science conference in Hanover which I intended to hear, but I was again prevented by illness.

I have talked about this Wiener case at some length because it was the most outstanding example of my being quite close to an important discovery and letting it slip by. There are several other cases; for instance the X-ray analysis of crystals for which von Kármán and I had all the material handy but which we missed while Max von Laue succeeded. And another case is Yukawa's discovery of the connection between the short range nuclear forces and a new type of particle, the mesons. I had often pondered about the force law $e^{-\kappa r}/r$ since my student days, when I found it mentioned in Minkowski's Encyclopaedia article on capillarity, and I knew the differential equation which it satisfied but I failed to translate the differential operators back into energy and momentum components and thus missed the particle interpretation. This is a kind of reverse of the Wiener case.

After having finished the lectures in January 1926, I set out on a big lecture tour of the West. Unfortunately, I had to go without Hedi, who had got food poisoning and was still not well enough for a long journey. She had to return to Germany and spend some time in a nursing home in Frankfurt. The new quantum mechanics drew big audiences, particularly in Pasadena and Berkeley. I tried meticulously to give the honour of the discovery of quantum mechanics to Heisenberg, with the effect that Jordan's and my contributions were hardly mentioned in the American literature of the following period. But a considerable number of young physicists from different places told me they intended to come to Göttingen to learn these new things at the source, and indeed they came; amongst them, to mention only some prominent ones, E. U. Condon, later President of the National Physical Laboratory, and R. Oppenheimer, the leader of the team which made the first atomic bomb. I shall speak of this American influx later.

In Göttingen I had to cope with a greatly increased number of research students attracted by quantum mechanics, amongst them those Americans just mentioned. Apart from our usual colloquium, I now had a private advanced one about theory alone, held in my large study at home in Planckstrasse. This was probably the most brilliant gathering of young talent then to be found anywhere. Apart from the Americans there were, for shorter or longer periods, my assistant Friedrich Hund (now my successor in the Göttingen chair), Enrico Fermi from Italy, Paul Dirac from England, Egil A. Hylleraas from Norway, A. Fock from Russia, Leon Rosenfeld from Belgium, E. Wigner and J. von Neumann from Hungary, and many others. I cannot remember any particular discussions at these gatherings, but I well remember the general impression, that it was soon very difficult for me to follow the arguments of these clever young chaps. Heisenberg was not among them, as he had been appointed a professor at Leipzig University. I tried to keep him in Göttingen by applying to the Faculty of Science for the foundation of a second chair in Theoretical Physics, but without success. Today Göttingen, like most larger universities, has two professors in our subject; but in 1926 this was regarded as a superfluous luxury, although a man like Heisenberg was available.

When Heisenberg left, he asked me to lend him my lecture notes on quantum mechanics as he had to start his course in Leipzig immediately. There is an amusing story connected with this. An Indian student in my department, named Sidiqi, a Moslem of splendid figure and fine features, therefore always referred to by his co-students as 'the beautiful Sidiqi', decided to go to Leipzig, in order to study under Heisenberg. Many years later, when I was in Edinburgh, the periodical *Nature* asked me to review a text-book on quantum mechanics by Sidiqi, and I accepted. It turned out to be essentially a presentation of my own lectures, a little evolved and extended; in the Introduction the author expressed his thanks to Heisenberg for the inspiration he had got from his lectures in Leipzig. I wrote in my review that I considered this book very satisfactory, as I could hardly criticize my own lectures at Göttingen, even when they appeared in a roundabout way, via Leipzig, in India under a different name.

I remember a few other stories about the young men in our brilliant group. The Americans were too numerous for me to have much time for all of them. Some of them, such as Condon, were therefore disgruntled. He complained about everything in Göttingen: the primitive digs without a proper bath, the food in the restaurants, the bad bus services, etc.; and last but not least the overworked professor who had so little time for him. The same complaint came from H. P. Robertson, who spent some time in Göttingen in one of the following years. I do not know why he came to me, as he worked on general relativity, a subject in which I had never done anything. One day he appeared with a big manuscript and insisted that I should read it, as it was important. I apologized for not being able to do so because of my ignorance of the subject and my being overburdened with other work. Then he said he would go to Berlin and discuss it with Einstein, and he asked me to give him a recommendation to Einstein. Now I had done this often, but just then Einstein had written to me, saying do not send people to me as I cannot cope with all these visitors. I showed this letter to Robertson, but he took no notice of it and went to Berlin. Some time later he reappeared, triumphant: Einstein had received him without a recommendation, had read his manuscript and declared it very important. And indeed it was, as I learned later; it was his celebrated work on the 'expanding universe'.

This affair finished me in Robertson's mind. He regarded me as rude and ignorant, and he did not conceal this opinion. Many years later I read in Leopold Infeld's autobiographical novel *The Quest*, which contains much gossip about physicists all over the world, how he had exchanged recollections with Robertson when they met at Princeton, about people and institutions. They came to talk about me, and how Infeld became my collaborator in Cambridge (1933). He said something like this: 'Born was first rude to me, but later angelic'; whereupon Robertson said: 'We started with a quarrel too. I did not give him a chance to be angelic.' Again many years later, it came to a reconciliation. During the second world war, I was in London for a few days and had arranged with my daughter Gritli (Margaret Pryce) to have lunch in the restaurant at Marylebone Station—of all places. When I got there, I found her sitting with Robertson and she succeeded very easily in making peace between us.

Oppenheimer caused me greater difficulty. He was a man of great talent, and he was conscious of his superiority in a way which was embarrassing and led to trouble. In my ordinary seminar on quantum mechanics, he used to interrupt the speaker, whoever it was, not excluding myself, and to step to the blackboard, taking the chalk and declaring: 'This can be done much better in the following manner. . . .' I felt that the other members did not like these perpetual interruptions and corrections. After a while they complained, but I was a little afraid of Oppenheimer, and my half-hearted attempts to stop him were unsuccessful. At last I received a written appeal. I think that Maria Göppert, then a very young student, now a well-known professor in California, was the driving force. They gave me a sheet of paper looking like parchment, in the style of a mediaeval document, containing the threat that they would boycott the seminar unless the interruptions were stopped. I did not know what to do. At last I decided to put the document on my desk in such a way that Oppenheimer could not help seeing it when he came to discuss the progress of his thesis with me. To make this more certain, I arranged to be called out of the room for a few minutes. This plot worked. When I returned I found him rather pale and not so voluble as usual. And the interruptions in the seminar ceased altogether. But I am afraid he was deeply offended, though he never showed it and gave me a splendid present before he left, a first edition copy of Lagrange's *Mécanique Analytique*, which I still possess. But I was never again invited to the U.S.A., in particular not to Princeton after the end of the second world war, when every theoretical physicist of any standing spent some time at the Institute of Advanced Study, whose president Oppenheimer had become. But this is only a conjecture, and it is quite possible that neither his influence nor his rancour was as great as I imagined. Maybe the reason for my being neglected was the well-known fact that I was opposed to atomic weapons and criticized those who made them.

A short time after the appearance of our three-man paper and the corresponding papers by Dirac, we were taken by surprise by Schrödinger's papers on wave mechanics which appeared in 1926 in the *Annalen der Physik*. This was a new approach to quantum mechanics which, at first glance, had no connection with our own work. It was of fascinating power and elegance, and as it used mathematical methods well known to every physicist (while our matrix method was not familiar to many of them), it quickly became the standard theory. The connection with matrix mechanics was soon demonstrated by Schrödinger himself and could also be gathered from Dirac's papers.

The superiority of wave mechanics as a mathematical tool could be seen by its application to the fundamental atomic problem, the hydrogen atom. This had been successfully treated by Wolfgang Pauli with the matrix method but it was a rather difficult and abstract work, while Schrödinger's treatment, in one of his first papers, used methods well known from the theory of differential equations and could be understood without difficulty. From this time on, wave mechanics was favoured by the overwhelming majority of physicists.

During the following years, the theory developed very quickly, and for almost inescapable reasons I was induced to write a text-book on it. Just before Heisenberg's discovery in 1925, I had published a book on the present situation in atomic theory, with the title *Vorlesungen über Atommechanik*, in collaboration with my assistant Friedrich Hund; I had had the audacity to add to the title: Volume I. In the Preface I explained this promise of a second volume by my confidence that the 'final theory' could be expected soon, 'in a few years'. This shows how certain I was that the decisive step was near, though it came even quicker than I had predicted, in fact the same year. This book is, I think, not only a final and satisfactory presentation of the older quantum theory, connected mainly with the names of Bohr and Sommerfeld, but has its intrinsic value as a condensed presentation of classical mechanics. Indeed, it has just been republished in America in the English version with the title *The Mechanics of the Atom* (translated by Fisher and Hartree; Frederick Ungar, New York, 1960), which shows that it is still regarded as useful.

Now, hardly a year after the publication of this Volume I, I was in the position to write the promised Volume II, and felt obliged to do so. As Heisenberg had left Göttingen, I did it in collaboration with Jordan. It took us several years; the book appeared in 1930 (published by Julius Springer, Berlin). And it was not a success. For we made a blunder: instigated by a kind of 'local patriotism' we decided to use only matrix methods, rejecting not only wave mechanics but also Dirac's intermediate methods. I think it contains many valuable parts, but it took no notice of the tremendous advance which was going on at that time in atomic physics with the help of wave and operator mechanics. Then Pauli published a destructive review of the book, in which he pointed out our mistakes pitilessly. I think now that he was completely right.

I still cannot understand this error of judgement. I remember dimly that Jordan insisted very strongly on this methodical restriction because he regarded the matrix approach as the deeper and more fundamental one; but that does not release me from my responsibility. The most astonishing thing, however, was that meanwhile, in the years between 1926 and 1930, I had successfully taken up wave mechanics and thus given, for the first time, a rational interpretation to quantum mechanics as a whole, which until then had been a successful and fertile formalism without any reasonable meaning.

It is well known that Schrödinger's work was a development of Prince Louis de Broglie's thesis, which is one of the most celebrated papers in theoretical physics. Prince Louis was the younger brother of a well-known experimental physicist, the Duke Maurice de Broglie, who worked mainly on X-rays. It was Louis de Broglie who first had the idea that with each particle there is associated a wave with a wavelength λ whose reciprocal, the wave number k (number of periods per unit length) is connected with the momentum p of the corresponding particle by the relation $p = hk$ (where h is the quantum constant introduced by Planck). De Broglie was led to this assumption by the relativistic generalization of Planck's equation $E = h\nu$ between the frequency ν of a light wave and the energy E of the associated particle (light quantum of photon). While in the case of light (electromagnetic radiation), the wave concept was generally accepted and Einstein's interpretation of Planck's equation with the help of 'photons', i.e. particles of light, was a new, revolutionary idea, the situation was reversed in the case of electrons, which were regarded as particles until de Broglie's introduction of 'waves of matter'. De Broglie also gave a non-relativistic interpretation of the connection between waves and particles, based on celebrated papers by Sir William R. Hamilton. Moreover, he indicated the reason why integral quantum numbers ought to appear for stationary periodic motions; e.g. for a circle the wavelength must be an integral fraction of the circumference. From these ideas Schrödinger developed his general differential equation for the 'waves of matter'.

A letter from Einstein directed my attention to de Broglie's thesis shortly after its publication, but I was too much involved in our own speculations to study it carefully. One day in 1925 I received a letter from the American physicist C. J. Davisson, who was, I think, at the Bell Telephone Laboratories and had visited us in Göttingen. This letter described experiments about the reflection of electrons from metal surfaces and contained some photographs and curves which showed very remarkable results, strange maxima and minima of reflection in certain directions. I showed these results to James Franck, and I think it was he who first mentioned the possibility that Davisson's maxima represented interference fringes produced by the de Broglie waves of the electrons reflected by the crystal lattice of the metal. I then made a preliminary estimate of the angles of maximum intensity, and as this was favourable to the idea we asked one of Franck's pupils, W. Elsasser (who later changed over to my department and worked on theory), to investigate the matter in detail. His work published in *Die Naturwissenschaften* gave the first confirmation of de Broglie's theory. Later Davisson, with his collaborator L. H. Germer, and quite independently G. P. Thomson, the son of J. J. Thomson, proved the existence of de Broglie's waves of matter by systematic experiments.

Thus the idea of electron waves was familiar to me when Schrödinger's papers on the structure of simple atoms appeared. At that time it was clear that the proper interpretation of quantum mechanics must be of a statistical type. There were several attempts to formulate this, all based on matrix mechanics, by Heisenberg, Bohr, Kramers, and others. I do not remember when and how the idea struck me that collision processes, i.e. aperiodic motions, must provide the clue for the solution; for the relative numbers of the incoming and outgoing particles could be counted and regarded as empirical probability values. I think this must have been already in my mind when I was at M.I.T. and tried, together with Norbert Wiener, to find a formulation of quantum mechanics for aperiodic motions. As soon as I had digested Schrödinger's papers I saw the right way to approach this, guided by a remark of Einstein's about the meaning of intensity of light (i.e. of an electromagnetic wave) in terms of photons: this intensity must represent the number of photons; but the latter was of course to be understood statistically, as the average over a certain photon distribution. Einstein had considered in depth the statistical nature of this distribution, particularly the fluctuations about the mean, which are closely connected with Planck's radiation formula. These investigations were well known to me, and they led immediately to the conjecture that the intensity of the de Broglie wave, i.e. the (absolute) square of Schrödinger's wave function, must be regarded as the probability density, which is the probability of finding a particle in a unit of volume.

This idea had now to be proved, and I was kept busy for a few years doing this. I shall not go into details concerning this work, which I pursued generally alone, but occasionally with the assistance of a collaborator; I shall mention only a few aspects and incidents.

The first line of attack was the theory of atomic collisions. The solution of the Schrödinger equation for this case is quite analogous to that of Maxwell's electromagnetic equations for molecular scattering of light. However, I did not use the rigorous methods well known for this optical problem, but invented an approximation method which was valid only under special conditions (high velocities), but which had the advantage of yielding rather simple and handy formulae. It is now known as the 'Born approximation' but instead of being pleased by the fact that my name appears in this term in almost all the innumerable papers on collision problems of atomic and nuclear physics, I am rather cross about it, for the following reason.

In 1933, about seven years after my first publication on the quantum mechanical treatment of collisions, a book appeared by Mott and Massey, a systematic text-book on this subject, in which, among different methods, my approximation was described and my name attached to it. But there is no mention of the fact—not even in the second edition of 1949—that the whole collision theory and its fundamental concept, the statistical interpretation of the Schrödinger function, was given in my papers. I think that this omission has done me a lot of harm; for, although I am not particularly ambitious, I was quite proud of this discovery, and I think rightly, for eventually I got the Nobel Prize for it, albeit twenty-eight years later.

In this work on collisions, a result which gave me special satisfaction was the derivation of the quantum mechanism of excitation of atoms by electronic collisions, because the laws of these, as divined by Bohr and experimentally confirmed by Franck and Hertz, were in complete contradiction with classical mechanics, while quantum mechanics combined with my statistical interpretation was able to account for them without any difficulty. One other obvious application was the famous formula derived and used by Rutherford for the scattering of α-particles in the electric field of nuclei. It was this formula on which Rutherford's nuclear model of atoms was largely based. This problem was on my programme, but I had to postpone it because of other urgent work, and fatigue. It was solved in a most satisfactory manner by Gregor Wentzel, a pupil of Sommerfeld's.

There were other ways of confirming the statistical interpretation. The simplest method seemed to me the action of an external force with given time dependence on an atomic system. A special feature of such processes, which interested me particularly, was the so-called adiabatic principle, which was formulated, during the Bohr-Sommerfeld period of quantum theory, by Paul Ehrenfest in the following manner. As an atomic system exchanges energy, according to Planck, only in quanta of finite energy $h\nu$, where ν is the frequency, h Planck's constant, a very slowly changing force (with only small frequency components) can produce no 'quantum jump' between stationary states at all, but only a continuous modification of the properties of these states. This principle had been used to determine (or should one rather say to guess?) the quantities which were characteristic of the stationary states (adiabatic invariants). Now the question arose as to how this adiabatic property could be derived from quantum mechanics.

This question was answered in several papers, the first by myself alone, later ones in collaboration with the very gifted Russian physicist V. Fock.

I then returned to the collisions, with the idea, not only to treat general features and derive general formulae, but to discuss a special, practical case in full detail. The simplest one was obviously the collision of a charged particle (electron, α-particle) with a hydrogen atom. At that time I was familiar with Schrödinger's eigenfunctions of the stationary states of the hydrogen atom. I tackled the problem using the approximation method which, since Mott and Massey's book, had carried my name. It led through lengthy calculations to definite formulae for the elastic scattering and for those inelastic collisions, which produce excitations to the lowest excited states. The paper was published in *Göttinger Nachrichten* and has hardly been noticed. The reason lies not only in the periodical itself, which few physicists outside Göttingen read, but also in the fact that I made no numerical calculations and gave no diagrams of the results—I think because of fatigue; one of my pupils, W. Elsasser, later developed my method and completed the results.

I shall now add a few anecdotes to this account of scientific work.

Heisenberg, then already in Leipzig, was at first not very happy about my using Schrödinger's wave mechanics. I got a letter from him in which he reproached me for deserting the matrix mechanics which he, like Jordan, regarded as more fundamental than wave mechanics. But about the same time Dirac independently dealt with the collision problem using a method formally different from mine, but exactly equivalent. Thus Heisenberg was very soon converted to my standpoint.

Another anecdote is concerned with Oppenheimer. When I had finished my paper on the collision of electrons and hydrogen atoms I gave it to Oppenheimer in order for him to check the involved calculations. He brought it back and said: 'I couldn't find any mistake—did you really do this alone?' The astonishment expressed by these words and visible in his face was rather excusable, for I was never good at long calculations and always made silly mistakes. All my pupils knew that, but Robert Oppenheimer was the only one frank and rude enough to say it without joking. I was not offended; it actually increased my esteem for his remarkable personality.

A few years later, my paper was completely superseded by the work of Hans Bethe, a pupil of Sommerfeld, now one of the leading theoretical physicists in America. After reading his paper, I wrote him an appreciative letter. His answer was at least as rude as Oppenheimer's remark. He wrote something like this: It is rather a pity that you did not see the connection between the collision problem of material particles and the well-known scattering of X-rays. It was a letter of the kind an angry teacher would write to a feeble pupil, not that of a young scholar to a much older one. Though it is difficult to suppress anger in such cases, I succeeded after a while. Experiences like this taught me to consider myself not to be a first class physicist like Einstein, Bohr, Heisenberg and Dirac. If I have advanced now (1961) in the opinion of the world into this class, it is solely due to my luck in having been born into a period when fundamental results were lying about waiting to be picked up, to my industry in picking them up and to my reaching a considerable age. To return to Bethe, I met him later and found him to be a most charming young man. My friendly feelings for him grew when he married one of the lovely daughters of my friends Paul and Ella Ewald, mentioned in Chapter XIII in the section describing 'El Bokarebo'. Later I learned to admire not only Bethe's scientific talent but also his political wisdom in the struggle concerning nuclear armaments.

During these first years of quantum mechanics Göttingen remained a centre for theoretical physics, where numerous excellent students from many countries gathered. On the budget of my department there was, as I said before, at first only one assistant. But the generosity of our friend Carl Still, a successful industrialist in Recklinghausen, Westphalia, provided me with the means to afford a private assistant—moreover to improve the library of my little department, and other such things. Later I got a second official assistant, but now I cannot exactly remember who occupied which post. After Heisenberg had left, F. Hund was with me for many years. For shorter periods E. Hückel (mentioned already in the last chapter) and F. Möglich were my assistants. Then for many years, until the beginning of the Nazi period, I had two excellent assistants, W. Heitler, now professor at Zürich University, and L. Nordheim, now in California; they soon became lecturers and relieved me of much of my teaching burden. Later the Belgian Leon Rosenfeld was my private assistant. He managed to combine two philosophies in his mind which are usually considered to be mutually exclusive or even contradictory: he was and is an ardent follower of Niels Bohr's ideas about the fundamental structure of scientific thinking (complementarity) and at the same time a Marxist. But he is such a charming, clever man, that these two aspects of his mind never seem to clash. Now he is the director of the Research Institute for Theoretical Physics of the Scandinavian States in Copenhagen and lives near his idol, Bohr.

I cannot recall all the other students from different countries who worked in my department, often in collaboration with me, some graduating Dr rer. nat. I shall mention only a few who later became well-known scientists.

Prominent among the German students was Maria Goeppert, the daughter of the professor of paediatrics at Göttingen University who had often attended to our children. Maria was a lovely and lively young girl, and when she appeared in my class I was rather astonished. She went through all my courses with great industry and conscientiousness, yet remained at the same time a gay and witty member of Göttingen society, fond of parties, of laughter, dancing and jokes. We became great friends. After she got her doctor's degree with a very good thesis on a problem of quantum mechanics, she married a young American, Joe Mayer, who worked with me on problems of crystal theory. Both had brilliant careers in the U.S.A., always remaining together, first at Johns Hopkins University at Baltimore, then at Chicago University, and now at La Jolla, California.

Another member of my group of research students was Victor Weisskopf, who came from Vienna. He was at first very timid, and several times came near to giving up theoretical physics when he made a blunder in his reasoning. But I encouraged him and succeeded in keeping him on his path. He became one of the leading theorists of America and is now head of the theoretical department of CERN, the gigantic European research institute for nuclear physics near Geneva.

An unusual case was Max Delbrück, son of a famous historian at Berlin University. His thesis, written under my direction, was excellent, and I regarded him as one of my best pupils. But in the oral examination for the Dr Phil., he failed. Pohl asked him some questions of practical physics which he could not answer. When Pohl insisted on this kind of question Delbrück became obstinate and refused to answer at all. So Pohl decided that he would not let him pass, and as the verdict of the examiners had to be unanimous, I could do nothing but protest. Delbrück, well aware of his intellectual superiority, was of a sensitive, delicate constitution of mind. Thus this blow was more than he could stand. I was actually afraid he might commit suicide and did not dare to leave him alone, but took him home. There it was Hedi who succeeded in re-establishing his self-confidence, showing him how silly the dependence on other people's estimation was. In fact, he passed the examination half a year later. However he did not remain a theoretical physicist but changed over to biology. He studied genetics, I think in Dahlem. When Hitler came he left Germany, although he was a pure Aryan, and went to America. Now he is one of the leading geneticists in the United States. When we lived in Edinburgh, he was invited by the University to give a series of lectures; he came and we renewed our friendship. He gave a fascinating account of his work in my department, and in my house we remembered the dramatic beginnings of his career.

Many foreign students came to work with me. To start with the exotics, there were some Indians and Japanese; of the latter several became well-known physicists, such as Sugiura.

Then there were quite a number of Russians. I have already mentioned V. Fock, who worked with me on the 'adiabatic' processes in quantum mechanics. A temporary visitor was J. E. Tamm, who later (1958) received the Nobel Prize, together with P. A. Cerenkov and J. M. Frank, for their work on the so-called Cerenkov Effect. I want to say a little more about George Rumer, who was a particularly attractive youngster, frank, a little boisterous and with a loose tongue. He was mainly interested in relativity, and as I did not work in this field, we wrote only one joint paper, which I now regard as rather fantastic and of little value. Later we lost sight of him, and when I went to Russia after the second world war (1945) with a British delegation to attend the 220th anniversary of the Soviet Academy, my enquiries after him were unsuccessful. I had the impression that he was a victim of Stalin's purges. This was confirmed when he wrote me a nice letter for my 75th birthday (1957). He had been sent to one of those terrible prison camps in arctic Siberia, but he survived, and when Stalin died he was suddenly called back and appointed head of a big physical institute in Novosibirsk where the Soviets are building up an enormous scientific centre. There he is living happily, married to a woman who shared his years of suffering, he has become a loyal Soviet citizen and has forgotten all the cruelties of the system under which he suffered himself. He now accuses the West of similar crimes against humanity and defends Eastern ways as being on a higher level of civilization. Well, such is human nature.

After the Russians, I wish to mention only one other group of foreigners, the Hungarians, or, to be precise, the Hungarian Jews. Concerning this distinction, I want first to tell a little story. In the twenties, Hungary had a very reactionary government, under the Regent Horty, if I remember rightly. One day during this period (about 1930), I was invited to a luncheon party by the University *Kurator* (the financial administrator) privy councillor Valentiner, in honour of the Hungarian Minister of Education, Count Konigswarth (or something like that), who was travelling in Germany to inspect the educational institutions. I was sitting opposite the honoured guest and talking to my neighbours, one of whom was James Franck, when suddenly the count addressed us across the table and asked us what we thought about the Hungarian mathematicians and physicists. I replied with a hymn of praise for my Hungarian colleagues, mentioning first my old friends Haar and Kármán (about whom I have written earlier in these recollections), then Polya in Zürich and others I cannot now remember, and finally the young generation who were at present in Gottingen: John von Neumann, Eugen Wigner and Edward Teller. At this point I got a fearful kick on the shin from Franck, whereupon I stopped and let him continue the discussion. I did not understand what he meant by this violent interruption until he explained it to me after lunch. All I had mentioned were Jews, and therefore, in the eyes of a representative of that anti-semitic government, not Hungarians at all. They had all left Hungary because there they had no chance of advancement. Most of them finally emigrated to America and were very successful. Kármán became the foremost expert in hydro- and aerodynamics; von Neumann was considered the leading mathematician in the United States (but he died young); Wigner and Teller are among the best theoretical physicists. All of them did important war work. Teller is known as the 'father of the hydrogen bomb' and an ardent advocate of nuclear armament of the United States. I once had a talk with Wigner, who shared these convictions (though to a lesser degree) and learned that one strong motion was their resentment of the suppression of the Hungarian people by the Russians, in particular after the revolt of 1956. In these questions our ways have parted. But that belongs to a later period.

I cannot claim to have been their teacher, for when they came to Göttingen they were fully fledged scientists. Wigner published papers with others of my collaborators. Teller, who was an expert in the theory of the Raman effect, helped me in writing the relevant chapter of my optics book. That was very kind of him, but the collaboration was not easy because his way of thinking was rather different from mine, and he insisted on his formulations with the same stubbornness as he now insists on his political concepts.

The list of my collaborators given here is quite impressive and shows that, after Heisenberg's departure, Göttingen was still a centre of theoretical physics. In the well-known book by Robert Jungk *Heller als tausend Sonnen* (*Brighter than a Thousand Suns*), there is a chapter (No 2 'The Good Years') containing a description of Göttingen during this time. Considering that he had not been there but, many years later, gathered information from interviewing survivors, it is a brilliant piece of work; certainly more entertaining than these recollections. However, it makes Göttingen appear too much as the centre for theoretical physics of the world. This would be wrong if only for the one reason that nuclear physics, which soon became so overwhelmingly important, was hardly discussed in Göttingen, except by a few pupils of Franck, Atkinson and Houtermans (to whom Robert Jungk gives a page or two). In the narrower field of quantum theory, we could perhaps claim to have been on top for the short time in 1925–26 when we were in possession of matrix mechanics, and nobody else was. But before and after that interval there were equally important places of research, in Copenhagen (Bohr), Paris (de Broglie, Langevin), Munich (Sommerfeld), Zurich (Schrödinger), Cambridge (Dirac). On the other hand Göttingen represented the most successful school, with which only Copenhagen competed. All these other places continued to flourish when the glamour of Göttingen suddenly vanished and the dark age of Hitler descended on Germany.

237

II. Approach of the Nazis

Physics was not the only branch of culture which flourished in Germany during the twenties. The theatre, for instance, particularly in Berlin, had rarely been better. Thus the inside pages of the newspapers, where the cultural life was reported, gave an impression of a high standard of civilization, while the front pages, with their leaders and political news, became more and more depressing. There were the endless negotiations about war reparations, leading to the French occupation of the industrial Ruhr district, and many other crises in foreign politics. But much more disquieting were the internal troubles. We hardly noticed the rise of Hitler and of his national-socialist movement; he seemed to us simply ridiculous, and we refused to believe that such a mean, low scoundrel could be taken seriously by the 'nation of poets and thinkers', as the Germans used to call themselves.

But soon horrible crimes began to be committed; statesmen were murdered, not only socialists like Rosa Luxemburg and Liebknecht, but bourgeois politicians like Erzberger, Rathenau and others, who had tried to resist the Nazis. The climax came with the financial crash in America. After that the situation in Germany deteriorated from day to day; no stable government was possible and the *Reichstag* was dissolved and re-elected several times. After 1930 terror reigned in the streets of the big cities, tussles between the Nazis and the Communists were frequent and the police were powerless or unwilling to stop them. However, very little of all this reached the peaceful town of Göttingen, and most professors did not seem to show interest in it. Yet this was a delusion, for after Hitler's rise to power it turned out that a considerable number of the university staff were supporters of the Nazi party.

As for myself, I had been worried about the political happenings all the time. I attended political meetings and, on one occasion, stepped in when a crowd of citizens and students became turbulent. This was when the well-known journalist Georg Bernhard, chief political leader writer of the old liberal newspaper *Vossische Zeitung*, had been invited by a left-wing student group to speak in the big Stadtpark hall. The meeting got out of hand when Bernhard attacked Hitler and the Nazis. The students made such a noise that he could not continue. Then I stepped on the platform and told the students that as one of their professors I was ashamed of their behaviour, and I asked them to listen to Bernhard even though he was expressing opinions they disliked. The noise actually stopped, but when Bernhard tried to speak again, the uproar broke out with greater fury, accompanied by tear gas. I was saved from violence only by a group of liberal students who helped Bernhard and me out of the hall.

In the summer of 1928 I was invited by the Russian physicist Joffe to take part in a congress of physicists. I travelled with a group of Western physicists, among them Pohl from Göttingen, and Dirac from Cambridge, by boat from Stettin to Leningrad. There the conference had already started, and we attended only the last of the Leningrad meetings. Now this was planned as a 'Travelling Congress', to visit several places right across Russia. We travelled by train to Moscow. In the meetings there, two things impressed me: the burning interest of the young Russian physicists and the incredible linguistic feat of Joffe, who translated not only every foreign talk into Russian, but also every Russian contribution into English, French or German. I remember very little of the city, apart from the Kremlin and even that is now mixed up with my second journey to Russia in 1945.

From Moscow we were taken to Nishni-Novgorod, where there was another meeting. The next day we embarked on a Volga steamer and began the long journey downstream. We stopped at many places, where meetings were held, often attended by large audiences. As our steamer was nearly always late, these people had to wait for hours, but they never became impatient and always greeted us with enthusiasm. In Kazan, for instance, we arrived, instead of at six o'clock, at about 10 p.m., but nevertheless were taken to the city hall where a large crowd received us with no sign of fatigue; we then returned to the steamer in the early hours of the morning.

This was very fatiguing. Through Joffe's solicitude I got a tiny cabin for myself. But sleeping in it was hardly possible because it was near the engine, and also because of the noise on deck. The same students, boys and girls, who showed such an enthusiasm for science in the meetings, were laughing, singing, dancing, flirting all the night, and even the Russian professors took part in these little festivities.

The long journey through the Volga steppes was very impressive and often very beautiful. Of the cities I remember only Kazan, with its picturesque churches and its 'Kremlin'. We were told that only a year earlier these streets had been strewn with dead people who had starved to death in the terrible famine (which Nansen had tried to alleviate).

Among the foreign guests were a number of old friends: Gilbert Lewis from the U.S.A., Robert Pohl and Paul Dirac, Léon Brillouin, Alfred Landé, Charles G. Darwin and his wife Katherine. These two were perhaps the most notorious among us, not only because the name of Darwin is one of the greatest in the materialistic Soviet world, but also because they insisted on having a bath every morning, and as there was no bathroom for passengers on the steamer they went into the shower-bath of the engine stokers!

In the lower reaches of the Volga we were told that one could get sterlet. So Landé and I ordered one for lunch. Our meals were rather perfunctory and slowly served, as the kitchen was much too small and we had only two or three untrained waiters for the crowd of more than a hundred. I had to wait till 3 or 4 o'clock till my sterlet came: an enormous fish, not a meal for two but for a whole company. So we invited Joffe and other friends, and we had a jolly party.

When we approached Saratov I heard that some people, Russians and others, were to leave the boat and take a train back to Moscow. As I was very tired, I joined them and returned home after a short stop in Moscow. Later I heard that the other West Europeans and Americans had had an adventurous return journey. From Astrakhan they were taken through the Caucasus to Tiflis and then to Batum on the Black Sea. But there they were left to themselves, nothing was provided for the return journey. Some went by rail through Southern Russia, others by boat via Istanbul. But all got safely home.

In the winter of the same year (1928) I fell ill and was off work for almost a year. It began with fatigue and insomnia. One day I had arranged to go skiing in the Harz mountains with a group of young people. I arrived at the station, but when I tried to enter the train I simply could not get up the steps and had to turn back home. There were several causes for this breakdown: the permanent political upset, intensified by the growing awareness of anti-semitism as the driving force behind the Nazis, their 'blood and soil' cry, but mainly the attempt to keep up with a group of keen and extremely gifted young men.

After this breakdown I went to a sanatorium on Lake Constance. This winter of 1928/29 was one of the hardest for a century; the whole lake was frozen over, and one could walk across it from Germany to Switzerland, therefore it was difficult to have an open-air cure. The patients in the Hydro were mostly middle-aged people, businessmen, lawyers, civil servants. They talked nothing but politics, and all of them were pro-Nazi. I had had no idea how this fanatical nationalism, this militarism and anti-semitism, had taken hold of the ordinary German citizen. They believed all the lies spread by Hitler: that the loss of the war was due only to treason and that the military leaders were not to blame for it; that socialism and liberalism ought to be suppressed; that Germany must fight again and enlarge her territory, and so on. When I tried once or twice to oppose these views I was treated as an outcast.

From Constance I was sent to the Black Forest to convalesce. I went to the small town of Königsfeld, a pretty place, with graceful eighteenth-century houses built according to a well conceived plan by Huguenot refugees.

There I had an experience which was a compensation for the depressing Nazi talk in Constance. Returning one evening from skiing on the hills, I passed the main church of the town and heard organ music. I went in and listened in the empty nave to a Bach fugue, played in so masterly a manner that I forgot the terrible cold. I was astonished that a first class organist should live in this little place and mentioned it in my next letter to Hedi. She answered that I must have heard Albert Schweitzer playing, for she had read in a newspaper that he was staying in Königsfeld. Then I went to the church several times at about the same hour, until one day there was the organ music again. I found the steps to the organ loft, climbed up and saw, sitting in front of the organ, beside a small iron stove, the great man whose face and appearance were so familiar to me from pictures. He soon noticed me and said he was pleased to have a one-man audience. He knew me by name and received me with great friendliness. After some more Bach, I accompanied him home and was invited to go for walks with him. Our conversations during these ranged over a wide field of human, religious, political and artistic subjects. This was a consolation and refreshment for my mind after the affronts suffered in Constance. It was in memory of this meeting that, after the rise of Hitler to power and the first atrocities committed by his government on Jews, liberals, socialists and communists, I wrote a letter to Lambaréné, asking Schweitzer to return to Europe and to make an attempt to arouse the civilized world against this barbarity. I remember saying that I believed that Nansen, were he still alive, would have taken up this cause, and that I regarded Schweitzer as his successor. But his answer was disappointing. He regarded the situation in Germany as hopeless, and felt unable to desert his African patients.

After my return to Göttingen in spring 1929, I slowly resumed my official duties. But I did not feel strong enough to do research work again and decided instead to write down some of my lecture courses in order to publish them later. I thought of starting with thermodynamics because my approach to this subject differed from the ordinary one; I used not the classical method but Carathéodory's, in the origins of which I had taken some active part. But as this demanded great concentration, I gave it up and turned to my optics lectures of which I had not only good notes, but even some parts worked out in detail.

This work, however, proved no less strenuous than original research; so I had underestimated my strength. I found much in the current presentations unsatisfactory or incomplete and began to work out new methods and results. Yet after having taken up my ordinary duties, lecturing, examining, supervising theses etc., there was little time left for the optics book. Therefore I began my day a little earlier and worked on the book every morning before my lecture. I dictated the text to two students, Lieb and Weppner, and they brought me a readable manuscript the next day; they also carried out and checked calculations and prepared the numerous figures. The book grew beyond the volume of my actual lectures and became a fairly complete text-book of the electromagnetic theory of light, as far as this can be treated by classical methods (i.e. without quantum effects). It appeared in 1933, just when Hitler came to power, and was of course classified as 'Jewish Physics' and destroyed wholesale. Therefore the financial results were practically nil. I got a small sum as the first instalment of royalties. Whether the book was sold in Germany in considerable numbers I do not know; I think that apart from some scientific libraries there were very few purchasers. I had only one copy, which I kept with the larger part of my library on the shelves of my Edinburgh department for the use of my students.

After the second world war, the Principal of Edinburgh University, the physicist Sir Edward Appleton, told me that an American photo-reprint of my book was available in the U.S.A. and was much used in radar research on the propagation of long-wavelength electromagnetic waves. Shortly afterwards I discovered my book with a somewhat different jacket, in the window of Thin's bookshop, near my department. I bought a copy for the department library and wrote to the American publishers, Edwards Brothers, Ann Arbor. They answered that the book had been confiscated as 'enemy' property (German) and that they had acquired the publication rights of it (as well as of many others) from the Custodian of Alien Property in Washington. So I wrote to this office to say that when the war broke out (1939) I was a British subject, and asked for compensation and the return of the copyright to me. They sent me an enormous bunch of application forms in triplicate which I had to fill in.

Nothing happened. After a year I wrote to them again enquiring about my application. The answer was that my case would be considered in due course, but that there were thousands of other applications of greater importance.

I waited several years, writing to the Custodian about once a year; without result.

One day we found in a newspaper a report that the eighty-year-old Finnish composer Sibelius had attended a concert in Washington where one of his symphonies was performed. President Truman, who was present, enjoyed the symphony so much that he asked the composer to his box. Questioned whether he like America, Sibelius answered, yes, very much but he resented it that no fees and royalties were paid to him because all his works were regarded as alien property and their copyright had been sold to an American firm. Truman became very angry and approached the Custodian to rectify this, which indeed was quickly done.

Thereupon I wrote a letter to the *Manchester Guardian*, referring to this story and claiming that numerous writers, artists and scientists were in the same position as Sibelius and felt wronged. In some cases, as in mine, though the publishing firm was enemy alien, the author no longer was, and some might be in real distress.

A number of letters were written to me, after mine had been published, by others who were in the same position as I, and also by a lawyer. At last the Custodian wrote to me that my case would be treated as urgent.

Still some years went by before I got compensation. Meanwhile some strange and amusing things happened which I shall report briefly.

One day a 'Scientific Attaché' of the American embassy in London appeared in my department. He told me that there was a project under the auspices of the American Navy to publish a modern book on optics written by a group of experts, because in parts my book was becoming obsolete. But they had heard that I was going to do this myself. He wanted to know whether this was true and if so how far I had proceeded. In fact, I had started on a new English version of the optics book. I was urged to do this by many people, but particularly by Sir Edward Appleton. I had succeeded in getting a collaborator, Dr E. Wolf, who was a specialist in modern optics. We had already collected much material, worked out the general plan and begun writing.

We showed all this to the American Attaché; he declared himself satisfied and departed with the promise to persuade his superiors in Washington to intervene on my behalf in order to accelerate the restoration of my rights in the old book.

The result of this interference was most surprising; the Custodian wrote to me he had heard that I was preparing a new optics book. He directed my attention to the fact that in my contract with Springer there was the (normal) clause that in the case of my writing a new book on the same subject I had to get the permission of the Springer firm. I had therefore to ask him, the Custodian of Alien Property, as the 'legal successor' of Springer, for permission to publish the new book as it would compete with my old optics. This time I got angry and answered that I could in no way agree to this demand, as the confiscation of the book was illegal; I should be pleased if he would bring the case before a British court. This speeded up the overdue final settlement.

The new optics book appeared in 1956 and soon became the source of new worries. Wolf and I are at present (November 1961) involved in an arbitration suit with the publishers about the royalties for the American sales and other things. Writing text-books on optics seems to bring nothing but trouble.

To return now to Göttingen, early in the twenties, I took up again my connections with England. It began with my attending a meeting of the British Association for the Advancement of Science in London. I stayed in the house of Sir Lawrence Jones, in the best part of South Kensington; his wife was a sister of the eminent physicist Schuster. It was the first time I had lived in that special atmosphere of wealthy English people, with a butler, footmen and maids, 'early tea' in bed, and so on. They were kind and deeply devoted to science, of which they understood nothing. I was not the only guest; I remember that among them was Siegbahn from Sweden, so I had interesting company.

A short time later a young English physicist, Lennard-Jones from Cambridge, came to Göttingen with his family, to study crystal theory with me. He continued and improved some investigations of my group concerning the forces which keep the atoms in ionic crystals together. We became good friends. His papers made him one of the central figures in theoretical work on crystal dynamics in England, and he was soon appointed to a professorship at Bristol University.

In 1927 this University offered me an honorary degree, the first honour of this kind given to me. Lennard-Jones wrote in his accompanying letter there would be a surprise for me when I came to Bristol. And indeed, there was: the Chancellor of the University who had to 'cap' me was Lord Haldane, former Minister of War, whom I knew well.

This had come about in the following way. In the 1920s Hedi used to visit an old Fräulein Schlote who was an invalid and confined to a ground-floor room in her little house in Kirchweg. In her younger years she had let rooms to students, preferably from Britain, and in her later years when she was quite lame, she gave English lessons. That is how Hedi got in touch with her. One day she found Fräulein Schlote in great excitement; an old English friend, one of her former boarders whose connection with her had only been stopped by the first world war, was coming to visit her—it was Lord Haldane. Soon he arrived and he heard from Fräulein Schlote that I was living in Göttingen. He had read my book on Einstein's theory of relativity, and this had impressed him so much that he himself wrote a book with the title *The Reign of Relativity*. It had actually nothing to do with the physical theory but was a philosophical discourse on the relativity of ideas and values. Now he wished to make my acquaintance and called on me in Planckstrasse. There was at once a friendly mutual understanding. We gave a little dinner party in his honour to which we invited a few of my distinguished colleagues, among them Hilbert, Runge and Franck; it was a jolly evening. From this time I had sporadic correspondence with him. For instance, one afternoon just before we were leaving for a holiday, I got a letter from him with a manuscript by his brother, the famous physiologist in Oxford, in which the physicists were attacked and a proof was given that the second law of thermodynamics was wrong. Our Haldane had doubts about this result and asked me whether his brother was right or wrong. As I did not want to spoil my holiday I sat at my desk till late in the night and of course found a mistake.

Some time later I came to Oxford and called on the physiologist Haldane. He was kind and charming, showed me many interesting experiments and avoided the subject of thermodynamics with considerable skill!

After my arrival at Bristol University I met the other graduands, all men of great fame; Sir Ernest Rutherford, Sir William Bragg, Sir Arthur Eddington, the spectroscopist Fowler and Paul Langevin from Paris. That they were all physicists was due to the fact that the celebration (on 20 October 1927) was connected with the opening of a new physics laboratory, given by the cigarette firm H. H. Wills to the university. We were photographed together with the head of the physics department, Professor Tyndall. I am quite proud of this picture where I, a relatively young fellow, was standing amidst some of the greatest men of my science, and it was always—and is still—hanging on the wall of my study. I must here mention an anecdote connected with this photo.

When we were in Edinburgh, many years later, we had a party in honour of some foreign guest—I have forgotten who it was, I think an American. He noticed this photo and was delighted to see the group of these great scientists, Rutherford, Bragg, Eddington, Langevin, together with some others. He turned to me and said: 'Isn't it a pity that this unknown youth is standing between the giants?' and pointed to my figure in the picture. I was not quick-minded enough to suppress: 'But that's me', producing thus an awkward situation.

Returning now to 1927 and Bristol, we honorary graduands were introduced to the Chancellor, Lord Haldane. Under normal conditions I would hardly have had an opportunity to exchange more than a few words with him. But Haldane had a bad cold, and so did I. So he arranged that we both were exempted from the ceremonial banquet and were served the meal in a private room. There we had a fascinating talk. He spoke about his negotiations in Berlin in 1914 about a restriction of the naval armaments race; these broke down because of Admiral Tirpitz's insistence on his programme of building war ships. Then he spoke of his resignation from the war office on the outbreak of the war in desperation, since he loved Germany. All this is now well known, but then it was new to me.

The situation today between the atomic powers in East and West is very similar, but the statesmen seem to have learned nothing from past experience.

In the following years I visited England several times, renewed old friendships and founded new ones. I remember a day's excursion by car with Lawrence Bragg and Charles Galton Darwin and their wives to the Downs near the South Coast which brought us close together; and another one with Lindemann and some others, among them the German physicist Willy Wien, to see some of the attractive old castles north of Oxford.

All these connections were later of the greatest importance to me when I was compelled to leave Germany and went to England.

The opening of the H. H. Wills Physics Laboratory, Bristol, 21 October 1927. A. M. Tyndall (rear) with honorary graduates (left to right) W. H. Bragg, A. S. Eddington, E. Rutherford, A. Fowler, M. Born and P. Langevin

III. *Arrival of the Nazis*

During the years 1930–33 Hitler's struggle for power was going on with increasing intensity. I do not remember how many times the *Reichstag* was dissolved and a new one elected. When the number of Nazis in parliament increased, we were dejected, when it fell we were hopeful. But though there was in my mind a permanent feeling of impending doom, our life went on quite normally and gaily. There are very few events which I remember. Our house was quieter than it had been, since both our daughters were at the boarding school at Salem (near Lake Constance) founded by Kurt Hahn, my friend from student days.

In order to counteract the increasing specialization and separation of departments, the university had introduced a system of general lectures for students of all faculties, given on Saturdays in rotation by members of different departments. During this period it was my turn to give these lectures. It was my first opportunity to speak to a large audience on my subject from a general philosophical point of view, and I enjoyed it very much. I regret having lost my lecture notes as I should like to see how I presented the philosophy of science at that time. Another course of lectures which I enjoyed, though it entailed a fatiguing journey to Berlin, was given at the Technische Hochschule, Charlottenburg, in March and April 1932, to an enormous audience which remained faithful even though I once had to interrupt the series because of illness, after I caught a cold in the train. On arrival in Berlin one half of my face was paralysed; I had no pain but could hardly speak, but with the help of some quickly installed loudspeakers I gave my lecture. The night journey home, where I arrived about 2 a.m., was not pleasant, nor were the next weeks. It turned out to be a paralysis of the facial nerve due to influenza. However, I recovered soon, though my face lost its symmetry for good. There was another strange consequence, which has lasted to this day: whenever I eat something with a strong taste, and in particular a pleasant taste, tears come out of my right eye. So my wife can always see whether I am enjoying a dish or not. The facial nerve seems to have become coupled with the nerve serving the tear gland.

After a few weeks I was able to take up and finish the lectures. They appeared as a book (Springer-Verlag) in 1933 under the title *Moderne Physik* and were, just like my optics book, financially a failure because of the Nazis. When I came to England as a refugee and had to live with my family in Cambridge on a relatively small income, I found a British publishing firm, Blackie and Son, Ltd in Glasgow, who were keen on having the book. The head of the firm, Mr Bisacre, and his family became very dear friends of ours with whom we spent several weekends in lovely Helensburgh, near Glasgow. Mr Bisacre arranged for the translation to be made by one of his trusted employees, since my English was not good enough at that time. This book, under the title *Atomic Physics*, was a great success in the Anglo-Saxon countries. The seventh edition appeared in January 1962; the original plan, though enlarged, was essentially preserved throughout.

But this is anticipating the future. During these years from 1930 until our emigration, we travelled quite a lot, partly in connection with lectures I had to give, partly for holidays. In March 1930 I gave a series of lectures in Paris. Then we went to Avignon but were driven away by the Mistral, a cold wind, and proceeded, via Marseilles, to Corsica where we had a most fascinating time, enjoying the lovely wild country and the marvellous wild flowers. On the way back, via Leghorn, we stopped at Florence. I had to give another set of lectures in October 1930 at Wittenberg, where we were the guests of the director of the Fixed Nitrogen Works, Herr Tauss, who showed us his very interesting factory. He and his wife also became very close friends, but they perished under the Nazis.

In the summer of 1930, we went with the children to Ehrwald, where my wife had gone often as a child with her parents. In 1931 we had our summer holiday in the Dolomites at Selva in Val Gardena. We took rooms in the house of a peasant, Perathoner, who was also a woodcarver, and we were enchanted by the country, the green valleys between stupendous mountains. This Perathoner house was soon to be our first refuge after our expulsion from Germany.

In the following year, I went with my friend Ladenburg to Schloss Elmau in Upper Bavaria, where a pastor Müller had installed a holiday resort of a peculiar kind. There were no waiters or chambermaids, but all the work was done by pretty young girls of good families. One could not choose the place where one sat at meals, but the management arranged the seating and changed it frequently. In the evenings there was folk dancing in which all those pretty girls took part, or philosophical and theological talks by Müller, or very good music. While we were there, a company of young musicians from Munich gave a very decent performance of Mozart's *Figaro*. Hedi came to Elmau too for a few days, but she did not like the atmosphere of the place, and we went on together to the South Tyrol.

I was elected Dean of the Faculty of Science for the academic year 1932. This was the only period in my long university life that I was in charge of an administrative post, and a very hard and difficult job it was. The economic slump and unemployment had reached their peak; Hitler's power was rapidly growing, and so was anti-semitism, and no steady administration was possible. At first things went smoothly, as there was only routine business in the faculty, with which I could easily cope, though my managerial gifts are poor. But then came the Emergency Decrees of Chancellor Brüning who tried to stem the economic depression and inflation by rigorously reducing government expenditure. One day all universities (which in Germany are State institutions) received the order to dismiss all those members of the staff who were not officials employed on a permanent basis. In this category were many young assistants, several in each of the bigger scientific laboratories. When this decree came before the faculty, the members were stunned. Many declared that they could not continue their work under these circumstances. As the dean responsible to the administration of the university, I had to insist that every head of a department should supply a list of all who came under the category hit by the decree. There followed a horrible wrangling between the various departments, each of them declaring that its work was extremely important and could not be continued if the number of collaborators was reduced in this way. In the end there was such turmoil in the usually well balanced faculty meeting that I had to adjourn it.

I then calculated the total sum of money saved by the dismissal of these young people and found it amazingly small. So I convened a conference of my friends Franck, Courant, Windaus (the chemist) and a few others and we decided that we would try to persuade the faculty to keep these young men, some of whom were married, as private assistants and to pay them out of a fund, collected from the permanent members of the staff, each contributing a certain percentage of his salary. It turned out that this percentage was not very high, between 5 and 10 per cent (I have forgotten the exact figure). I was rather shocked by the difference between the income of a young assistant and a full professor and I expected that the directors of the departments who had so vehemently declared the indispensability of these men would be very willing to make a personal sacrifice in order to keep them.

However, that was not the case. The faculty regarded the proposal as quite irregular, against all tradition. But our little group who had hatched the plan pleaded for it so energetically that it was eventually accepted, of course not as an official faculty decision, but as a private agreement of its members. It turned out that our plan was opposed mainly by those who, a few months later, came out openly as members of the Nazi party. Yet in the end, even some of these voted for our proposal when they saw that the majority was on our side; we had decided that all departments would get their share of the fund collected, whether they voted for it or not, and this acted as a moral compulsion to vote in favour.

When I informed the *Kurator,* privy councillor Valentiner, a distinguished civil servant and a very kind man, of the faculty's decision, he said with tears in his eyes: 'If every public body showed the same spirit of sacrifice as your faculty has done, there would be no need to worry about the future of our country.'

Another remarkable event during my service as Dean of the Faculty of Science was the graduation of Lord Rutherford of Nelson, the great physicist who, born in New Zealand, was now professor at Cambridge in England. I knew him from several previous meetings, and we had come into closer contact during the Congress in Como, in 1927, in honour of Volta. One morning I had found the lecture dull and the air in the big hall intolerable. So I went out and collided at the outer door with Rutherford and Aston, the inventor of the mass spectrograph and discoverer of numerous isotopes. Rutherford said: 'We cannot stand this any longer—nor can you, I assume. Let's go for a drive around the lake.' So we took a car and spent a lovely day, first driving along the eastern shore up to the Swiss frontier, then crossing the lake on a ferry and, after a nice lunch at a hotel by the lake, returning along the western shore. A day spent in this way brings people together. We spoke about everything, physics, politics, and very intensely about the language problem in science: the Tower of Babel situation which had already developed and was to become worse. Rutherford rejected all artificial languages and declared that English was already in international use and ought to become more so.

Now in 1932 Rutherford came to Göttingen to receive the honorary doctor's degree of our faculty, and I, the Dean, had to confer it on him. He was accompanied by his son-in-law, Professor Fowler of Cambridge, who was the leading theoretical physicist in Great Britain at that time. The graduation ceremony took place in the big lecture-room of the Physics Institute. I had to address the guest and confer the degree on him. I spoke first in German and repeated it in English. After the graduation Rutherford gave a splendid lecture on his latest research into nuclear processes. Pohl had secretly arranged microphones and recorded the whole ceremony and the lecture on gramophone records, and he gave me the part which contained my address and Rutherford's reply. About two years later, when we were living in Cambridge, one of Rutherford's pupils and collaborators, Professor Oliphant, told me that he had heard of the existence of these records and wished to collect them in order to make them available to all interested people. I played my record to him, and he was enchanted. He then wrote to Pohl to obtain the names of the other men who possessed such records. They were then edited by 'His Master's Voice' and published. I got the whole set and have it still. It is a great pleasure to hear Rutherford's voice just as it was in life. My own voice seems to me strange and hardly recognizable, and my English is very bad. Still I am proud to be on the same record as Rutherford, who was certainly the greatest experimental physicist of our period, and in addition a wonderful man and character.

The University of Frankfurt, where I had been professor for two years (1919–20), invited me once to give a lecture at the physics department. I stayed at the industrialist Arthur von Weinberg's palatial house Buchenrode, in the suburb of Sachsenhausen, which I described in Chapter XVII. I did not feel very comfortable in these luxurious surroundings, though the Weinbergs were kind and hospitable. I walked through the narrow streets of old Frankfurt, and I remember that I was haunted by the feeling I would never see them again. Indeed they were almost completely destroyed during the war.

The same sensation of final farewell, still more vivid and distinct, came upon me in Nuremberg. I was invited by the directors of a branch of the Siemens–Schuckert Electrical Industry to give a lecture on modern atomic theory to their engineers. After the lecture, the firm gave me a dinner at a hotel to which a number of local people of distinction were invited. I was placed between the head of the branch firm and a general, the commanding officer of the garrison. The talk with these men soon turned to politics, as always in those days, and I heard opinions like those in the sanatorium in Constance, which depressed me deeply. Next morning when I walked through the lovely mediaeval town and visited the ancient churches and the museum, I could hardly enjoy the beauty of the architecture, the paintings, sculpture, and other testimonies of Germany's cultural heritage, for I had the distinct feeling that all this was condemned to destruction. Indeed, Nuremberg, the centre of Nazism, where the big annual rallies of the party were held, was efficiently destroyed by British and American bombs during the war.

I returned from Nuremberg the day before my fiftieth birthday (10 December 1932) and in our flat found Heisenberg, who had come over from Leipzig to celebrate the day with us. It was a very pleasant day, with innumerable visits from friends, colleagues, collaborators, and pupils. The highlight was the performance of a funny play by my closer collaborators, amongst whom I remember Edward Teller as particularly grotesque—who would have thought then that he would become the inventor of the American hydrogen bomb, the most terrible instrument of mass destruction? They gave me also a *Festschrift*, a fake number of the *Göttinger Nachrichten*, which contains that play and moreover some very amusing contributions by friends and colleagues, such as Brillouin, Ehrenfest and Fock. I still have a copy of this number, and the periodical *Physikalische Blätter* has (1961) reprinted a few of its items as historical reminiscences.

Meanwhile the situation in Germany deteriorated, the Nazis gained power, and even in Göttingen we began to feel the tension. To give an example, once the telephone rang in the middle of the night, and a violent shout came from the receiver: '*Juden raus, Juda verrecke*' (out with the Jews, may Judah perish), followed by the singing of the Nazi 'hymn', the 'Horst Wessel Lied'.

From now on I can give more detail of the events and their exact date, because Hedi began to write a little diary in which she jotted down the daily happenings in a few words; she has continued to do this up to this day.

Hedi spent the month of January 1933 in Braunlage in the Harz mountains, in order to improve her health and, I visited her there at weekends. During this time some most distressing news reached us: Paul Ehrenfest had committed suicide after having killed one of his children, who was mentally ill and lived in a mental hospital in Germany. Obviously Ehrenfest was desperate because he saw no hope for Germany and Europe. This tragedy was a horrible shock for us.

On 30 January 1933, Hitler became Chancellor, and on the following day there was a torchlight procession of the Nazis in Göttingen. Then one evil event followed the other. On 26 February there was the fire in the *Reichstag* building. On the next day the right of free expression of opinion, guaranteed in the Weimar constitution, was cancelled by decree. On 3 March there was a new election for the *Reichstag* which brought Hitler no majority, but with the help of the German Nationalist party he was able to govern. We were at that time back in Braunlage, but the worries about the future of the country and ourselves frustrated our hopes of relaxation. On 5 March the newly elected *Reichstag* was opened at Potsdam and gave Hitler all the powers he demanded for ruling by decree. The socialist members were not admitted, and all the others, even men regarded before as genuine liberals, voted for Hitler. It was a depressing spectacle. A few days after our return to Göttingen, a general boycott against the Jews was declared.

We had innumerable discussions, not only with our friends, Franck, Courant, Weyl and others, but also with Herr Lange, the editor of the liberal newspaper of Göttingen. But nobody could give any hope or advise what to do. We began to discuss emigration, and Hedi decided to go to Switzerland, to see how the situation was regarded there and to consult our friends, such as the Edgar Meyers and the Scherrers. She returned on 16 April and reported that all of them advised us to leave Germany in time, before it was too late; they promised us their help, even if it came to illegally crossing the border, should all other ways be barred.

A day after her return, on 17 April, Franck wrote to the University administration informing them of his resignation. As he had served at the front during the war of 1914–18 he was exempted from the racial measures against civil servants. Franck declared that he could not make use of such a privilege. A few days later the conservative newspaper *Göttinger Nachrichten* published a declaration by forty professors denouncing Franck's resignation as sabotage of the ideal aims of the national-socialist government. The following day, 25 April 1933, the newspaper published a list of civil servants dismissed, amongst them some professors and lecturers of the University. On this list were my name and those of Courant and other men of Jewish origin.

Though we had expected this, it hit us hard. All I had built up in Göttingen, during twelve years' hard work, was shattered. It seemed to me like the end of the world.

I went for a walk in the woods, in despair, brooding on how to save my family. When I came home nothing seemed to have changed. Hedi kept her head and showed no sign of desperation. Then visitors began to drop in to express their sympathy. Most of them showed sincere regret. But there were others who got on our nerves. I remember especially the zoologist Kühn, who tried to console me with the consideration that we were in a crisis like war; just as one man was killed in battle, while the other survived, so it happened that I belonged to the casualties. That idea might have been a comfort to him, but not to me, and it made me irrationally angry—though I did not show it. But I remember that I tried to avoid receiving these condolence visits by going for long walks.

Another incident which made me furious was this. Dr Otto Heckmann, Professor Kienle's assistant at the astronomical observatory, asked me to give him my lecture notes as he had been given the task of continuing my course. I

251

regarded this as terribly tactless and refused. Twenty-five years later I met Heckmann again who was then head of the big observatory at Hamburg-Bergedorf. He apologized for his behaviour and explained it by the pressure exerted on him by the new Nazi administration. Well, after so many years, I considered it best to forget these minor weaknesses shown by my colleagues during Hitler's régime.

The new head of the University, the Rector, was Neumann, the professor of German studies. We knew him quite well, had been guests in his house, and vice versa, and he seemed to us a harmless, friendly, unpolitical fellow. But that was a deception. He had been a Nazi all the time and used his harmless exterior to spy on his colleagues and prepare 'proscription lists' and so on.

Our feeling of insecurity increased when we read in the papers of the establishment of concentration camps and other methods of terrorization. I remember one incident which must have happened in this period of suspense. We were told, or we read, that house searches for socialist and communist literature were going on; people who were found to own such things were sent to the camps. Now I had very few books or papers of this kind, but we thought it safer to destroy what I had, for example a copy of Karl Marx's celebrated book *Capital*. I had tried to read it but never got very far as it appeared to me incredibly dull. I think now that the reason for my instinctive revolt against these speculations was their claim to be a science and certain as a science, which seemed to me, as a physicist, and natural philosopher, simply absurd. However, it was certainly on the 'Index' of the Nazis, and we thought it better to get rid of it. So we burned it and some other printed matter in the central heating boiler. But some neighbours observed our unusual activity at the furnace in the cellar, so we learned from our maid, and as denunciations were quite common we felt our situation more threatened than ever.

Thus our plan to leave the country became more and more definite. There was no immediate danger of destitution since my salary had not been stopped. When I said above that I, and my colleagues, were dismissed, this was not altogether correct. I was forbidden to teach and given permanent leave of absence.

I do not remember how the final decision to go away was made. We had clever and reliable advice from a friend, Erich Rosenberg. He had been a successful merchant, but when he had earned sufficient money he gave up his job and came to Göttingen in order to study science and philosophy—without any practical aim, just for 'fun'. His brother had married Hedi's friend, Elli Husserl, the daughter of my teacher in philosophy, of whom I have spoken in Chapter VI. Thus he came to our house and was a most charming, inspiring addition to our circle. He was a man of practical experience, kindness and wisdom. During these critical days, he was a great help.

If it was hard for me to leave the place of my work, it was much harder for Hedi. She was attached with all her soul to the country, to her home town and to the German language. Just in those years she had begun to write verses, very beautiful verses. Einstein later wrote to her that they reminded him of Goethe's late poems. I still read them now (1962) again and again. For Hedi, leaving the country was a much more total break than for me. But she never raised any difficulty when we decided to go. The summer semester of 1933 began, the students assembled in the lecture-rooms, and I was excluded from all that. It was unbearable, indeed.

We had again booked the little flat in the Perathoner house at Selva in Val Gardena, South Tyrol, for the summer holiday. We wrote to the owner and asked him whether we might come earlier, in May. He agreed, so we packed our trunks, dismissed our faithful housekeeper Fräulein Schütz and our maid, shut our flat and went off. Three of us: Hedi, Gustav (then a boy of twelve) and myself. Irene had just got her leaving certificate at Salem and was at a housekeeping school near the Harz. We left her there and Gritli at Salem.

IV.   Selva. Val Gardena

On 10 May 1933 we left Göttingen. I have only a vague recollection of this journey. We were sad, and yet glad to be out of the horrible atmosphere of Göttingen. Looking down from the train on a crowded market place of a small town in Bavaria we saw the burning of a pile of books. This made me furious, and Hedi had some trouble preventing me from expressing my rage; there was a Nazi-looking fellow in the compartment. We spent the night at Munich, and the following day we took a train across the Brenner Pass. After passing the Austrian frontier, we felt relieved and could breathe freely again. In Bolzano we left the train and hired a car which took us in a few hours to Selva.

There it was still winter, with deep snow and a sharp frost. But we had our cosy rooms in the Perathoner house and settled down to our lonely life. Hedi tackled the domestic arrangements with her usual skill. At first she did all the housework alone; later we had a woman from the village who was quite a good cook but unfortunately given to drinking. Once when we returned very hungry from a long walk in the mountains we found her unconscious, and our meat dish in the oven burnt to a cinder. We dismissed her and presumably got another help, whom I cannot remember. Our flat had a nice verandah with a beautiful view, where I used to sit and work as soon as the weather permitted. The centre of the small, scattered village was about ten minutes' walk from our house. There were a few little shops, a church and a hotel which at that time of the year was still closed. Our house was surrounded by wide meadows, and the narrow gauge railway which connected Val Gardena with the Brenner main line ended there with a large loop and a station consisting of one small hut.

I suppose we tried to continue Gustav's schooling as well as we could. He was certainly not bored and had a lot of entertainment, first with a sledge in the snow, then playing in the meadows and by the little stream.

I soon began to miss my accustomed work. But I had neither books nor periodicals. So I looked for a subject for which no literature was needed, something in quite a new line. I started from my old favourite problem, the electromagnetic mass of the electron, which I had treated in my thesis for admission as a lecturer. According to Coulomb's law, the energy of a charged

254

particle becomes infinite when the radius shrinks to nothing. One has therefore to assume that the electron has a finite size. But if it is a rigid sphere one gets into trouble with the theory of relativity because there, in general, rigidity does not exist. In my thesis (see Part 1, Chapter XI), I had discovered that there are special movements for which rigidity can be defined, and I had obtained for these a definite expression for the self-energy (electromagnetic mass). But for arbitrary movements this was not possible. The only way out of this difficulty seemed to be the assumption that the ordinary Maxwellian theory of the electromagnetic field holds only approximately for dimensions large compared with the radius of the electron, while for smaller distances it should be replaced by another, more subtle theory. A quite general frame for modifying Maxwell's equations had been given, many years before, by Gustav Mie. I remembered these and tried to fill out this frame by special, plausible assumptions. But I was not successful, though I spent many hours on my verandah pondering about it. At last I succeeded by abandoning Mie's formalism and inventing another one. The usual linear theory of the electromagnetic field was replaced by a non-linear one, in which point charges existed which carried a finite electrostatic energy.

I had no means (tables, etc.) to work it out numerically, and I remember the tension with which I waited for the letter from a colleague whom I had asked to evaluate the decisive elliptic integral. When it came I was very happy and proud; for I was filled with the intense desire to discover something fundamental to regain my self-confidence after the loss of my job in Göttingen, and to 'show the Germans what they had lost'. Alas, these ambitions were not fully satisfied. My non-linear field theory made some stir, but did not lead to the solution of the problem of the structure of elementary particles, although I spent a great deal of time and effort on it during the following years, together with a number of collaborators, in the first place the Pole Infeld. I shall return to this later. The so-called Born–Infeld theory was one of the first attempts at a non-linear field theory and worked perfectly in the 'classical' (non-quantum) domain, but it failed in the quantum domain. Today one has a wide empirical knowledge about elementary particles, and it is clear that the electron cannot be treated separately. The most promising theory, that of Heisenberg, also uses non-linear field equations and follows thus the direction indicated by my work, but rests on quantum theory right from the beginning. In any case, this work helped me to overcome the bitter feelings produced by the loss of my position and my expulsion. But there were other factors helping us to accept our fate and to look hopefully into the future.

The first, and a powerful one, was the beauty of spring in the mountains. Shortly after our arrival the snow began to melt, and on each patch of ground there appeared the most lovely alpine flowers we had ever seen. And these patches spread day by day, climbing up the mountain slopes. The meadows around our house became almost uniformly yellow from these troll flowers. Higher up we found specimens we had never seen before, in abundance. On a clear day when the Dolomite peaks stood yellow before a dark blue sky, the beauty of the scene was overwhelming. Hedi said she longed to send a telegram to Hitler or to his minister of education, Rust, thanking him for giving us this spring in the Alps.

Concerning our means for living, there was no immediate worry because my salary was still being paid, and part of it could be transferred to Italy, just sufficient for us. Still it was a considerable relief when the first invitation from a foreign university arrived. It came already on 25 May from my former pupil Alfred Landé, who had left Germany earlier and was now professor at the University in Columbus, Ohio. He had studied in Göttingen, but had got his degree, as far as I remember, somewhere else. When I was professor in Frankfurt, he joined my department for some months and had a place at my own big desk, opposite me, quietly working on problems of his own. I have not mentioned him in the narrative of my Frankfurt period as he was shy and unobtrusive, and hardly joined in the discussions between Stern, Gerlach and myself, nor in other activities of the department. Actually he made an important discovery during this time which contributed much to the development of quantum theory: by sheer incredible patience, using numerical calculations from the empirical facts, he found the laws which govern the magnetic splitting of spectral lines, the so-called anomalous Zeeman effect. They are expressed in some formulae which rightly are named Landé's g-factors. This empirical approach was alien to me, and though I admired Landé's ingenuity in dealing with numbers, I had no actual interest in his methods. Later, Landé's factors were derived from quantum mechanics and formed one of its most decisive confirmations. He became professor in Tübingen, but then accepted a job in America. Now came a letter from him to Selva inviting me to come to Columbus and join his department.

And then one invitation followed another. On 26 May came a letter from my friend Léon Brillouin with an invitation to Paris, and on 31 May one from Belgrade offering me a professorship under very favourable conditions. I knew nothing of Yugoslavia and wrote a letter to the physical chemist Professor Mark in Vienna as I had heard that he knew the Balkan world. His answer was something of this kind: if you arrive at Belgrade you will find none of what they promised you; but if you are prepared to sit every night in a restaurant drinking wine with the minister of education or the minister of finance, and if you have a gift for telling funny stories, and keep them amused, you might in due course get money, buildings, books and whatever else you want. This did not sound very promising. On 15 June came a more attractive proposition of a lecturership at Cambridge, England, and on the following day the information that I would get an offer from Brussels. Now Cambridge, which I knew from my young days, and which was one of the greatest centres of physics, attracted me of course most strongly. Hedi on the other hand had never been to Britain before and had a strange prejudice against England and English life. Yet she understood the advantages of this job for my scientific work and raised no objections. At the end of June there was to be a congress of physicists in Zurich, and I was informed that Blackett would be there. I knew him well from the year he had spent with his charming wife and his friend Dymond in Franck's department in Göttingen. Now he was one of Rutherford's most brilliant collaborators and was authorized to negotiate with me. Thus I set off a few days before this conference, went to Paris on some business I have forgotten, and then to Zurich.

The discussions with Blackett went on very smoothly, and on 30 June I definitely accepted his offer. I met in Zurich a number of German Jewish colleagues who were still in Germany and were keen to learn how the situation was regarded in Switzerland. Among them was Otto Stern, my former lecturer at Frankfurt, who had already decided to leave Germany and to go to America. Another was Franz Simon, a physical chemist in Breslau, who had made use of the privilege granted to Jewish front fighters of the first world war which Franck had rejected. He still had his position and was very unwilling to give it up. I tried to convince him that he would not be able to keep his job for good and that it was much better to go now, while there were still possibilities and no huge competing crowd of scholars. I had the impression that I had made him waver but had not persuaded him. Yet some weeks later he accepted an offer from Professor Lindemann to go to Oxford. The situation had quickly deteriorated. For instance, in this period the open conflict began between Hitler and the Evangelical Church, and the 'purging' of the universities increased.

I returned to Selva, and straight away on 2 July Hedi travelled to England to find accommodation for us. On the way she stopped at Amsterdam and spent a few days with our friend Erich Rosenberg, who was doing a marvellous job there in helping Jewish people to emigrate; he stayed in Europe as long as possible and went to America only when the occupation of the Netherlands by Hitler's armies was imminent. In Cambridge Hedi stayed in student digs and was well received and helped, in particular by the Dean of the Faculty of Mathematics, Max Newman, who took her around in his car to inspect houses. She rented the house at 246 Hills Road, where we lived for the next three years, and also arranged for Gustav to be accepted at the Perse School. Lord and Lady Rutherford invited her to tea.

During Hedi's absence I received an invitation to Istanbul where the Turkish Government was building up a modern university with the help of displaced German scholars. I had of course to decline.

Hedi returned to Selva on 12 July, and ten days later our daughters Irene and Gritli arrived, with our dog Trixi. Trixi was a beautiful Airedale bitch whom Hedi had acquired a few years earlier from an English breeder near Merano. Now Trixi was returning to her home country, the South Tyrol. We were all glad to have her back.

Our quiet life came now to an end. Not only did our daughters cause an uproar by their vivacity and desire for entertainments, excursions, dances, and so on, which Hedi rightly tried to curb, but the holiday season brought numerous friends and acquaintances to the South Tyrol. On 1 August I met my half-brother Wolfgang on the Brenner Pass. He had been living in Vienna as a painter and student of art history and was clever enough to foresee that Hitler would not allow Austria to remain independent. So he decided to emigrate, and now he was on his way to America.

The lecture-hall of the University of Selva. Myself, lecturing to Thompson

There appeared two English students, Blackman from London and Thompson from Oxford, who had intended to come to Göttingen; when they heard that I had left, they came to Selva to study with me there. Thus the University of Selva was founded, consisting of a bench in the forest, one professor and two undergraduates. I cannot remember what I tried to teach them; but they were a welcome addition to our circle and accompanied the girls on walks and climbing excursions.

Blackman later became well known by his work on specific heat of crystals where he showed definitely that the theory of Kármán and myself was, at low temperatures, superior to Debye's. Thompson also did some research work but then concentrated on teaching; he became a fellow and later a tutor at one of the Oxford colleges.

Other scientific friends appeared: P. W. Bridgman and his wife from Harvard; Wolfgang Pauli and his sister; Hermann Weyl, my colleague from Göttingen; and Anny Schrödinger, Erwin's wife.

The most unexpected visitor was F. A. Lindemann, professor of physics at Oxford. I had known him since the time he had worked in Nernst's laboratory in Berlin, but I was never very intimate with him. Now he had the idea of using the opportunity of many good scientists being expelled from Germany to improve the standard of science at Oxford, where theology, law, and the arts dominated. So he travelled in his big Rolls-Royce with a chauffeur from one German university to another and offered jobs to dismissed physicists. When he learned at Göttingen that I was in Selva he did not mind the long journey and came to see me. But I had just confirmed my arrangement with Cambridge and could not accept Lindemann's offer, though I think it was financially better. He stayed a day with us, and we talked about scientific problems and the political situation. Lindemann's strange personality has been

The Dolomites, 1933
Above: Hedi and myself (right) with Planck (left) and Annie
Schrödinger and Herman Weyl (centre), in the Fulpmes Tal

Below: Gustav with Annie Schrödinger and Weyl at the
Grodner-Joch

described since his death (1957) in several biographies. He was, in my opinion, a mediocre physicist, but a remarkable personality and a remarkable snob. He was devoted to the top British aristocracy, and he boasted about his noble friends in a peculiar manner. Once, some years later, I had to give a lecture at the Clarendon laboratory at Oxford whose director he was; he wrote to me that to his great regret he could not attend as he was invited to a party at Blenheim, the enormous palace of the Duke of Marlborough in Woodstock near Oxford. I thought this mildly snobbish. But when I entered the lecture-room, there was Lindemann in full evening dress and he told me he could not resist the desire to listen at least to the beginning of my lecture, but would have to leave in half an hour. I thought this was nice but even more snobbish than his letter. His connection with the Duke of Marlborough and the Churchill family came through his friendship with Winston Churchill. One can read in Lindemann's biographies how this started. During World War I there were frequent fatal accidents of British aeroplanes; they got into a spin and crashed. Lindemann found out by calculation the cause of this phenomenon and invented a remedy for it. To prove that this worked he learned to fly and demonstrated to an assembly of air officers that he could pull the plane out of the spin and land safely.

Churchill was impressed by this feat of intellect and courage, and he made Lindemann his adviser in all matters scientific. As these played a dominant part in the second world war, Lindemann became one of the most powerful men in Churchill's war cabinet. The dramatic events of this period can be read in C. P. Snow's book *Science and Government* (Oxford University Press, 1961), which deals mainly with the struggle between him and Tizard, about the air war against Hitler's Germany.

I knew that Lindemann was in a leading political position during the war, but I had no idea of his actual part in the events. I admired his energy, his wit, sometimes his judgement, not so much in regard to scientific facts and theories but to personalities. After he had visited me in Selva, he succeeded in bringing a number of German scientists to Oxford, in the first line Franz Simon (as mentioned above) and several of his collaborators including Kurti and Mendelssohn; also Franck's gifted pupil Kuhn from Göttingen.

Lindemann was elevated into the ranks of the high aristocracy he adored, by being given the title Lord Cherwell. During all these years, before, during and after the war, I was on very good terms with Lindemann. Whenever I visited Oxford I dropped in at the Clarendon and had a talk with him. Once when I stayed only a day, he came to Simon's house, assuming he would find me there. Discussions on physics often exasperated me, as he believed himself to have a deeper insight into the structure of the physical world than any of us, Einstein not excluded. I did not take him seriously but always managed to remain on very friendly terms with him.

This might not have been so if I had known what he had done as scientific adviser to the Government during the war. It was he who recommended and insisted on trying to break the resistance of the German people by bombing and destroying their homes. We believed at the time that a certain Air-

Marshall, 'Bomber Harris', was the driving force behind this strategy; he may well have played this rôle among the military but it is now clear beyond doubt that the idea was Lindemann's and that he was fully responsible for this kind of warfare, which I regard as abominable, mean and detestable. In the account of my childhood I have said that I learned the true character of war from my father's descriptions of his service in the military hospitals during the Franco–Prussian war of 1870–71. I have also told the story of my inglorious military service in two Prussian cavalry regiments and of my activities during the first world war. I was deeply convinced that war was a relic of primitive times, repulsive and barbarous, and that with increasing 'civilization', e.g. increasing technical development, it was becoming more and more obsolete. I have reported how I had a conflict with Fritz Haber because of his invention of poison gas warfare. Later I had the satisfaction of finding that Rutherford shared this opinion. When I came to Cambridge as a refugee, Haber appeared too, ill, depressed, lonely, a shadow of his former self. I felt pity for him and intended to bring him into contact with scientific people, but when I invited Rutherford to tea in my house to meet Haber, he refused, saying frankly that he did not want to shake hands with the inventor of poison gas warfare.

Now this was still a kind of warfare aiming at the destruction of the armed forces. Even in World War I, the idea that war was to be fought between armies and that civilians had to be protected was still alive—though the blockade with its consequences of famine was allowed by international law and practised by both sides, according to their power (surface navy against submarines).

But the systematic destruction of inhabited areas of cities with the purpose of breaking the will to resist of the civilian population is quite another matter. The development of aeroplanes as instruments of war was inevitable. Still, there is a difference between using these arms against military forces, even against factories working for these forces, and using them against helpless civilians and their homes, to say nothing of the churches, museums, historic buildings, and other irreplaceable treasures. Of course, Hitler was the first to use this method of terrorization; Warsaw, Antwerp, Coventry are a few examples. But Britain was in the war precisely to fight this kind of barbarity. Can one rid the world of an evil by committing the same evil, planned and amplified? I think one cannot. The bombing war as practised by the allied air forces appeared to me, right from the beginning, as morally wrong. That it was also strategically wrong has now been proved without doubt. The predictions made by Lindemann about the efficiency of destruction by air attacks have proved to be wrong by a large factor (I think about ten). Other methods of using the allied air power could have led to victory much faster. During the war I did not know all this, but I felt the strongest disgust when I read day by day about the destruction of German cities. Yet I was not a pacifist at that time. If I had been, at least in principle, before the appearance of Hitler, I suspended this attitude during the anti-Hitler war, since I regarded the Nazi régime as the greatest evil that had ever befallen the human race. I wished for its destruction with all my heart, and it was obvious that only crude power could bring it down. Thus I suppressed the objections against war which I had learned during my former life. But still there were limits. I knew that the German people had been stupid and tricked into letting Hitler obtain power; but I knew also that once he was in power there was no way of getting rid of him without defeat in war. The idea of breaking Hitler by killing ordinary people, women and children, and by destroying their homes seemed to me absurd and detestable. But it was just this idea which Lindemann suggested and forced through against the objections of better men.

I had believed in science not only as a way to obtain knowledge of Nature and to apply it to a better material life, but as a way to wisdom, to distinguishing between sense and nonsense; and as most wickedness is based on nonsense, on irrational thinking, I had thought that a real scientist could not do base deeds. The scientists I admired and loved, like Franck, Einstein, Rutherford, Planck, von Laue, seemed to confirm my belief. But Lindemann did base things and opened the gates of hell for other men of his type, men efficient and clever, but not profound and wise, who later became leaders in science and its applications to politics and war. I need not tell names here. I hope to take up this theme later.

After this long interlude I have now to return to my narrative of our stay in Selva. There are some events, sad ones and pleasant ones, which have stuck in my memory.

One day (12 August) a group of Boy Scouts appeared at our door. One of them was Otto Still, the younger son of our friends in Recklinghausen; he knew our address and had directed his friends to it. We improvised a meal for them—there were about twenty, I think—we had a lot of fun with them and helped them to find quarters for the night in the neighbouring farms. It was a jolly evening. The next morning Hedi left for Bolzano to do some shopping. Late in the evening an almost exhausted boy appeared in our house and told me that they, the Boy Scouts from yesterday, had had a terrible accident, and that one was dead: it was our friends' son Otto. Some of the boys had experimented with sliding down a steep meadow on their knapsacks; Otto Still had lost control of his 'vehicle', gone faster and faster and fallen, at the lower end of the meadow, over a sheer precipice of some hundred metres. I was terribly shaken, for I knew he was the favourite son of his parents. I was asked to inform them and I remember that in my desperation, I ran to the hotel and roused Hermann Weyl from his bed in the hope of getting some help and advice. We formulated a telegram to Recklinghausen. Next morning I took the train to Bolzano to meet Hedi according to plan; I found her in high spirits, enchanted by the lovely town and the surrounding valley. I could not bring myself to break the sad news. We walked to a little lake, enjoyed a swim in the clear water and I kept a serene appearance. I remember that it was a horrible experience. Only when night fell did I tell her what had happened and had to relive with her my agonies. The body was taken to Bressanone where his elder brother fetched it home. We could not bring ourselves to meet him there.

We heard from Annie Schrödinger that Max Planck and his wife were at St Peter in the neighbouring Dolomite valley of Villnös. On the last day of

August we all went there to visit them. There I had my last talk with Planck for a long time; I saw him again more than twelve years later after the war in Burlington House, London, at the Newton celebrations of the Royal Society, and then he was an old, broken man. In St Peter in 1933, he was still active and lively. I had always wondered that he had done nothing for his Jewish colleagues. He told me that during the critical weeks in spring 1933, he was in Italy on holiday and heard little of the events in Germany, until he was roused by a message from Lise Meitner, who informed him that Fritz Haber had been dismissed from his position as director of the Kaiser Wilhelm Institute for Physical Chemistry in Dahlem and all his other posts—Haber, whose invention of methods to produce explosives with the help of fixed nitrogen had prevented the complete breakdown of the German army after the first months of the war of 1914–18; who had, together with Walter Rathenau, mobilized and organized Germany industry for war production; who had invented the poison gas warfare which gave the German army a considerable advantage for some time—not to mention the fact that he was one of the greatest chemists of his period. When Planck heard of his dismissal he hurried home and asked Hitler for an audience. Planck has described this to me. After he had said a few words Hitler began to speak and soon to shout, abusing the Jews in general and the Jewish scholars in particular, and Planck could never get a word in. After this experience, he gave up all hope of helping people through his influence. Soon after, as the President of the Berlin Academy, he was compelled to sign the document through which this society expelled Einstein. I suppose he also signed the corresponding document in my case, though I have never seen it. But Planck was not a weakling. He knew that any resistance would be useless and dangerous for German science. So he yielded to pressure. I wonder whether Max von Laue would have given his signature in such cases, for he had more of the hero in him, and was not, like Planck, conservative to the degree of obeying orders of the State, even if they were bad.

I had already, on 9 August, written to the Minister of Education in Berlin asking him for my final dismissal from my professorship in Göttingen. About a fortnight later I received a reply in which my demand was refused and an extended leave of absence granted. In fact, I got my due payment until 1938, when I was expatriated and my property confiscated, but my salary had been paid by them into a closed bank account and could not be transferred to foreign countries. I used it for helping relatives and Jewish friends in Germany, and for several journeys by members of our family to Germany as long as these trips could be made without great danger.

Now I shall report about another human contact which was thoroughly enjoyable and free from political connections.

One afternoon in August, Hedi and I were sitting in the garden of the Selva hotel drinking coffee. We could not help overhearing the talk of two young men at the next table, both smartly and a little loudly dressed and looking like artists. They were complaining that they had not succeeded in booking a room in the hotel. At that instant a car drove up to the entrance of the hotel, an elderly couple emerged, entered the house and returned soon to join the two young men, both very angry because they also had been told that no rooms were available. I recognized them at once: they were the great pianist Artur Schnabel and his wife Therese Behr, the equally celebrated singer whom I had met as a youngster in the Neisser house in Breslau, more than twenty years earlier, and the young men were their sons. I went to the hotel and said to the manager, whom I knew: 'Do you know to whom you refused a room just now? Schnabel, one of the greatest living pianists. Can't you find something for them, and possibly also for their sons?' The man at once changed his attitude and said, yes, there were special rooms for such distinguished guests; if he had only known, etc., etc. Then I returned to the garden, addressed the Schnabels, introduced them to Hedi and told them that I had secured rooms for them. They were delighted and asserted that they remembered me (which I doubted). We had a nice talk about the old days in Breslau and the appalling contemporary events. Schnabel was Jewish, his wife not; but she had broken with her family, in particular with her brother, who in my time was deputy conductor of the Breslau Symphony Orchestra and, as she told us, an ardent Nazi.

From this day on we became great friends and met often. We introduced them to Hermann Weyl and Annie Schrödinger, and we went for walks and had lively discussions.

This friendship with the Schnabels lasted. When we were in Edinburgh they came often to take part in the Festival and once spent an evening with us. Hedi has written down a collection of Schnabeliana, clever and amusing bits of talk in which he excelled. After our return to Germany, we did not see them again and were very shocked and sad when we heard of Arthur's death in 1951.

The time of our departure from Val Gardena was approaching. We knew by now all the delightful walks and little climbs around Selva, and we loved them very much. Parting from them for a foreign country and an uncertain future was not easy.

The girls returned to their schools in Germany. Then Hedi left (2 September) for Germany, to arrange the transfer of our furniture and household goods and of some money. She stopped in Munich and there visited Heisenberg and his mother. Then she went to Cologne to get advice from Erich Rosenberg. In Göttingen she worked hard and successfully. I cannot relate all her exertions and adventures, among which the most exciting and dangerous one was the loss of her passport by the travel agent—the passport which, for an emigrant, was the most precious piece of property. She invited Grete Heller, who had in previous years helped her in the house and with the children, to come to Cambridge with her.

Gustav and I stayed with Trixi in Selva until the end of September. By that time all our friends had left, we were rather lonely, and the first snow fell, so we were quite glad to leave. We went to Zurich, stayed a day with the Edgar Meyers and took a night train through France to Calais. We had no sleeper and were sitting in a crowded compartment, which Trixi shared. The little dog slept with his head on my lap. I could not sleep much but pondered on what was in store for us. That night I shall never forget. But the future was not so dark as it then appeared to me.

264

V. Cambridge

From here on again there is no help from Hedi's diary and I have to rely on my memory. Hedi's suspension of her diary is itself a reflection of our life in Cambridge: she was not happy there, so unhappy that she cannot bear digging up the events of those years. She felt uprooted, cut off from her language and her poetry, from her country and her friends; she was overworked with domestic duties, burdened by the responsibility for two grown-up and temperamental daughters, and she had no compensations, such as I got from the scientific atmosphere of the Cavendish Laboratory and from my connection with the common rooms of several colleges. But she hardly ever showed her state of mind to us and devoted herself to her duties. Only now, after so many years (1962), has all this come to the surface.

When Gustav and I arrived at Dover we had to deliver Trixi to the health authorities. We watched her being put into a small cage and lifted by a crane from the deck to the quay. We were both very sad when our good doggie disappeared, and little Gustav cried. Trixi was taken to a veterinary surgeon who had a kennel for 'holiday dogs', a few miles from Cambridge. Imported dogs had to undergo a quarantine of six months. We visited her once during this period, and this was due to the kindness of Lord and Lady Rutherford. At the first tea-party we attended in their house we told them all about Trixi and complained that we knew nothing about her. The Rutherfords said: 'But why don't you go and visit her? We shall take you there in our car.' And indeed, a few days later they came, drove us to the kennels and told us not to hurry: they had taken books along and would read comfortably until we had assured ourselves about Trixi's well-being in prison. I tell this story because it throws a bright light on the deep humanity and kindness of Rutherford and his wife.

In London, Gustav and I met Hedi and the two Heller girls, mentioned above. I have forgotten where we stayed and what we did in London. On our arrival in Cambridge, we had a rather frightening reception. On all the big hoardings was written in large letters: 'The Man Born to be Hanged'! But soon we found out that it did not refer to me, and that it was quite harmless: the title of a cinema film, made from a crime novel. I found it later in a bookshop, read and enjoyed it; it is still on my shelf.

Many Cambridge families took refugees into their houses until they found some home of their own. Thus we came to 1, Chaucer Road, the home of Professor Robert Hutton and his wife Sybil, a daughter of the physicist Schuster, whom I have mentioned before in connection with the Jones with whom I stayed once in London. The Huttons offered us the kindest hospitality and taught us the elements of the art of living in Cambridge society. He was professor of metallurgy, but also interested in many other things, for instance in the problem of how libraries could cope with the ever-increasing flood of periodicals and books. Sybil was a most energetic, almost frightening little person, but in fact very good-natured, kind and helpful. We became great friends.

When Hedi and her helpers had got our house in order, Gustav and I moved in. I remember very little of its rooms, just that my study was tiny and overlooked our garden. This was rather large and contained a lawn tennis court. Beyond it there were the playing fields of the Perse school, to which we had access through a little gate in our fence.

Right from the beginning we had guests, friends who had to leave Germany like ourselves and stayed with us for some days. That did not simplify matters for Hedi. Irene and Gritli arrived in the spring of 1934, and our house became livelier still, with young people coming and going, among them the two young men to whom the girls became attached and soon engaged.

I was invited by the University to receive the degree of Master of Arts, and as I had no British degree I accepted with pleasure. The Vice-Chancellor of the University was at that time Mr Cameron, Master of Gonville and Caius College, the same college to which I had belonged years earlier as an 'advanced student'. He invited the honorary graduates to a luncheon party at his lodge before the ceremony. When I entered the dining room of the lodge I remembered it well, for I had been the guest of Mr Cameron's predecessor, Mr Richards, at several little parties in 1913. During the meal I mentioned this to Mr Cameron, and he asked me: 'So you have been here before; have you been immatriculated?' When I answered 'yes' he jumped up, ran out of the room to the consternation of all, and returned a few minutes later, obviously much relieved. Then he explained that they had prepared to immatriculate me before the graduation since nobody had remembered that I had been a student in Cambridge before. But it is a rule that a person immatriculated once remains a member of the University all his life. Never has anybody been immatriculated twice through all the centuries the University has existed. But then it had nearly happened.

I remember this story vividly because it illuminates the power of tradition at a place where at the same time the most modern research in science is flourishing. Caius College gave me dining rights, and I made use of these about once a week. A short time later St John's College, to which Dirac, Max Newman, and other scientists belonged, offered me the same honour. Thus I had dinner at high table about twice a week and enjoyed it very much. I took part in the 'Feasts' of these two colleges and was invited to many others. I remember one occasion when Rutherford gave a dinner party at Trinity for a group of about twenty men. I was sitting next to him. When the port was circulating some people discussed the question of whether they were satisfied with their profession or would have preferred another one. Rutherford became quite angry and suggested that everybody should declare what he would have preferred to be and why. I had to start and said that I did not think I should be good at anything else but mathematical physics. But many of the others declared they were dissatisfied and would have preferred another profession, art, music, literature, army or navy, even business. When the circle closed with Rutherford he hit the table with his fist and shouted: 'I shall be damned if I ever thought of being anything but what I am.' That was Rutherford in his glory.

I was given a small room in the Cavendish Laboratory as a study, and I started a course of lectures, I think two hours a week, on my new non-linear electrodynamics. It was quite well attended, not only by graduate students but also by fellows of different colleges. Among the students were Maurice Pryce, who later married our younger daughter Margaret (Gritli), Paul Weiss, a German refugee who had been assistant to Fritz Haber, N. Kemmer, born in Russia but educated mainly in Germany, who had already attended lectures of mine in Göttingen (he is now my successor in Edinburgh) and Leopold Infeld from Poland. Infeld soon became my main collaborator in the unified field theory, as I called the non-linear theory of the electromagnetic field; for it enabled us to deal with point charges, without the difficulty of infinite self-energy (which occurs in Maxwell's theory) so that the field and its sources could be combined in the same mathematical structure. The beginning of our collaboration was not very auspicious. Infeld has described it in his autobiographical book *The Quest* (London, Victor Gollancz Ltd., 1941), and I think his account is quite accurate. He asked me to lend him a copy of my manuscript as he had not understood my calculation of the self-energy of a point charge.

The next day he addressed me after my lecture and said: 'Your calculation of the self-energy of the electron is wrong,' and showed me what was wrong. I listened to him but I was tired and anxious to get home. So I did not follow his arguments properly but interrupted him a little angrily, saying: 'I shall think it over.' Infeld reports that he was disgusted with my behaviour and, for one afternoon, with Cambridge altogether, because one professor (myself) had treated him rudely and another (Dirac, known for this taciturnity) had not talked to him at all. But during the afternoon I considered Infeld's objection to my energy calculation and his proposed alteration and found that he was right. The next day I waited for him at the door of the lecture-room and addressed him when he arrived, apologizing for my rudeness, and asked him to collaborate with me.

In his book he gives a short account of my life and development from what I had told him during our daily conferences and discussions. He was impressed by the amount of work, correspondence and other activities which I undertook, and he describes my mood changing between high enthusiasm for an idea when it seemed to work, to deep depression when it failed. It is rather odd to read about such intimate details of one's character. But I think his account is quite correct.

I have already indicated that the work we started with so many great hopes was in the end not successful, as it turned out that it was not possible to 'quantize' the new field equations. Nevertheless, I still think there is something in it. In my first larger paper (*Proceedings of the Royal Society*, Vol. 143, 1931; p. 410), a method is indicated by which the non-linear field equations can be again linearized with the help of Dirac matrices. If the latter could also be regarded as field quantities, one would have a bilinear set of field equations (linear in the Maxwell field tensor and linear in a set of spinor quantities) which seems to me of a similar structure but more natural than Heisenberg's non-linear field equations in his theory of elementary particles. Dirac has recently (1962) taken up this idea and tried to work it out—so far without success. But I still think it is promising and I spoke about it to the large audience at the colloquium in Göttingen on the occasion of my eightieth birthday where Heisenberg gave a fascinating lecture on his theory.

I took part in the weekly colloquia in the Cavendish and learned a little about nuclear physics, which I had previously rather neglected. I came in touch with Sir Arthur Eddington, the great astronomer, who had confirmed Einstein's prediction of the deflection of light by the sun. He was working with the same group of spinor matrices which I used in order to linearize the non-linear field equations, for a quite different purpose, namely in an attempt to derive the ultimate laws of physics *a priori*, in particular the fine structure constant (or its reciprocal $hc/e^2 = 137$). I have never had much confidence in such a philosophical approach to science and later explained my standpoint in a little book, *Experiment and Theory in Physics*, published by Cambridge University Press (1943) and reprinted (1956) by Dover Publications, New York.

The professor of physical chemistry, Lowry, came to see me because he was interested in my work on optical activity of asymmetric molecules and crystal lattices. He became very much attached to us, took us in his car for little excursions and showed us the Gog and Magog Hills just outside Cambridge where there was a tiny 'wood', so small that one could look right through it—it did not mitigate our homesickness for German forests but even increased it. He was a most friendly and kind man, a little philistine and with narrow interests. We were deeply shocked when we later learned that he had committed suicide. It is still a riddle for me what dark experience may have driven this quiet, even-tempered, kind man to such a desperate step—probably the knowledge that he was suffering from an incurable disease (cancer).

Gustav was sent to the Perse School and had a few very difficult months until he learned the language. I remember that he once came home crying, and when we urged him to tell us what had happened, he said that in the mathematics lecture he could not follow because they were talking about 'eggs' all the time, and when he asked the master what an egg had to do with maths, the class broke out in a roar of laughter. But soon he learned to use x and y in maths as well as the English tongue in general; he became one of the best pupils of his form and was given a scholarship, which was a great relief for our very restricted purse.

We had much trouble in finding suitable training opportunities for the girls, as we had not the means to send them to university. Irene learned dressmaking and Gritli graphic art, first in London, later in Vienna. They enjoyed life in England, where they were a 'social success' because they were 'different'.

A great deal of my time was absorbed by correspondence connected with the plight of scholars, artists and other people who had lost their jobs through the Nazis and were compelled to emigrate. I was quite successful in some cases. For instance, I heard that the Academy of Science in Lima, Peru, was willing to employ half a dozen scientists; I wrote to Lima and proposed names. All these men obtained jobs. Then the neighbouring state of Colombia decided not to be outdone and offered even more jobs to refugee scientists. I was appointed an honorary member of the Lima Academy in acknowledgement of my advice.

But there were some cases where I dismally failed to save men of distinction because I could not find a place for them. The worst was that of the brilliant violinist Alfred Wittenberg, once a member of the famous Klingler Quartet in Berlin, whom I had met in the house of my stepmother's sister, Helene Mühsam, at her husband Ismar's most enjoyable chamber music evenings in the years before the first world war (see page 26). The imploring letters from this great artist shook me deeply, but I could do nothing, as the British musicians were not as magnanimous as the scientists and tried to keep competition out. Wittenberg has perished; I never heard of him again.

As long as I was a 'foreigner' I could not become a Fellow of the Royal Society, and my scientific status was not high enough to be elected a Foreign Member. But I attended many meetings of the Royal Society in London. Sometimes Dirac took me in his car to London; he was known as a fast driver, and indeed, I was always glad when we arrived safely. I found the meetings rather dull. Somebody lectured about a piece of special research which only a minority of the audience understood. But on occasions there were brilliant lectures summarizing whole subjects. Altogether the meetings of the Göttingen Academy had been more attractive.

Once, I think in 1934, Heisenberg was invited to give a lecture at the Cavendish. We offered him our guest-room and all would have been pleasant except for one unfortunate event. One afternoon walking with him in our garden he told me that he had to bring me an official message from the German Physical Society. They had obtained from the Nazi Government the permission to call me back to Germany. I would not be allowed to teach but could do research. I was rather bewildered by this sudden offer, which to a homesick emigrant sounded quite attractive. After a few moments of consideration I asked, 'And this offer would include my wife and children?' Then Heisenberg became rather embarrassed and answered: 'No, I think your family is not included in the invitation.' This made me very angry, I broke off the discussion and went into the house, where I told Hedi. We could not understand how Heisenberg, whom we knew as a decent, humane fellow, could have agreed to convey such a message to me. I have never talked to him about it. But later he, and others, described to me the situation in Germany at that time, the pressure under which they lived; thus I came to a kind of understanding, though not of forgiving this lack of feeling and tact—to use the mildest expressions.

After that experience I avoided contact with visiting scientists from Germany, even with Walter Gerlach, whom I had liked very much since my Frankfurt period. On the other hand, I regard it as a great privilege that I came into closer contact with several emigré scientists of very high distinction: Erwin Schrödinger and Franz Simon. I knew both of them quite well; but now that they lived in Oxford we visited one another frequently, stayed in each other's houses and, feeling we were in the same boat, came to know each other more intimately. Schrödinger was one of the most fascinating men I have ever met. I can give here only a very brief outline of his life. After his great discovery of wave mechanics, he became Max Planck's successor in Berlin. When Hitler seized power, he emigrated, although he was a pure Aryan and not forced to; he did it in protest against the Nazi barbarism or, as he expressed it with an understatement in the British manner: 'I could not stand being bothered with politics.' He went to his home country, Austria, and became a professor in Graz. When the Nazis invaded Austria he fled over the mountains, carrying a rucksack, to Italy. In Rome he met de Valera, the Prime Minister of Eire, who was enchanted by him and promised to found for him a special research institute in Dublin. But this project took some years to accomplish. Meanwhile Schrödinger went first, I think, to Belgium and then to England, where he was attached to one of the great Oxford colleges. He bought a house and was greatly honoured and respected by the scientific community of Oxford, but never became a real Oxford don. He disliked college life and high table dinners. He regarded a society without women as detestable and barbaric. Once he said: 'These colleges are academies of homosexuality. What a queer type of men they produce!' When I asked him why he did not enjoy the college dinners he said: 'You never know who your neighbour may be. You talk to him in your natural manner, and then it turns out that he is an archbishop or a general, or even a former prime minister—huh!' His private life seemed strange to bourgeois people like ourselves. But all this did not matter. He was a most lovable person, independent, amusing, temperamental, kind and generous, and he had a most perfect and efficient brain. There was never a steady flow of letters between us. But if one of us informed the other about some new scientific idea, there followed an explosion of correspondence; often I got a long letter every day, full of mathematics and critical remarks about colleagues and the general situation in theoretical physics.

Schrödinger became interested in the 'unified field theory' and wrote several papers about it. They appeared in the *Transactions of the Royal Irish Academy* when he was already in the new Institute for Advanced Studies in Dublin. He invented a new mathematical method to deal with the theory and applied it to different problems, e.g. the superposition and the mutual scattering of electromagnetic waves, the derivation of the equations of motion of a point charge from the field equation, and others. All this was done 'classically', without the use of quantum theory. In this line he did not succeed any better than Infeld and myself.

Franz Simon was a very different type. He was solidly bourgeois, matter-of-fact, efficient in his experimental work and in administration and finance as well. He brought with him to Oxford a number of excellent collaborators, Mendelssohn, Kurti, and others, and built up a brilliant research team, who worked mainly on low-temperature thermodynamics. Through their efforts the Clarendon Laboratory at Oxford became one of the leading institutions in this field.

Simon adapted himself very well to the life of an Oxford don and became in many aspects thoroughly English. The one thing he detested was the way of heating houses by open fires; he regarded this not only as a disagreeable nuisance but as a serious defect in the economic system of Great Britain, and he wrote articles about it, in which he showed that the coal shortage was mainly due to the waste of coal for domestic heating. He also wrote about the insufficiency of the technical education in Britain and recommended the foundation of schools corresponding to the German *Technische Hochschulen.* During the war, he took part in the development of nuclear energy and worked out methods for the separation of isotopes. In acknowledgement of all these public and patriotic efforts he was knighted and became Sir Francis. When his chief, Lindemann (Lord Cherwell), died in 1954, he became his successor as director of the Clarendon. But he was already suffering from an incurable disease and died a short time later.

Our scientific discussion dealt with experimental problems of the laboratory, and I could help him occasionally. I enjoyed his dry humour, a mixture of his native Berlin and his acquired English background.

During these years in Cambridge we travelled quite a lot, but I cannot remember when and where. I visited Paris several times, lecturing at the Institut Henri Poincaré. We spent some summer holidays at Mittelberg in the Walser Tal near Oberstdorf. There was a financial reason for going to this particular Austrian valley. It was surrounded by high mountain ranges, and there was no railway and no road to the Tyrol, to which it politically belonged. But a good road and a railway connected it with Oberstdorf in Germany. Therefore there existed an old treaty according to which the Walser Tal belonged economically to Germany; one could pay there with German money but was under Austrian rule, which meant, at that time, still free from the Nazis. Now I was still receiving my salary as a professor, although it could not be transferred to England but was paid into a closed account at a bank in Göttingen. It enabled me to assist relatives and friends, and occasionally to purchase things which could be transported to England. Now we discovered that we could pay for our holiday in the Walser Tal with the help of this money. The only risk was the journey across Germany, but during this early period of Hitler's rule this was not too dangerous.

VI. Bangalore

One day I received a letter from Sir Chandrasekhar Venkata Raman, the only Nobel Laureate in India, discoverer of the scattering of light with change of frequency, called after him the 'Raman Effect'. He asked me whether I could recommend to him a young, efficient theoretical physicist for an appointment at the Indian Institute of Science at Bangalore, whose president Raman was: preferably a non-British man. I answered that I knew well only the German physicists but, having been forced to emigrate, I did not wish to have anything to do with them. Moreover, how could I persuade a man to go to Bangalore when I did not know the place and the conditions there? After a short time I received another letter from Raman in which he expressed his understanding of my first argument; but the second difficulty could easily be overcome: Why could I not come myself for say, half a year, and have a look at the place? The journey and a decent salary would be paid by the Institute.

Hedi and I considered this suggestion, and as my appointment at Cambridge was soon coming to an end anyhow, we decided to accept.

We arranged for the children to live with families of friends, and also found a place for our domestic help, Emmy Witteschnik from Vienna, who had joined our household not long before, and in the autumn of 1935 we set out for India on the steamer *Staffordshire* of the Bibby Line.

Hedi has described our life in India in charming letters which have been typed and collected in a little volume. I shall give here a short account mainly of those matters which she does not report.

I must mention an acquaintance we made on the boat, the Reverend Alexander Fraser and his wife. He had spent a good deal of his life in Africa and founded the Achimota College on the Gold Coast. As far as I know this was the root from which the new University of Ghana developed. He had also done educational work in Ceylon and was just on his way to Colombo where he was to receive some academic honour. Both Frasers were the most charming people, quite different from the average British parson; lively, amusing, full of fun.

The boat was full of forestry experts from Burma with their rather elegant wives, who were returning from a year's leave to live in the jungle for another

272

few years. They were a noisy lot, drinking and dancing all the time. There were also some British officials, amongst them a Captain Harvey, tutor to the children of the Maharaja of Travancore, one of the South Indian states. Our boat was to call at Cochin, the capital of a little state near Travancore—by the way the first big passenger ship bound for this harbour, and we intended to leave the ship there and travel by rail to Bangalore. Harvey persuaded us to postpone our departure from Cochin for a day and arranged for us to stay a night in the house of his friend, the British Resident, who was just then absent.

This house was on a small island in the lagoon, most beautiful and romantic. I remember that first night in India, the large bedroom with small lizards running up and down the walls, the narrow beds with mosquito nets, the polite servants clad in white. We had a motorboat at our disposal to bring us to the town. On the afternoon of our arrival and the next morning we went sightseeing in rickshaws. We saw the bazaar streets, crowded not only with human beings but also with cows and goats, we saw palaces and mosques and a very ancient synagogue which was said to have been founded long before the Christian era. The whole place was indescribably beautiful, built on islands covered by palm trees and separated by canals with picturesque barges.

We paid a visit to the British Club where we had our first impression of the separation between the ruling British and the Indians. It was a perfectly English atmosphere; everyone knew everyone else even if he had just arrived from a distant corner of the world, and they talked about school-mates and what had become of them. In spite of the heat, a considerable amount of alcohol was consumed. We rather felt ourselves foreigners and outsiders, though they were all very friendly to us.

At the station we found Joseph, the butler, who had been engaged, together with our whole domestic staff, by Mrs Metcalf, the wife of the Vice-Chancellor of Mysore University. Joseph had collected our luggage at the Residency and showed us our comfortable compartment. He was a Christian and a thoroughly honest, efficient and faithful man, who ran our household at Bangalore smoothly and accompanied us on journeys. He still writes to us every year at Christmas.

In Bangalore we were received by Lady Raman, who took us to our 'bungalow' which was actually a big two-storey house with numerous rooms, electric light, and two W.C.s, at that time a rare convenience in India. A few rooms had been furnished for us by Mrs Metcalf, quite cosy and comfortable.

We had a large garden with beautiful trees and flowers and two tennis courts which were screened off by marvellous bougainvillea shrubs. The Raman family lived in a similar house just across the road.

We liked Lady Raman right from the beginning. Her husband was absent and appeared a few days later. We were fascinated by his appearance and talk. Hedi said he looked in his Indian dress and turban like a prince from the *Arabian Nights.*

At first we were the only Europeans at the Institute. Later an English professor of engineering, Aston by name, appeared, with his wife and baby.

We invited them to stay with us until their bungalow was ready; later we had little contact with them.

We lived mainly among Indians and enjoyed it. We had occasional invitations from British officials, such as the Resident; there a stiff ceremonial was observed as if this colonel, representative of the British Raj, was actually a sovereign. The highest medical officer, Colonel Hance, received us amicably, with cold beer, 'genuine Pilsener', and efficiently treated me when I had strained my heart in a tennis match between Hedi and me and two Indian students.

During the first weeks we received an invitation from the Maharaja of Mysore to attend the Dasara Festival in his capital. There we stayed a week in a big, luxurious tent in the park of the Guest House and saw the old, now extinct system of the small, half-independent states in all its splendour: colourful processions, illuminations, a reception by the Maharaja (Durbar) in the big hall of the palace for Indians, and one for Europeans, sporting events such as polo. Unfortunately, Hedi caught a germ and was rather ill after our return to Bangalore. Later she had a more serious illness, due to a slight sunstroke. One of Raman's sons wanted to take snapshots of us and made us take off our topees in the full sunshine after lunch. Immediately afterwards Hedi complained of disturbances in one of her eyes. She was compelled to lie in bed in a darkened room for some weeks. The Indian professor of ophthalmology treated and cured her. She was so attracted by him that she wished to see some of his work, and he allowed her to attend some of his cataract operations in the hospital.

This illness spoiled our plans for a longer journey during the mid-term holiday. We had intended to go as far as Darjeeling, to see the Himalayas. Now we had to restrict our journey, and Hedi had to wear glasses covered by cardboard with a small hole in the centre in order to restrict the movement of the eyes. Nevertheless we saw many places: Bombay, Agra, with the Taj Mahal, Aligarh with its Muslim University, whose president, Ziu Uddin Ahmad, a co-student of mine from Göttingen, had invited me to give a lecture. Finally Old and New Delhi. There we found Raman waiting for us in the hotel; he was there on some administrative business. I found also an invitation to give a talk on the radio the same night. There was only about an hour in which to prepare a script for a thirty-minute talk, and I had written only half of it when I had to begin talking. So I had to improvize the other half, a most exciting and fatiguing experience. Later I learned that the radio system had just been inaugurated a few days before, so there were not many listeners, and it did not matter much what I said.

We attended a dinner given by the Minister of Education—I do not remember whether in my honour, or whether I was just invited on Raman's recommendation. All I remember of this dinner party is that it almost put an end to my career. I got a pea into my windpipe which all but choked me. I had to leave the table and the room. But it came out just as I was getting blue and desperate.

Later in the year we spent a week in the Nilgiri Hills at a health resort 6000

274

feet above sea-level, beautifully situated amidst hills covered by forests of rhododendrons in full flower.

Now I come to my teaching and research work, which had been the motive for our Indian journey. Actually there is not much to report. I gave a series of lectures which were attended by Raman and his staff, by the heads and assistants of some related departments and by a few post-graduate students. I cannot remember exactly what I talked about, presumably the theory of crystal lattices, in particular crystal optics, which includes Raman's famous discovery, the Raman Effect. There was only one young scholar with whom I came into closer contact, Nagendra Nath. We took up some work initiated by Jordan and by Kronig, the 'neutrino theory of light'; the idea was to consider the photon (with spin 1 and Bose statistics) as a kind of combination of two neutrinos (with spin $\frac{1}{2}$ and Fermi statistics). I was never very enthusiastic about this suggestion, and nothing has come out of it. Nagendra Nath later published another paper on the subject. I had the impression that he was gifted and keen on research, but after I had left India, I heard nothing more about him or his work.

I gave a lecture to the South Indian Science Association (9 November 1935) with the title 'The Mysterious Number 137', a survey of the situation at that time of quantum electrodynamics and the dimensionless constant $hc/e^2 = 137$ which determines the coupling of the electromagnetic field and the charges.

I had innumerable discussions with Raman and his collaborators on their experimental work, mostly optical problems connected with the Raman effect. They seemed to me not very interesting; in any case I was not very useful to the experimenters. On the other hand there were several violent disputes between Raman and me about his theoretical ideas. But on the whole we were on very friendly terms.

We had some social contact with the students. Once they invited us to a 'gymkhana' with all kinds of sporting competitions. They had noticed that we often played tennis in the early morning, before it became hot, and so they asked us to take part in a tennis tournament. This took place in the afternoon when the sunshine was still strong, and had a bad after-effect on my heart, as mentioned already. There was also a performance of Indian dancing by a young student, Ramgopal. We were delighted by the beauty of his body and movements and asked him later to dance at a tea-party given by us to the members of the staff of the Institute. We invited the ladies to this party as well; quite a number of them came, but while the men all spoke English and could converse with each another, the ladies, who had their own separate table, spoke only their dialects, and as there were almost as many different tongues as persons, they could hardly communicate. We had the impression that our Indian guests did not admire the dancer Ramgopal as much as we did because he was too 'Europeanized'. In fact he later became an international celebrity and danced in all the major cities of the world.

Events like these have stuck more firmly in my memory than the scientific work I tried to do on a verandah of the Institute. I do not remember what this was about, but I do remember that monkeys used to come and to watch me, and that they once stole my glue and ate it!

I tried to learn something about the government of Mysore. We did not meet the Maharaja, nor his brother the Yuvaraja (apart from seeing them at the Dasara Festival) but we came to know the real ruler of the state, the Divan (prime minister), Sir Mirza Ismael. He was a Mohammedan, while the majority of the inhabitants of Mysore and the family of the Maharaja were Hindus. But this seemed to cause no friction. Once Sir Mirza invited me to a discussion in his office. He asked me whether I could recommend to him a young architect, preferably not British, to help in the extended building programme of the state. Now by a strange coincidence I could. My nephew Otto, eldest son of my sister Käthe Königsberger, had graduated, some years earlier, at the *Techniasche Hochschule* in Berlin-Charlottenburg, and a short time later won a highly valued prize for which big architectural firms had competed. But when Hitler came to power in 1933, Otto had no chance in Germany. He found a job as architectural assistant to the well-known archaeologist Professor Borchardt in Egypt, and there he was at that time. I could recommend him with a good conscience, and a few years later he actually got the appointment to Mysore and had a very successful career in India, first in Mysore, later in New Delhi.

Sir Mirza explained to me his ideas about raising the standard of living amongst the peasants. He was sceptical about big projects and preferred things such as to induce—if necessary by law—every peasant to have one banana tree, whose fruits should not be sold but eaten by the family. I was impressed by this simple kind of social legislation.

Life in India was very pleasant for us. Hedi enjoyed it even more than myself. She met a Swami of the Ramakrishna order, and they became great friends. He told her that she had been an Indian woman in an earlier incarnation because she understood Indian spiritual life so well and looked quite Indian when wearing a sari.

So when Raman told me he wished to offer me a permanent position at his Institute, Hedi was at once in favour of accepting it. I was not so very keen on this idea. I thoroughly disliked the social conditions, the poverty of the peasants, the wealth and luxurious life of the Maharajas and their high officials, the gulf which divided the Indian people from the ruling British, the superstitions and antiquated customs of daily life in all sections of Indian society, and the caste system, in particular the treatment of the untouchables. But as I had no other job, I was willing to accept Raman's offer, if he could obtain the consent of the Council of the Institute.

To get this consent he applied all his energy and cleverness; but, alas, he showed also a considerable lack of tact and consideration. He sent out invitations for an extraordinary Council meeting and—as he explained to me proudly and confidentially—he arranged the mailing in such a way that the Bengali members of the council with whom he was not on good terms would receive it too late. Then he insisted on my attending the next faculty meeting which had to decide on bringing my appointment before the Council. I strongly objected but he overcame my resistance. Now this faculty meeting was one of the most horrid experiences of my life. I had to listen to a talk by Raman about my merits as a scientist, a teacher and a man. I felt that it was not well received by some members of the faculty. The English professor Aston went up and spoke in a most unpleasant way against Raman's motion, declaring that a second-rank foreigner driven out from his own country was not good enough for them. This was particularly disappointing since we had been kind to the Astons, as I mentioned before. I was so shaken that when I returned to Hedi I simply cried.

After that meeting there was of course no question of our staying at the Indian Institute, and we began preparing for our journey home. Later we heard that Raman's trick with the Bengali members of the Council had been disclosed and led to his resignation as head of the Indian Institute of Science. He succeeded in collecting money for the building of a private laboratory and has worked there ever since. I presume that this regrettable result of my visit rankled in his mind, for later a difference of opinion about scientific questions developed between us. Otherwise the bitterness of his attacks against me and his personal behaviour are hard to understand. The scientific dispute was concerned with lattice dynamics. The number of frequencies of any quasi-elastic system, no matter whether treated by classical or by quantum mechanics, is equal to the number of 'degrees of freedom', i.e. three times the number of particles. A system of four particles has $4 \times 3 = 12$ frequencies and a crystal of about 10^{24} atoms has the corresponding enormous number of frequencies which can be regarded as a practically continuous distribution. Now when experiments on a so-called second-order Raman effect became possible, a spectrum was observed where distinct peaks rose from a diffuse background. Raman interpreted this as a proof that the vibrational spectrum is not quasi-continuous but consists of a finite (rather small) number of frequencies, and he developed a theory of lattice vibrations on this basis. We had a few exchanges of opinion about this, mostly in *Nature* notes. Raman attacked me and my pupils also in his larger publications but we hardly answered him, as we were sure that most physicists were on our side. Today the question has been experimentally decided in my favour, as the whole spectrum of lattice vibrations is now directly observable with the help of neutron scattering. I have just (August 1963) attended an international meeting on lattice dynamics in Copenhagen where this subject was thoroughly treated. Raman was not even invited to attend.

We have met Raman twice since we left India. The first occasion was a conference in Bordeaux to celebrate the twenty-fifth anniversary of the discovery of the Raman effect. He received an honorary doctor's degree, but the same degree was also conferred upon me. I am sure that the French colleagues did this to demonstrate that in the dispute about lattice vibrations not Raman but I was right. At the first reception in Bordeaux we greeted each other very cordially and had a lively talk. Then Raman abused some theoretical physicist (I have forgotten whom) because he had done experiments which Raman regarded as poor; I replied: 'But, my dear Raman, what about the other way round, when experimentalists venture to make theories?' or something like that. Though he first remained quite friendly he later became furious and said to Hedi, his neighbour at the banquet, that I had given him deadly offence and that he would leave the conference. She had great trouble in appeasing him, but during the whole congress he was nervous, excitable and aggressive.

The second time, we met Raman at one of the Lindau meetings of Nobel Laureates. He was sitting at the next table in the dining room of the Schachen Hotel, greeted us in a very friendly manner and talked with us in his lively manner from one table to the next. But the next day his attitude had changed. He avoided us and went out of the way when we met in the house or garden. He must have suddenly remembered that I was his 'enemy'.

Actually I never was. I still admire his fascinating personality, his devotion to science and research. It makes me sad to think that by inviting me to India and trying to keep me there permanently he has brought himself into a precarious situation, and had to give up this leading position at the Institute of Science. But I cannot see that I am to blame for this misfortune. Nor can I accept a scientific theory which I regard as wrong. Hedi and I regret all this and particularly the split between us and Lady Raman, whom we loved dearly.

When we left Bangalore there was a crowd of Indian friends at the station to see us off. They gave us many lovely garlands and the scent was so strong that we had to throw a great many out of the window of the compartment.

When our ship entered the Canal at Suez, my nephew Otto, mentioned above, came on board and travelled with us to Port Said in order to discuss with me the offer of a job by Sir Mirza Ismael. That is all I remember of our return journey.

VII. Edinburgh. The Department of Applied Mathematics

Back in Cambridge in spring 1936, I had a rather uncomfortable time as I did not know whether my position there would be prolonged. My friends, primarily the mathematician Max Newman, Dean of the Faculty of Science, made great efforts to persuade the university authorities to provide me with a more permanent job, if possible a readership; but it was quite uncertain, and we considered other possibilities, even emigration to America.

Just then, in the summer of 1936, came a surprising offer from Moscow. Peter Kapitza wrote to me and offered me a professorship at Moscow University.

Kapitza was well known to us since our Göttingen days. He had left Russia during the Bolshevik revolution, studied in Cambridge and there made a brilliant career. He became a fellow of Trinity College and worked at the Cavendish Laboratory with such success that a special building, the Mond Laboratory, was erected under the auspices of the Royal Society, for his investigations on strong magnetic fields and other subjects. When we came to Cambridge, he and his charming wife Anna received us very kindly. At that period he had made his peace with the communists and often went to Russia during the long vacations. It happened, I think in 1936, that he was retained there. The reason was this. The Russian physicist Gamow, later well known through his research as well as his amusing popular books, used to visit the centres of physics in the western world. But one year—it must have been 1932—he did not get a permit to leave Russia. He made several attempts to escape but each time was caught. Then he induced some distinguished physicists, amongst them Einstein, Bohr and Langevin, to send a letter to the Soviet Government in which they guaranteed his return with their signatures. Thus he got the permit; but he did not return. After having visited the European centres of physics he settled in the United States.

The Soviet Government was of course very angry, and in retaliation, when Kapitza appeared in Moscow, they did not permit him to return to England. They offered him a good position and a laboratory. He first refused, but after some time changed his mind and accepted. The Royal Society behaved very generously towards the Russians by sending the whole of Kapitza's very expensive equipment at the Mond Laboratory to Moscow.

Now he was settled in Moscow and invited me to join him there. He promised me a rather grand position. I remember in particular a sentence in one of his letters which showed that at that time he was already converted to communism and believed in its victory; he wrote: 'Now, Born, is the time to make your decision whether you will be on the right or the wrong side in the coming political struggle.'

We were of course not very keen on going to Russia, which would mean learning a new, very complicated language, uprooting the children a second time and starting an entirely new life. Still, as we believed ourselves to have hardly a chance in England, we thought we might have a look at Moscow and its university. So we applied for Russian visas and made enquiries about the journey.

But we also considered the case that we might have to remain in Cambridge on the small income of a lecturer. In this case the rent of our house in Hills Road was much too high. Hedi, as usual in such emergencies, took the initiative. One day she came home and said to the family assembled for tea: 'I think I have bought a house.' It turned out that, walking along Hills Road, she saw a small house with a 'For Sale' signboard. So she approached a lady in the next-door house, who showed her the empty house and told her the name of the solicitor in charge of it. And she added: 'If you tell him you are a friend of mine he will give you the house for a much cheaper price.' As Hedi liked the house, she went at once into town, called on the lawyer and asked him whether she, as a friend of Mrs What's-her-name, could have the house for a reduced price. His answer was: 'Yes, if you sign the contract here and now.' This she did, and now she was a little overwhelmed by her own audacity. The whole family went at once to inspect the house, and we all found it nice and suitable. But we never moved in.

During this period I received several invitations to lecture at other universities. One of these was Edinburgh, where the chair of theoretical physics—the official title was Tait Chair of Natural Philosophy—was held by my old friend Charles Galton Darwin. My lecture on non-linear electrodynamics was received well, and I had a pleasant time in the house of Charles and his wife Catherine. They drove me to see the sights of Edinburgh and the surrounding country and I remember how impressed I was by the rocky hill called Arthur's Seat just inside the city, by the Forth Bridge and by the bare, steep Pentland Hills. I had no inkling that we would soon be citizens of this beautiful city.

Indeed, Darwin's letter in which he informed me that he had resigned his chair to become Master of one of the Cambridge colleges and that I was chosen to be his successor, was a complete surprise and a tremendous joy for us all. It solved at once all our difficulties and promised a secure and stable future. Although it is the practice at British universities that the resigning professor does not belong to the committee which chooses his successor, I am sure that in this case Darwin's opinion was requested and followed, and I am deeply grateful for this.

We dropped all other projects and accepted the offer from Edinburgh. The house in Hills Road purchased by Hedi was soon sold again without loss of money. The journey to Moscow was abandoned. The events of the following years in Russia showed that accepting Kapitza's offer would have been a catastrophe; Stalin's 'purges' eliminated not only politicians and soldiers whom he regarded as possible rivals for power, but also many of the foreigners who lived in Russia either as refugees from Nazi terror or as experts called in by the Soviet Government.

We went to Edinburgh and looked for a house, assisted by Mrs Whittaker, wife of the mathematician. Hedi chose one of the first houses we saw, 84 Grange Loan; it was perhaps a hundred years old, had a well designed front of smooth stone, while the other walls consisted of rough, unworked stones in the Scottish manner; its main advantage was that it had only two storeys, connected by a not too high and steep staircase. There was not much to do in the line of repairs and improvements, apart from the kitchen and the installation of two iron stoves instead of open fireplaces. There was a small garden in front and a larger one at the back with many apple trees and opportunities for Hedi to work and improve it; and indeed, it soon became very beautiful—and later during the war very useful as a producer of fruits and vegetables. There was also a garage, but as we had no car, it was used as a store-room for gardening tools, superfluous furniture, empty boxes etc., replacing the cellar and attic of German houses.

This house in Edinburgh was our home for the next seventeen years, the longest period we have ever been in the same place. It was also a very happy period. We learned to love the town, the country and the Scottish people, and when the war broke out we hardly expected ever to leave Scotland. Yet we did leave in the end. It is not easy to understand why, even for ourselves. Germany had been in the grip of the most horrible gang of criminals which ever came to power in the whole of recorded history; the German people had submitted to them without resistance and fought a cruel war for them; we had lost numerous relatives and friends through the Nazi terror, heard the horrible tales of the concentration camps and the murder of millions of Jewish people and we witnessed the overthrow of peaceful nations. Yet in spite of all this, there remained an extinguishable homesickness for the German language and landscape. On the other hand, Scotland had invited and accepted us, given us nothing but kindness, opened our minds to the ways of democracy and political fairness, and widened our horizon by making us members of the great British community of nations, the Commonwealth. But still we were not Scots and would never be. Germany meant for us a struggle between hatred and love, Scotland one between love and strangeness. Thus the choice was difficult. In the end the decision was helped by rather trivial factors of a practical, financial nature.

Our life in Edinburgh consisted of three distinct periods: before, during, and after the second world war. In the first period our life was externally peaceful but we felt a constant tension due to the growing Nazi threat. We had the impression that the British did not grasp the danger and did nothing to counter it. We hated war, but could not help agreeing with Churchill, who

281

incessantly warned against Hitler's plans of aggression and pleaded the necessity to arm, and to arm quickly. We watched, with increasing apprehension, Hitler's progress: his attacks against Czechoslovakia and Austria, the Munich negotiations ending with Chamberlain's foolish declaration that he had succeeded in securing 'peace in our time'. However, I do not want to write about politics, but to describe our life and the people with whom we came in contact.

My department in Drummond Street was quite near to the main University building, Old College, in one of the many dark, depressing slum districts. The building had been a hospital in which Lord Lister, the great surgeon of antiseptic fame, had once worked. Now it contained two scientific departments, that of experimental physics or 'natural philosophy', as physics is traditionally called in Scotland, and that of 'applied mathematics', attached to my 'Tait Chair of Natural Philosophy'. The latter was part of the basement and consisted of one enormous, high room, a small room used as office and a lecture-room with steeply raked benches such as are used in medical 'theatres' but quite pointless for a theoretical subject. It was not much, but more than I ever had in Göttingen, and I was quite content with it.

My staff consisted of one lecturer, Dr Robert Schlapp, called Robin by his friends, among whom we soon belonged, and a half-time typist/secretary. Schlapp's father had been professor of German and himself had emigrated from Germany in his youth. He and his wife were most charming and cultivated people who lived in Peel's Terrace, not very far from us. Robin himself was a great help. I had to run the department according to a syllabus printed in the annual University Calendar, prescribing lectures on elementary statics and dynamics and a little electromagnetic theory. It at no point reached the level of modern knowledge and research. The students were trained in solving problems of a type which was a residue of an ancient and—in my opinion—quite out-moded tradition. In Göttingen we used to make fun of this kind of problem when we found them in English text-books. There was one on dynamics, I think by Rouse, which contained the most abstruse examples. It was only a slight exaggeration when we made fun of it by quoting the following (invented) example: 'On an elastic bridge stands an elephant of negligible mass; on his trunk sits a mosquito of mass m. Calculate the vibrations of the bridge when the elephant moves the mosquito by rotating his trunk.' It took me quite a time and a considerable effort to learn how to lecture about these things using the traditional terminology. There was a tutorial hour each week where such questions were practised. The students then occupied only every second bench, so that the teacher could pass between the rows and look at their work. As the steps between the rows were steep this meant a quite considerable amount of climbing, and as the same errors were repeated by many students, the same thing had to be explained many a time. This seemed to me a waste of physical and mental effort, and I suggested applying the method used in Germany: the students were allowed some time to find a solution and then one of them was asked to explain it on the blackboard. But this was a complete failure. No student came forward, and no amount of persuasion could overcome their shyness. That was particularly astonishing in that the same students showed not the slightest shyness as actors in their theatrical society, to which they invited us. We enjoyed the performance of a play where the same boys who refused to address their class acted and spoke in the most natural manner.

A great difficulty for me was the preparation of questions for the tutorials and examinations. At the end of each set of lectures was a class examination, and there were final examinations at the end of the three-year courses. To find suitable questions was not difficult as English text-books contain a lot of them. There were also collections of examination questions. As I was not accustomed to this kind of problem, it took me more time to solve them than was available to the students in the examination. Therefore I had the feeling that I myself would not be able to pass the examination with a decent mark. In the course of time I improved a bit, but never felt happy in this part of my work.

Another difficulty was the reading and judging of the examination papers. English or Scottish handwriting is a little different from the continental, and the papers written down in a hurry were often hard to decipher.

Then there was the question of marking. This was traditionally done by points. Each paper consisted of ten or twelve questions; each question of two parts, one asking for a definition or a theorem and its proof, so-called 'book work'; the other putting a numerical example. The perfect solution of all papers was valued 100 points, and each question had a maximum figure, about 10, according to its difficulty. Now at the beginning it seemed to me quite impossible to decide which percentage should be given to a student's performance. But Robin Schlapp insisted it could be done. So we marked a set of papers independently, and to my astonishment the results were in good agreement. I changed my mind and am now convinced that the British way of written examinations and marking gives reliable results. Compared with the methods of oral examinations as practised in Germany, it has the advantage that it is free from personal prejudices and the temper of the examiner, particularly if there is an external examiner who does not know the candidates.

Soon a few research students appeared. In the course of time there came Australians. Indians, Egyptians, Americans, Argentinians, Frenchmen, Chinese, and other nationalities, but very few Scots. During the whole seventeen years of my teaching in Edinburgh I had only two Scottish students, and both deserted our subject after graduation and chose other professions. One, named Begbie, was a Quaker and pacifist; he refused to join the army, came before a tribunal for conscientious objectors and was ordered to do agricultural work. He had a job in one of Edinburgh's big gardening firms; it allowed him time enough to join my department and work on a doctoral thesis. After having got his degree, he was sent to the front with an ambulance unit and was so attracted by this kind of work that he decided—after persuading his parents to help with the finance—to study medicine. Today he is a lecturer in the Faculty of Medicine in Edinburgh.

The other deserter was a girl, Miss Helen Smith, very pretty, clever and gifted. About the time of her graduation the dropping of atomic bombs on

Japanese cities became known. As a result of this, she decided that she would never have anything more to do with physics, began to study law, passed all the examinations and is now a flourishing lawyer in Aberdeen. I wish scientists would take her as an example of conscientious behaviour and refuse all work connected with mass destruction.

The reason why so few Scotsmen joined my department to do research and get a higher degree was that there were fellowships available for first class students to go to Cambridge. The fame of Cambridge guaranteed a good career. I tried several times to persuade my faculty that this arrangement was injurious to its interest, but in vain. Many of my colleagues were themselves Cambridge men and proud of it.

I had very little help from Schlapp in teaching research students. Although he had started with excellent research work and possessed a wide and thorough knowledge of physics he had no wish to contribute to it. He was too modest and did not believe in his ability to do research and he loved teaching and talking to students. He did not even regularly take part in my seminar, which I started as soon as some research students had arrived.

One of the first of these was Klaus Fuchs, later so well known through the spy affair in which he was involved. One day I received a letter from Professor N. Mott in Bristol, telling me that he had an excellent German refugee student and wished him to study in Edinburgh to learn my special methods. I thought this rather strange, as Mott had been in Göttingen and knew my ways of thinking very well; further, I had no funds available to support Fuchs. Mott answered that he would take care of the financial side of the matter. Fuchs arrived, and we found him a very nice, quiet fellow with sad eyes; he was extremely gifted, on quite a different level from all my other pupils in Edinburgh. He had suffered from the Nazis more than the ordinary refugees, though he was not Jewish, but the son of an evangelical pastor. Some of his relatives had been killed, and he had only narrowly escaped. He never concealed that he was a convinced communist. But he did not speak about politics very much, except during the time of the Soviet–Finnish war. All of us in the department, including Indians and Chinese, were sympathetic to the Finnish side, while Fuchs was passionately pro-Russian.

I enjoyed working with Fuchs so much that I wondered why Mott had sent him away. This was explained when I encountered Mott at a meeting in London. He asked me how I was getting on with Fuchs, and when I answered 'splendidly', and praised his talent, Mott said 'What a pity I had to get rid of him. He spread communist propaganda among the undergradu: :es.' Mott told me that he had arranged for his own contribution to the general refugee fund to be directed to Fuchs, a generous gesture which possibly also showed how much he was afraid of communist propaganda.*

---

*Sir Nevill Mott comments: 'I must have made a remark which Born misunderstood or took more seriously than I intended. I do not remember believing that Fuchs spread communist propaganda among the students, and at a time when Hitler was the enemy I would not have worried unduly if he had. What happened was this. In Bristol we had research funds from the

Fuchs and I collaborated on several different problems. We improved the paper by Joseph Mayer (husband of my brilliant pupil Maria Goeppert who has just in 1963 been awarded the Nobel Prize) on the kinetic theory of condensed systems, which aims at generalizing the equation of state of Van der Waals. We worked on the fluctuations in the black-body radiation but discovered later that Heisenberg had done the same, and better. Fuchs assisted me also in working out an idea of mine, which I called the 'Principle of Reciprocity'. It seemed to me that the symmetry (or more precisely: antisymmetry) between coordinates and momenta which appears in the commutation law $pq - qp = h/2\pi i$ might have a deeper meaning, that a general principle of physics was behind it. At that time physicists were already convinced that the next generalization concerned with the explanation of elementary particles would need the introduction of an 'absolute length' a of the order of 10^{-13} cm (nuclear dimensions). Then Planck's constant h could be split into a product, $h = ab$, where b is an absolute momentum.

Thus I was led to the postulate that all laws of physics should have a reciprocal counterpart, and I have attacked this in several ways. One way which I followed at that time with Fuchs and some Chinese students led to quite interesting mathematical results about reciprocal properties of wave functions, but to no physical consequences.

Altogether I have spent a lot of time pondering over this reciprocity idea, and I think it was wasted time. Actually this approach to physics did not correspond to my general philosophical conviction that all progress in physics is due to the rational interpretation of facts and not to speculation. But though I was sceptical about speculative theories like those of Eddington and Milne, and also about Einstein's late work on a unified field theory, I succumbed to the attention of this reciprocity idea and even induced some of my pupils, amongst them Fuchs, to work in this line. We published several papers in the *Proceedings of the Royal Society of Edinburgh*, but as this periodical is not widely read, there was no response, not even a contradictory one.

Apart from this, Fuchs studied nuclear physics, where I could teach him nothing, and he published a number of papers on this subject.

Now I shall report on the 'espionage' case of Fuchs as far as it touched our life in Edinburgh.

Almost at the same time as Fuchs appeared in my department, another German refugee, named Kellermann, arrived. He and Fuchs soon became attached to one another. They were rather different in body and in mind: Fuchs slim, with delicate limbs, weak in appearance but with a powerful brain, taciturn, with a veiled expression which disclosed nothing of his thoughts; Kellermann tall, robust, with open, frank, almost child-like face, not highly gifted but careful and industrious. Kellermann's thesis on a problem of lattice

---

generous gifts of the Wills family, and with these and help from the Academic Assistance Council we built up a very strong group of physicists who had left Germany in 1933. Some we wished to keep; but established positions then as now were few and far between and for others we helped as far as we could to find jobs elsewhere. This is how we acted about Fuchs.'

dynamics (vibrations of the sodium chloride lattice), which appeared in the *Transactions of the Royal Society*, was not a work of genius but very solid and reliable; it was the first example of the calculation of a complete frequency spectrum of an ionic lattice and has frequently been quoted.

In spite of their differences, Fuchs and Kellermann became inseparable. But I do not think that Kellermann penetrated deeper into Fuchs's private thoughts than any of us.

Fuchs graduated first with a Ph.D. and about two years later gained a Sc.D. The subject of his Ph.D. thesis was something in nuclear theory, not in my line, therefore I have forgotten it. For his Sc.D. he submitted a whole little volume of papers, most of them written without my assistance.

When the war broke out, we refugees from Germany of course stuck together to discuss the situation. It seemed to us serious because we knew the efficiency of Hitler's *Wehrmacht* and the fanaticism of the Nazis, and we saw no evidence of war preparations in Britain.

We thought it was stupid not to attack Hitler before his army was fully developed; the quiet period of the 'phoney war' seemed to us ominous. It was a strange time of expectation and tension, and it ended as we had foreseen: Holland, Belgium and France were overrun. I think it was after Dunkirk that all the refugees from Germany were interned. Fortunately Hedi and I had become British subjects by naturalization just a few weeks before the declaration of war and so remained free.

One morning when I came to my department I found Fuchs and Kellermann not at their usual places. The other research students had heard rumours of the internment of Germans. So I picked up the telephone and soon found out what had happened and where my men were, namely in a provisional camp established in one of the big schools which were empty during vacations. I hurried there and found a great number of 'aliens' in very poor circumstances. There had been no time for preparations, and the major responsible had lost his head; not even straw bags were available for the prisoners. I soon discovered not only Fuchs and Kellermann, but also other friends and acquaintances, 'Aryan' and Jewish alike. The whole affair was due to panicking by the authorities. Here Fuchs's ability became evident. He took the initiative of approaching the officer in command and not only insisted on getting the most urgent commodities for the prisoners, but helped the staff to get them.

The men were soon transported to different camps in England and the Isle of Man, and later taken to Canada. I was told that all the time Fuchs played the role of a leader who took care of his comrades efficiently and prudently.

The detention did not last very long, especially for those who could be used for war work. Thus Fuchs was back in my department one day in the summer of 1941, quite unchanged by the adventure of deportation. He was disinclined to speak about it, and we took up some work. Then he got a letter from Peierls, who had become professor of mathematical physics at Birmingham University. Fuchs showed the letter to me and asked my advice. It contained an invitation to join a group of physicists on important war work. I do not remember whether this was specified, but we both knew what it meant. Just before the outbreak of the war, I had been at a conference in Cambridge and met there Leo Szilard, a refugee physicist from Hungary who was in a state of utmost excitement because he had convinced himself that an atom bomb could be made. We had heard rumours of the discovery of nuclear fission by Hahn and Strassmann, and I think that we soon discussed the possibility of exploiting the gigantic energy involved. But we had not worried about it. Szilard however had considered the chain reaction, which leads to a nuclear explosion. He saw a formidable danger in the possibility that Hitler might concentrate on the construction of this super-weapon and thus conquer the world. I think that I had discussed this with Fuchs after my return from Cambridge. In any case Fuchs was much more expert in these matters than I was, as he knew much more about nuclear physics. He was inclined to accept Peierls's offer. I was doubtful and warned him. Already in the first world war I had disliked doing war work, though sound-ranging was relatively harmless; I considered it as a welcome way to pull students and young scientists out of the front and save their lives (as I have described in Chapter XV). Though my recollections of these discussions with Fuchs are dim, I believe I had a strong feeling that an atomic super-bomb would be a devilish invention and I wanted nothing to do with it. For though I hated Hitler and the Nazis more than I can express, and though I despised the German people because they had brought him to power and fought for him like lunatics, I could never bring myself to consent to actions by which not only Nazis and Hitler's soldiers were killed but also innocent children and people who shared my feelings. But Fuchs thought otherwise. He hated Hitler and his gang so violently that he was willing to use any weapon to destroy them and to prevent the world from getting into their grip. So he accepted Peierls's offer and disappeared.

We heard nothing from him for a long time, and little about the progress of the fission project. We knew that the British team had been transferred to the U.S.A. and Canada. Later, I have forgotten whether during the war or after the armistice and Hiroshima, when food was still scarce in Great Britain, we received a few parcels with very welcome edibles from America, sent by Fuchs.

I ought to speak now about our reactions to the news of the nuclear attack on Japan. But this is a big subject, and I prefer to continue the tale of Fuchs.

Some time after the armistice, the British nuclear team returned to England. The Atomic Energy Establishment at Harwell, near Oxford, was founded, with Sir John Cockcroft as its chief and Klaus Fuchs head of one big department.

Cockcroft invited me to give a lecture to his staff on our work in Edinburgh. When I arrived in Harwell, he and Fuchs received me at the station and Fuchs was attached to me as a guide to show me the laboratories. I spent a very pleasant day with him, not only in the cyclotron hall and other scientific places, but also in his tiny and tidy prefabricated house where he gave me a pleasant lunch. He talked freely about problems of science, organization and technology but said nothing about his personal experiences and plans.

It turned out that his father, Emil Fuchs, originally an evangelical pastor, later a Quaker and remarkably at the same time a convinced communist, was in Oxford. So I invited both of them to dinner for the next evening, in an Oxford hotel. I remember very little of our talk, but one thing has stuck. Klaus complained that the routine work on nuclear technology—construction of reactors and such things—was rather dull. So I asked him why he remained there and did not return to academic work, and I offered him my help in getting a good job at a university. His reply was: 'Thank you, but it is too late.' This struck me as strange; I could not understand what he might mean, for he was still a young fellow.

I met Fuchs once more at a meeting on elementary particles held in Edinburgh during 14–16 November 1949. I have a photo of the members grouped in front of the portico in the court of the natural philosophy building. It shows many of my colleagues and friends, such as Darwin, Peierls, Feather, Fröhlich, Powell, Proca, Bopp, Möller, Rosenfeld, Kemmer and Pryce. In the front row Klaus Fuchs can be seen, sitting beside Janossy, who was and is a communist (now in Budapest) and, just two rows above, Pontecorvo, who a short time later was also found to have given away secrets to the Soviets and vanished to Moscow. I wonder whether this grouping was accidental.

One day soon after, when coming home for lunch, I found Hedi in a state of great agitation: 'Have you heard about Fuchs? They say he has been arrested for being a spy and a traitor.' A journalist had rung her up wishing to interview me, and when she asked why, told her of Fuchs's arrest. She simply could not believe him; nor could I, as Fuchs had always appeared to us a most likable, kind, harmless fellow. We had to stand some interviews by reporters and could only repeat that the whole affair had taken us completely by surprise, that we knew nothing and had always regarded Fuchs as perfectly honest. I still think in a way he was. He believed in communism and considered it his duty to supply the Soviets with secret information which could save the world from being helpless in the hands of American capitalists. However he broke his promise to keep secret his knowledge of nuclear technology, and he did so without considering the awkward and dangerous position into which he brought other refugee scientists, among them friends of his like ourselves. He has suffered for this by a long imprisonment.

Concerning ourselves, we encountered no serious difficulties through the Fuchs affair. My colleagues behaved in the most decent manner; they refused to give any information about me to journalists and asked them to interview me directly. But none came.

It was a little different during the Pontecorvo affair, which happened a short time later. Then some newspapers had on the top of their front page a kind of 'rogues' gallery', containing the portraits of a number of refugee scientists who had obtained professorships, Peierls in Birmingham, Fröhlich in Liverpool, Frisch in Cambridge, Freundlich in St Andrews, myself in Edinburgh, and perhaps one or two more, with a caption such as: 'These scientists of foreign origin have delivered their passports to the authorities to show their allegiance and their intention to stay in Great Britain.' This was of course untrue. A young reporter, after having tried to interview my experimental colleague Feather without success, came to our house and wanted to question me. But it was not difficult to get the better of him by showing him the meanness of this kind of pictorial journalism, and no interview with me was published.

All this happened after the war. I return now to the pre-war time and to my department in the basement of the physical building in Drummond Street.

We had several guests there who were above the level of research students. One was Reinhold Fürth, professor of theoretical physics at Prague University. He came with his wife to Great Britain, when Czechoslovakia was overrun by Hitler's armies. I knew him from German and Austrian physics congresses and respected him as a scholar of high rank. He had worked mainly in the field of statistical mechanics, in particular about thermal fluctuations, and had published several books, apart from many papers. He continued to work in this line, and we joined forces on several problems, for instance in considerations about the stability of crystal lattices. I started these with a paper in the *Proceedings of the Cambridge Philosophical Society* (1940), and some of my collaborators, amongst them Fürth, followed with a series of papers which eventually ran to about ten, some of them quite good. Then I had an idea on how to construct a photoelectric Fourier transformer which would show the transformation of a function (cut out of cardboard) on the screen of an oscillograph. Fürth and a young man from the experimental department, named Pringle, constructed the instrument, and when it worked, the well known firm Ferranti took it over and tried to improve it; but without great success.

The Fürths took a house in our road quite near to our own; we saw one another very often and played bridge together. They were a very welcome addition to our circle. Later he became a reader at Birkbeck College, London.

The first of my Chinese students was a small, sturdy fellow named Peng. His gifts were quite outstanding; he published a few papers and got a Ph.D. degree. I remember that once he made a mistake in a theoretical consideration, and when it was discovered, he was so depressed that he decided to give up scientific research and to write instead a big *Encyclopaedia of Science* for the Chinese people which would contain all important discoveries and technological methods used in the West. When I said this seemed to me a task far too big for a single person, he answered that one Chinese could do as much work as ten Europeans. Asked about his attitude to the communists, whose final victory seemed very likely at that time, he said he hated politics and would retire to a quiet village in innermost China to work quietly and without interference. Events proved otherwise. He was appointed to a professorship at Schrödinger's Institute of Advanced Studies in Dublin, Eire, as successor to Heitler, who went to Zurich. I think Peng was the first Chinese to get a European professorship. After a few years he decided to return to China; before leaving he visited us and came all the way up to Ullapool in the Highlands where we were spending our holiday. He had left London in hot weather and taken no coat; but it rained heavily on the Scottish west coast, and we had to give him dry clothes and a mackintosh. We spent a few very nice days together and showed him a little of the Scottish landscape. Then he left us, and we have not seen him again. Nor did he ever write. But I heard about him through letters from my other Chinese friends. He did not go into isolation far away in the country but became a professor in Peking and a member of the Chinese Academy. A few years ago, during the Jubilee celebrations of the Royal Society, I received a letter from him written in London. It turned out that he was one of the Chinese delegates, perhaps their head (I do not know). Hence he had become a great man in the communist state which once he had despised. I answered at once and asked why he had not written to me all these years, and I got a second letter from him saying he did not wish to cause trouble for me through receiving a letter from communist China. Apparently the Chinese were told that it was dangerous for a European or American to have any Chinese correspondence. The strange thing is that a clever man like Peng believed this; he could have just asked one of my other Chinese pupils, most of whom corresponded with me.

Peng was followed in Edinburgh by two of his compatriots, Cheng and Yang, men of a very different type. While Peng was a very simple soul, apart from his uncanny scientific talent, and looked like a sturdy peasant, these two were refined, elegant, highly educated gentlemen, both very well trained in mathematics and also gifted in physics, though perhaps not on Peng's level.

The last of my Chinese, Kun Huang, cannot be called my pupil, as he was already a competent theoretical physicist when he came to me. He had studied as an I.C.I. Fellow in Liverpool under Fröhlich, who advised him to spend part of the holiday in my department in order to learn my methods, mainly dynamics of crystal lattices. Now I had a manuscript of a new book on this subject on which I had worked intermittently for a long time. My older books on the subject were written during the period before the discovery of quantum mechanics, and nobody had tackled systematically the crystal theory from the new standpoint. My idea was to derive the whole theory of crystals from first principles and deal with observable phenomena at the end of the deduction. I had worked out quite a large part of this project, but other duties and current work had prohibited me from continuing it. I gave this voluminous manuscript to Kun Huang who became deeply interested in it, discussed it with me and wrote one or two papers on special problems.

The idea occurred to me that he might be a suitable collaborator to finish it, and I suggested it to him. He did not accept my proposal at once but went back to Liverpool. Some months later he wrote me that he had obtained permission and financial assistance from the Chinese government to stay in Great Britain, and he intended to come to Edinburgh and finish that crystal book with me.

He came, and we had a most pleasant and interesting time. Huang did not approve of my plan of constructing the book in a strictly deductive manner. He was a convinced communist and materialist, and he had no use for abstract thinking; he regarded science as a way to improve the life of people. So he suggested that my systematic deduction of lattice dynamics should be preceded by some chapters containing elementary considerations which demonstrated the practical usefulness of the theory. We struggled about this, but as he made it his condition of collaboration without which the book would never be finished, I agreed. So the book now consists of two parts, an elementary one containing all the fundamental experimental applications, and a higher one on the general theory; the latter leads to some more involved experimental facts and predicts a series of new, mainly optical phenomena, e.g. about the fine structure of infra-red absorption (so-called 'residual rays') in crystals. Experimental investigations of these theoretical results have since (1963) been made at different places, e.g. Freiburg im Breisgau, with satisfactory results.

The text of the book in its final form is mainly due to Kun Huang. It is amazing that a Chinese was able to write a European language so fluently and correctly. In the end I had to do the final work before publishing. A short time before my retirement from my Edinburgh chair at the end of 1953, Kun Huang decided suddenly to return to China to take part in the communist reconstruction of his country. He married an English girl, secretary of Professor Fröhlich in Liverpool (as far as I know against the wish of her parents) and disappeared with her in the direction of China, leaving about three quarters of the manuscript with me and promising to send the rest soon.

I had to wait a long time for these final chapters. The end of my time in Edinburgh came and our removal back to Germany. Letters to Kun Huang were of no avail. I cannot quite remember how I succeeded eventually in inducing Kun Huang to finish the work, but I think it was through a colleague who visited Peking and called on him. Eventually the manuscript arrived, and I had to do the final editing work, proof-reading etc. The book was printed by the Clarendon Press, Oxford, in the most efficient manner. But as I was then over 70, it was one of the hardest and most gruelling jobs I have ever done. I had to compare every line, every formula, with Huang's not always easily readable handwriting and check all the calculations. In the end the result was satisfactory for the book looks attractive and not a single error has been found, either by myself or anybody else.

Recently (August 1963) I had the satisfaction of learning that all this effort was not in vain, for lattice dynamics has become high fashion in physics. It had already been shown in 1925, by the Swede Ivar Waller, that just as the Laue spots in the scattered X-rays determine the geometry of the lattice, the diffuse background determines the quasi-elastic lattice forces and hence the vibrations. During the last years of my time in Edinburgh my old Indian friend C. V. Raman directed violent attacks against the systematic theory of lattice vibrations as developed by me and tried to replace it by primitive ideas of his own. Thus I was led to occupy myself again with these matters, which I had considered as completely settled. My pupil Miss K. Sarginson wrote her thesis (published in *Proceedings of the Royal Society*, 1941) on this subject, and I wrote a big article for the annual *Reports of Progress of Physics* (1942–43), where I derived Waller's results in a modern form; in the same volume Kathleen Lonsdale published a comprehensive article on the experimental side of the subject, which at that time consisted in the observation and interpretation of the diffuse background of X-ray scattering.

Max Born with lifelong friend James Franck in Göttingen, in 1964

Then I forgot all about these things—until in the summer of this year (1963) I was invited to attend an international conference on lattice dynamics in Copenhagen. There I learned that neutron scattering has proved to be much more efficient than X-ray scattering, and that it was now possible to check the theory of lattice vibrations in detail. The old investigations started by Kármán and myself fifty years ago have thus become fashionable and my book written with Huang is now much used.

After this digression I return to Edinburgh and my collaborators. I wish to mention only two of them. One, Edward Corson, was an American who excelled as much by his eccentricity as by his scientific enthusiasm. In the space of less than a year while he was in my department, he published a considerable book on an abstract aspect of quantum mechanics and wrote the greater part of another book which appeared soon after (both with Blackie and Son, Glasgow). He kept my secretary busy and all members of the department in a permanent state of tension through his eccentricities. When the news of Fuchs's arrest came he sent a telegram to the authorities vouching for Fuchs's honesty. Later I heard that he had a hard time in the U.S.A., but he has now a decent teaching position.

Herbert S. Green was of a different stamp: very quiet, shy, retiring and highly gifted. The main work we did together was on the kinetic theory of condensed gases and liquids. I had followed the work done by my former pupil and collaborator Joseph Mayer, on the equation of state of a dense gas, by a careful elaboration of collision processes. Assisted by Klaus Fuchs, I published a paper (*Physica*, vol. 4, 1937) on it in which I declared it the most important step since Van der Waals, and suggested some mathematical simplifications. But in the end I was sceptical about Mayer's conclusions concerning condensation. Then I pondered about a new approach to the condensed, disordered (fluid) state. With Green's help, a new method was developed which starts in the phase space of $6N$ dimensions for N molecules and then reduces the continuity equation in this multi-dimensional space step by step down to the six-dimensional phase space of one molecule or the ordinary three-dimensional space for its coordinates. We later found that the French physicist Ivon had taken about the same line. Green and I published a series of papers on the subject which later appeared as a little volume (Cambridge University Press, 1949), and Green wrote a comprehensive book about it. Meanwhile the theory has developed, in the hands of Ivon, Mayer, Uhlenbeck, Kirkwood and others into a major branch of science, which I have not followed.

Postscript

Gustav Born

My father wrote these recollections, as their text shows, at various times before and after his retirement from the Edinburgh Chair in 1953. Although my parents had been happy in Scotland there was no question of staying on there, mainly because they found the climate too harsh. At first they considered moving to the South of England to be near children and grandchildren. However, as my father had worked in Britain for a comparatively short period, his British pension was far too small to live on whereas under the German restitution laws he was now receiving his full professorial salary from the University of Göttingen. That money was not transferable to Britain at the time although it became so some years later. This financial situation was one reason for a return to Germany. Another reason, to my parents more important, was their wish to contribute what they could to a democratic rehabilitation of Germany and the hope that my father's special position as a universally esteemed leader and teacher in physics might be put to use to help prevent any possibility of German rearmament and specifically to prevent atom bombs getting into German hands. In this, as it turned out, he achieved as much as any private individual could expect to do. Nevertheless, my parents' return to Germany understandably offended many other refugees, particularly those whose families had suffered more than ours, and who saw this move as too conciliatory.

My parents had a house built in Bad Pyrmont, a gracious little spa town embedded in rolling fields and forests not far from Hamelin on the Weser river. The reasons for going there were partly sentimental, partly practical: sentimental because they had spent part of their honeymoon there; practical because the town was near Göttingen but not so near as to have their envisaged existence of writing, reading, music and walking made impossible by too many visitors. Pyrmont is also the centre for north Germany of the Society of Friends which made it attractive to my mother who had become a Quaker in Edinburgh.

My father did not give up physics entirely but continued to work on its philosophical implications. He has described this as follows:

295

'On retiring from my Edinburgh chair I was awarded a *Festschrift* [a publication in his honour]. . . . containing a paper by Einstein, in which he presented a brief and clear argument for rejecting the statistical interpretation of quantum mechanics based on his concept of physical reality. I was unable to agree and even regarded his mathematical treatment of an example as insufficient. I wrote a reply in which I tried to justify my statistical standpoint by showing that the claim of classical mechanics to be deterministic is not justified because it depends on the assumption that absolutely precise data have a physical meaning, and this I regarded as absurd. So I developed a statistical formulation of classical mechanics. Then I offered a forthright quantum-mechanical treatment of Einstein's example and showed that, in the classical limit, it passed over exactly into the result previously attained from my statistical formulation of classical mechanics.

Einstein replied that I had misunderstood him, as his objections had to do with the concept of reality, not with determinism. A correspondence ensued which was full of mutual misunderstandings. Pauli, who was at Princeton at the time, tried to mediate and told me frankly that I was not a good listener to other people's opinions. I suppose he was right. But he helped me to reformulate my paper, and the final version, which appeared in a number of the *Transactions* of the Danish Academy in honor of Niels Bohr's seventieth birthday, met Pauli's complete approval. Although the dispute with Einstein was rather sharp, it did not in the least affect our friendship.

In Pyrmont I continued working along this line. In the end it led to the publication of a paper (with W. Ludwig) in which a formula was developed which represents the movement of a freely moving (e.g. rotating) body from the extreme quantum domain with discrete states up to the classical domain of continuity.'

(from *My Life and My Views*, pp. 44–45; New York: Scribner, 1968)

Towards the end of their first year in the little house in Pyrmont came the award of the Nobel Prize to my father (jointly with Professor Bothe). This was at just the right time to add weight to his main retirement occupation, which was to educate thinking people in Germany and elsewhere in the social, economic and political consequences of science. He was, of course, principally concerned with the implications of nuclear physics for war and peace but also with other, what he called pathological features of our scientific age such as rocketry and space travel. There seemed hardly any public interest in these matters at the time of my father's return to Germany. That changed rapidly when he and like-minded friends, notably Otto Hahn inside Germany and Einstein, Bertrand Russell and many others elsewhere began to challenge public opinion with the revolutionary implications of scientific discoveries and their technological exploitation. In the 'Göttingen Manifesto', eighteen leading atomic physicists including Born and Hahn declared that they would never take part in work that could lead to the possession of atomic weapons by Germany. My father was also a founding member of the Pugwash movement, and remained indefatigable in promoting his views through lectures, broadcasts and newspaper articles, and in conversations with people at all levels from his postman to Government ministers. I know from personal experience that the clarity and sincerity of his urgings made a profound and lasting impression on many people throughout the world and particularly so on the younger generations in Germany. These writings and speeches have been published in several collections (e.g. *My Life and My Views*). They remain well worth reading for prophetic insights and for original conclusions so well reasoned that they seem self-evident. It says much for the regeneration of Germany as a liberal democracy that, although his public utterances were often opposed to the policies of the Federal Government, there was never any question about his freedom to say what he liked, despite remaining a British citizen. On the contrary, he received from the Government the highest distinction in its gift, the Grosse Verdienstkreuz, which is the German equivalent of the Order of Merit. The purposefulness and fearlessness of his public utterances was appreciated not only by like-minded politicians but even by opponents such as Franz-Josef Strauss. For the dangerous posturings of the East-West protagonists he expressed disdain and detestation. Because of their mental similarity he referred to them as 'Khrushless' and 'Dullchev' in correspondence with Bertrand Russell (see Russell's Autobiography, Volume 3, page 136). With Russell my father shared an abiding interest in the impact of science, particularly of theoretical physics, on philosophy. Unlike most professional philosophers, Russell was also a trained mathematician. It was to this that my father attributed the singular logical and critical clarity of Russell's writings. He agreed with many of Russell's positions but by no means with all, especially in matters of personal ethics. About a week before he died I read aloud to him a letter just received from Bertrand Russell. He laughed about something Russell had written and said; 'That clever old rascal will outlive me by many years'. This particular prophesy was not fulfilled: Russell died only a few weeks later.

My father retained to the end his wonderful clarity of mind but in his eighties began to weaken physically. He was a fastidious man and detested the bodily failures which accompanied his last illness. It was probably these that made him say to me repeatedly and dispassionately that he had had enough. My father appeared to have little interest in biology and medicine, even in their strictly scientific aspects, which was strange in view of his curiosity about almost everything else. This made him rather naïve about his own condition and as my mother was similar in this respect, though for different reasons, it was no surprise that their bed-side drawers were stocked with a large assortment of drugs, many of them useless or inappropriate. He was not religious and regarded beliefs in individual survival as absurd, apart from the very limited continuity achieved through one's genes and the consequences of one's work. He certainly appeared unafraid during his last days, although somewhat quizzically thoughtful, as if he were trying to come to terms with the biological events that were dominating him and which he did not understand in the same way as he understood so much else.

These recollections give away little about my father's inner life. He was not in the least secretive in the ordinary sense, but his deepest feelings were hidden from everyone until he began to open up a little towards the end of his life, mainly to his children. Out of the hundreds of letters I had from him, I treasure one above all in which he calls me his closest friend and rather diffidently begins to say something about emotional relationships. The wealth of feelings inside him found resonance in the varieties of classical music he loved, from Bach to Brahms, but his emotions remained largely inaccessible through verbal expression. Why that was so I have, of course, asked myself innumerable times without finding a convincing answer. It may be that the loss of his mother when he was still so young made a naturally shy child even more so; but this begs all the important questions. It is interesting that my mother was exactly the opposite in her ability to show her emotions to others, both verbally and in an untold number of letters which were often so frank as to be deeply disturbing to the recipients. The sometimes stormy relationship of my father with his wife settled down in their later years to a symbiotic situation in which they came to depend completely on each other. After his death she deteriorated rapidly and died two years later. Her papers include a collection of poems which he composed annually for her birthday. They express a loving appreciation of their long life together which makes one glad that any difficulties between them had faded from his mind.

He obtained great enjoyment from poetry of all kinds, profound, profane or playful. What moved or amused him he rapidly learnt by heart. Thus he was able to recite much of the Iliad in school Greek, as well as Silesian doggerel in his native dialect. He clearly derived comfort from the wonderful Goethe poem which ends:

> Und solang du dies nicht hast
> Dieses: Stirb und werde!
> Bist du nur ein trüber Gast
> Auf der dunklen Erde.

> (And until you grasp this—'die and be transformed!'—you are naught but a sorry guest on the dark Earth.)

because he recalled it so often. During retirement he indulged his poetical whimsy by translating Wilhelm Busch into English; so well indeed that *Klecksel the Painter* became a successful publication in the United States (New York: Frederick Ungar Publishing Co., 1965). He was also very fond of *Alice in Wonderland* which he used as a quarry of quotations for lectures. The playfulness that appeared in his own poems also came out in other ways: in the funny faces he pulled whenever anyone wanted to photograph him and in gentle teasing of his grandchildren.

My parents' tombstone in the city cemetery in Göttingen bears the strange equation which he clearly considered his main single contribution to science, viz. $pq - qp = h/2\pi i$. But that the great mind of Max Born encompassed far more than science is well shown by a quotation from his Nobel Lecture, entitled 'Statistical interpretation of quantum mechanics', which seems to me so profound and important that it stands framed on my desk and makes an appropriate end to this postscript: 'I believe that ideas such as absolute certitude, absolute exactness, final truth, etc. are figments of the imagination which should not be admissible in any field of science. On the other hand, any assertion of probability is either right or wrong from the the standpoint of the theory on which it is based. This *loosening of thinking* seems to me to be the greatest blessing which modern science has given to us. For the belief in a single truth and in being the possessor thereof is the root cause of all evil in the world'.

Index of persons

Valera, de 270
Van der Waals 293
Vogt, W. 110, 111
Voigt, W. 81, 86, 87, 95, 98, 100, 105, 125, 135, 136, 153, 154, 156, 163, 199, 200

Wachsmuth 190
Wagner, C. 144
Wagner, R. 101, 143, 144
Wagner, S. 144
Waller, I. 291
Walter 154
Wätzmann 56, 123, 173
Weber 71
Weber (Mrs) 71
Weber, E. 71
Weber, H. 71
Weber, L. 70–72, 75, 77
Wedekind 8, 9, 16
Weierstrass 56, 66
Weigert, K. 46, 47
Weinberg, A. von 192, 249
Weiss, P. 267
Weissenborn 6–8, 18, 24, 43
Weisskopf, V. 235
Wende 200

Wentzel, G. 233
Weppner 241
Wertheimer, M. 173, 174, 184
Weyl, H. 135, 201, 206, 251, 258, 259, 262, 264
Whitehead 54
Whittaker 281
Wiechert, E. 88, 135, 154
Wien, M. 123, 168, 198
Wien, W. 198, 212, 213
Wiener, N. 226, 227, 232
Wigner, E. 214, 227, 236, 237
Wilhelm II of Prussia 7, 16, 111, 112
Windaus, A. 207, 248
Wittenberg, A. 26, 269
Witteschnik, E. 272
Wolf, E. 242, 243
Wolfer 73
Wolfskehl 146

Yang 290
Yukawa 227

Zeiss 118
Zermelo 86, 90, 135, 139
Zsigmondy, R. A. 154, 207